T0253469

ETHNOBOTANY
Local Knowledge and Traditions

Editors

José L. Martinez

Vice Rectory of Research
Development and Innovation
Universidad de Santiago de Chile
Santiago, Chile

Amner Muñoz-Acevedo

Chemistry and Biology Research Group
Department of Chemistry and Biology
Universidad del Norte
Barranquilla, Colombia

Mahendra Rai

Department of Biotechnology
SGB Amravati University
Amravati, Maharashtra, India

CRC Press
Taylor & Francis Group
Boca Raton London New York

CRC Press is an imprint of the
Taylor & Francis Group, an **informa** business
A SCIENCE PUBLISHERS BOOK

Cover credit: Reproduced by kind courtesy of Prof. Amner Muñoz-Acevedo (co-editor)

CRC Press
Taylor & Francis Group
6000 Broken Sound Parkway NW, Suite 300
Boca Raton, FL 33487-2742

First issued in paperback 2021

© 2019 by Taylor & Francis Group, LLC
CRC Press is an imprint of Taylor & Francis Group, an Informa business

No claim to original U.S. Government works

Version Date: 20181026

ISBN-13: 978-0-367-78046-3 (pbk)
ISBN-13: 978-1-138-38898-7 (hbk)

This book contains information obtained from authentic and highly regarded sources. Reasonable efforts have been made to publish reliable data and information, but the author and publisher cannot assume responsibility for the validity of all materials or the consequences of their use. The authors and publishers have attempted to trace the copyright holders of all material reproduced in this publication and apologize to copyright holders if permission to publish in this form has not been obtained. If any copyright material has not been acknowledged please write and let us know so we may rectify in any future reprint.

Except as permitted under U.S. Copyright Law, no part of this book may be reprinted, reproduced, transmitted, or utilized in any form by any electronic, mechanical, or other means, now known or hereafter invented, including photocopying, microfilming, and recording, or in any information storage or retrieval system, without written permission from the publishers.

For permission to photocopy or use material electronically from this work, please access www.copyright.com (http://www.copyright.com/) or contact the Copyright Clearance Center, Inc. (CCC), 222 Rosewood Drive, Danvers, MA 01923, 978-750-8400. CCC is a not-for-profit organization that provides licenses and registration for a variety of users. For organizations that have been granted a photocopy license by the CCC, a separate system of payment has been arranged.

Trademark Notice: Product or corporate names may be trademarks or registered trademarks, and are used only for identification and explanation without intent to infringe.

Library of Congress Cataloging-in-Publication Data
Names: Martinez, Jose L. (Jose Luis), 1959- editor.	Munoz-Acevedo, Amner, editor.	Rai, Mahendra, editor.
Title: Ethnobotany : local knowledge and traditions / editors: Josâe L. Martinez (Universidad de Santiago de Chile), Amner Munoz-Acevedo (Universidad del Norte Barranquilla), Mahendra Rai (SGB Amravati University).
Description: Boca Raton, FL : CRC Press, [2018]	Includes bibliographical references and index.
Identifiers: LCCN 2018041061	ISBN 9781138388987 (hardback)
Subjects: LCSH: Ethnobotany.	Medicinal plants.
Classification: LCC GN476.73 .E845 2018	DDC 581.6/36--dc23
LC record available at https://lccn.loc.gov/2018041061

Visit the Taylor & Francis Web site at
http://www.taylorandfrancis.com

and the CRC Press Web site at
http://www.crcpress.com

Preface

The local knowledge or indigenous knowledge of the use of plants is based on the wisdom and experience of the local community since ancient time. In due course of time, this knowledge becomes the tradition of the local community or tribes. Such knowledge is passed from one generation to another generation and is applied in agriculture, medicines and other household activities. In the past decade, the botanists have made tremendous efforts to gather this knowledge, and eventually, a new branch of botany known as 'Ethnobotany' has originated.

Ethnobotany has always been used as primary source to search biologically active molecules (ca. 60%) from nature and at the same time, it has contributed to the development of drugs (pharmaceutical industry) that have improved the quality of human life. However, some active constituents found within those plants are not the result of the scientific validations based on traditional medicinal uses of plants containing them. On the other hand, there are a few documents/books on ethnobotany (e.g., WHO monographs) containing information about plant-based treatment for specific chronic diseases such as hypertension, cancer, CNS disorders, psoriasis, and urolithiasis, etc.

Most of the plants described in the present book have been used in traditional medicine by local community for specific ailments and the biological activities related to each "sickness" have been scientifically verified.

We hope that the present book will be an appropriate consultation tool for scientists/professionals/experts such as ethnobotanists, botanists, cell/molecular biologists, chemists, pharmacists, pharmacologists, and environmentalists/ecologists. It will also be useful for undergraduate and postgraduate students from the same areas and for the pharmaceutical industries.

José L. Martinez
Amner Muñoz-Acevedo
Mahendra Rai

Contents

CHAPTER 1

Mururé (*Brosimum acutifolium* Huber) in the Treatment of Syphilis in Colonial Amazonia

From Historical Data to the Actual Contribution in Treatment

Erika Fernanda de Matos Vieira,[1] *Maria da Graça Ribeiro Campos*[2,3,*] and *Flávia Cristina Araújo Lucas*[4]

Introduction

Studying the historical ethnobotany of medicinal plants is a new method of obtaining information about plants and people derived from historical records, including documents about botany, anthropology, ecology, and history that are still used and relevant today (Prance 2000, Medeiros 2009). Living and learning with diverse ethnic groups have provided valuable contributions to the development of research about natural products, as well as knowledge about the close relationship

[1] Universidade do Estado do Pará, Belém, Pará, Brasil.
[2] Observatory of Herb-Drug Interactions/Faculty of Pharmacy, University of Coimbra, Heath Sciences Campus, Azinhaga de Santa Comba, Coimbra Portugal.
[3] Coimbra Chemistry Centre (CQC, FCT Unit 313) (FCTUC) University of Coimbra, Rua Larga, Coimbra, Portugal.
[4] Herbário MFS, Universidade do Estado do Pará, Belém, Pará, Brasil, Rua do Una, n°156, Telégrafo.
* Corresponding author: mgcampos@ff.uc.pt

between the chemical structures of compounds and their biological properties (Viegas Júnior et al. 2006).

Medications considered from natural origin, many of which were used by native groups, began to spread during the time of colonial Brazil when knowledge and practices were taken by missionaries, naturalists, and travelers that were experimenting with new plants and methods to cure diseases using plant chemistry (Costa 2007, Veríssimo and Pereira 2014). This period was known for the use of natural resources in the synthesis of toxic and medicinal substances, as well as for the identification of biologically active molecules (Jain and Sklani 1991, Viegas Júnior et al. 2006, *apud* Medeiros 2009). In this historical context, the Jesuit João Daniel, who lived in Amazonia for 16 years, experimented with a vast repertoire of herbal recipes. His main work, "The Treasure Discovered on the Great Amazon River (original title: *O Tesouro Descoberto no Máximo rio Amazonas"*), is impressive for a publication produced almost 200 years ago, especially because of how precise it is in relation to the flora and forms of use (Pinto 2005, Val 2014). His writings are technical descriptions of healing techniques for the most common diseases of the time, which often became plagues or pests and were treated with medicinal plants (Sousa 2013).

In Amazonia, Father João Daniel commented about the seriousness of venereal diseases among natives and settlers, highlighting the numerous cases of syphilis that severely affected a large number of people and sometimes became epidemics that generated innumerable social impacts (Veloso 2001). The description of his treatment for syphilis mentions the frequent use of a plant popularly known as mururé, which was considered an effective remedy.

Until the discovery of penicillin, other treatments with plants or chemical elements, such as mercury, bismuth and arsenic salts, were used to treat syphilis; however, they were ineffective and caused side effects (Carrara 1996, Veloso 2001). Syphilis is a contagious, systematic, chronic disease, caused by the bacteria *Treponema pallidum*, which can be transmitted sexually (called acquired syphilis) or vertically when a pregnant woman infects the fetus (congenital syphilis) (Neves and Araújo 2013).

As far back as the 15th century, there were references to serious syphilis epidemics in Europe; however, the origin of the disease was unknown (Geraldes Neto et al. 2009). The WHO (2016) estimated that there are approximately 18 million prevalent cases of syphilis in the world and around 5.6 million new cases per year. In addition, the WHO warned that this disease is becoming more resistant to penicillin due to its overuse. According to Anvisa (2016c), this is a problem because penicillin is also effective at treating other bacterial diseases.

Validating the use of plants to treat diseases is recommended by the Ministry of Health (Ministério da Saúde 2012) and more investigations involving traditional knowledge and modern technologies to produce medicines should be carried out. WHO (2011) indicated that approximately 25% of modern medicines originate directly or indirectly from medicinal plants. In addition, the strong curative efficacy of plants has been observed since ancient times.

The aim of this chapter is to share a case study that can be used as an example of the tools and the various steps that should be followed in ethnobotanical research. Based on this, the ethnopharmacology of mururé since the 18th century for the treatment of syphilis will be described. A correlation of the past-present therapeutic use and contribution of historical records in the use of natural products will also be discussed.

Historical Reports of Syphilis, Ethnopharmacology, and Reactions in the Organism

João Daniel designated syphilis as the Gallic disease. Gallic evil was one of the common terms used to refer to the disease (Geraldes Neto et al. 2009), as was *morbus gallicus* or French evil, due to a famous internal European conflict where the French army was defeated by the disease (not by weapons) (Sousa 1996).

The proven therapeutic efficacy of mururé was associated with using its exudate, called Milk of Morure (*"Leite do Morure"*), which was extracted by cutting the bark of the tree. The action of this liquid to combat syphilis was also reported by Pinagé (2011). Due to the abundance of exudate produced, João Daniel called the tree as the King of the Milks (*"Rei dos Leites"*). In the transcription below, the curative and rapid effect of the exudate on syphilis after ingestion has been mentioned.

> "Milk of morure: is the king of the milks, because its effective virtue is superior to all. It is distilled from the tree of the same name, which we have already described. It serves to cure the Gallic disease in a marvelous way because no matter how severe the disease is in the sick, it heals, and cures in 24 hours, or less, without further preparation"
>
> (*"Leite morure: é o rei dos leites, porque na sua eficaz virtude a todos é superior. É destilado da árvore do mesmo nome, que já descrevemos no seu lugar. Os seus préstimos são para curar as doenças do gálico por modo maravilhoso, porque por mais arraigado que esteja este mal nos doentes, o tira, e os cura em 24 horas, ou menos, sem mais preparos ou preâmbulos"*) (Daniel 2004, p. 556).

The administration of the milk to patients with the disease was very simple. It was directly ingested after collecting it from the tree, in the shortest possible time after exposure to air, as indicated here: "[...] *duas colheres ordinárias ou três colheres não tão cheias,*" two full spoons or three moderate spoons. After the first dose, João Daniel wrote that a few additional daily doses are necessary, said the treatment is effective in 24 hours or less, and highlights the reactions of the body of the patient after ingestion.

Of the reactions observed after ingesting the milk, it was the urine of the patients very concentrated with sediment ("[...] *bastante materiada*"), which resulted from overloading the kidneys with the ingested medicine and reflected the curative action.

This was interpreted as an indicator of the effect of the medicinal plant. Similar urinary tract laden with concentrated sediments also occurs when syphilis patients use current medicines for this pathology (Neves and Araújo 2013, Anvisa 2016b). Other effects noted were neurological disorders described by the priest as chills, tremors, and seizures that arise immediately after ingestion, which were treated by placing a warm cloth on the stomach.

João Daniel also verified the performance of the plant for cases of neurosyphilis, which could appear in 12 to 18 months after infection and could be asymptomatic. He noted a small amount is enough to get rid of the disease and bad mood from the bones and nerves in one night ("[...] *e basta esta pequena quantidade para dos mesmos ossos e nervos arrancar em uma noute toda doença, e mau humor*"). For neurosyphilis, Vargas et al. (2000) affirmed the strong psychological effect on the patient, and Barros et al. (2005) characterized it as an infection that affects the nervous system and can cause lesions on the brain and spinal cord. These clinical observations were similar to what the Father reported. Neurosyphilis had great repercussions before the use of penicillin and resurged in the late 20th century, mainly in HIV patients who were more susceptible to the disease (Neves and Araújo 2013, Caixeta et al. 2014).

Mururé was also prescribed for other illnesses: it treated *alporcas*, and all the Gallic diseases "[...] *Da mesma sorte cura alporcas, e todas as doenças que pecam de gálico*" (Daniel 2004). *Alporca* is a term that signifies scrofula, where there is swelling in the lymph nodes, and is associated with tuberculosis (Sayahi and Thomas 2005, Capone 2006, Almeida 2012). Carrara (1996) reported that in the past syphilis was considered by the medical community as an aggravating agent of tuberculosis. Evidence of this was reported after Father started administering the milk when there were clinical symptoms of tuberculosis and considered the treatment very effective.

The experience and effectiveness of using mururé in the missions, on both Indians and Caucasians, impressed João Daniel who soon saw the export potential of the species to use as a medicine in Europe. Introducing this plant to Europe would be the beginning of its wide dissemination throughout the world. The articulation of knowledge, healing practices, and market views by the Jesuits was discussed by Pinto (2005), who analyzed the transfer of natural resources and traditional knowledge to other parts of the world and promoted advances in the sciences (Calainho 2005, Medeiros 2009, Laws 2013). In relation to this, João Daniel was a visionary by promoting the value and use of the riches in the region. Studying mururé and other plants did not simply occur because people were searching for knowledge about specific plant species, but was instead driven by the fact that these plants were potentially useful to the colony and cities abroad (Costa 2007, Santos 2013, São Bento and Santos 2015).

Based on tests using mururé to treat syphilis in the 20th century, in the year 1918 there was an elevated number of reported cases (about 6 million people) of the disease throughout the country. Due to this health crisis, drugstores popularized therapeutic treatments using mururé, as seen in the advertisements written by the dermatologist Dr. Zopyro Goulart (Figs. 1.1 and 1.2).

In addition to the historical indications of this plant, the species is still used. In Amazonian communities on the Maruepaua River, the sap of *Brosimum acutifolium* is still used in the treatment of rheumatism, dislocations, and swollen body parts (Baptista 2007). Also, in 2017, you can find mururé bark for sale as

Fig. 1.1. Advertisement for syphilis treatment using mururé in a magazine published in 1918. Source: Revista Careta, 1918, edition 502.

Fig. 1.2. Advertisement for syphilis treatment using mururé in a magazine published in 1918. Source: Revista Careta, 1918, edition 501.

a medicinal product in the largest open fair in Latin America, Ver-o-Peso Fair in Belém, Pará (Iphan 2017), with similar indications as the historical records from the 18th century and the 1918 advertisements (Fig. 1.3). The herbalists at Ver-o-Peso, who are considered holders and transmitters of traditional knowledge in an urban environment (Dantas and Ferreira 2013), continue to promote mururé as a curative and therapeutic product for syphilis.

Fig. 1.3. Mururé bark for sale by herbalists at the Ver-o-Peso Fair in 2017.

Brosimum acutifolium Huber. (mururé) in the Treatment of Syphilis in Colonial Amazonia: A Case Study

Documentary research

The main method of this work involved collecting and analyzing documents (Godoy 1995). The data obtained was then used to conduct a comparative study between the past and present (Medeiros 2009).

Documentary research and bibliographic works were consulted in this study. Despite their similarity, Sá-Silva et al. (2009) differentiated them mainly by their source because the former comprises records that were not analytically treated (primary source) or insufficiently analyzed, and the latter includes material with contributions from different authors about a theme (secondary sources).

The book *Tesouro Descoberto no máximo rio Amazonas*, written between 1757 and 1776 by Father João Daniel (1722–1776), was selected as the main literature source of this study. Father Daniel was a naturalist, who wrote this work in Lisboa while in prison (due to Pombaline orders), which became a massive codex that is considered by many historians as a significant document about Amazonia during the colonial period (Salles 2004). It is the most comprehensive record and most complete source of knowledge about Amazonia during the 18th century because the information in it is the result of extensive fieldwork and vast cataloging of data, especially about the flora of the region (Siewierski 2014). According to Tocantins (1976), although there are other publications in the same genus, this book is the "Ecological bible of Amazonia," and is a complete, well-researched work about the region produced by a member of the Society of Jesus.

Research sources: consulting works, databases, and files

To study the original works of Father Daniel, the following databases were searched: Biblioteca Digital Luso Brasileira; Catálogo Biblioteca da Universidade de Coimbra; Biblioteca Nacional do Rio de Janeiro; the digital collection from Biblioteca Nacional; and the virtual library at the Universidade Federal do Pará ("UFPA 2.0"), that has a free PDF version of *Tesouro Descoberto no Máximo rio Amazonas*. For all searches, the key words "*jesuítas*" (Jesuits), "*Amazônia colonial*" (colonial Amazonia) "*plantas medicinais*" (medicinal plants), and "João Daniel" were used.

Other works and catalogs in libraries were also examined based on searches for material from the 16th, 17th and 18th centuries related to medicinal plants, as well as Amazonia and people (missionaries, travelers, and naturalists), who wrote about colonial Amazonia. The libraries consulted included the following: Museu Paraense Emílio Goeldi, Belém, Pará, where the Rare Works Section and the *Catálogo Biblioteca Universitatis* were searched; University of São Paulo database; and the Coleção *História da Companhia de Jesus no Brasil* and *Ensaios de história das ciências no Brasil: às luzes à razão independente*. The use of several bibliographic sources, which considered the Jesuits, medicinal botany, Amazonia and the way of life in the sociocultural context of the 18th century, allowed different narratives and interpretations of the writings of João Daniel.

In addition, at the University of Coimbra in Portugal, the general library and Faculdade de Letras were visited. Some of the notable original documents and facsimiles consulted were: *Florae Lusitanicae et Brasiliensis Specimen Plantae exoticae Brasilienses*; annual report of the feats made by the Society of Jesus in parts of East India during the years of 604 and 605, by Padre Fernam Guerreiro; and *História da Província Santa Cruz*, by Pero de Magalhães de Gândavo, 1605.

Species definition

The species was identified using the taxonomic clues method in *Identificação de termos oitocentistas relacionados às plantas medicinais usadas no Mosteiro de São Bento do Rio de Janeiro, Brasil*, by Medeiros (2010). To properly confirm the scientific nomenclature, the following criteria were used: (1) text search of the common name "moruré" and "mururé"; (2) detailed examination of morphological descriptions and life form of mururé, both by the Father and other authors; (3) confirmation of the botanical identification using scientific articles and specimens listed in virtual herbaria (www.tropicos.org) and the "List of Species of Flora of Brazil" in "Flora do Brasil 2020 Project" (floradobrasil.jbrj.gov.br)".

For the writing in older works, spelling and punctuations were modified based on the current norms for Portuguese. Transcription into Portuguese and identifying characteristic linguistic terms of the period were based on *Estratégia e Táctica da Transcrição* (Castro and Ramos 1986). For quotes by some authors, the vocabulary used was absolutely respected.

Past-present comparative study

The past-present study compares how the use of plant resources in the past was related to local ethnoknowledge (Medeiros 2009). The indicated therapeutic history was analyzed using current scientific directories of plant research that provide information about compounds, isolated substances and pharmacology, which included the following: The Plant List (National Center for Biotechnology Information), PubMed, Biological Abstract, Jstor, Biblioteca Regional de Medicina da Unesp (Bireme), Medline, Phytotherapy Research, and Pubchem Compound.

Medical-pharmaceutical concepts during colonization

During encounters between Europeans and people who would eventually be colonized, various plant species were collected and introduced to different parts of the world where they were cultivated in the gardens of royalty and published in floras and catalogs, influenced the art world and, most importantly, were included as basic ingredients in European pharmacopeias (especially Lusitanian) (Vigier 1718, Almeida 1976). Missionaries, apothecaries, druggists and traders combined their knowledge from other places with the knowledge of indigenous and African populations to promote American ingredients, medicines, and recipe books (Domingues 1992).

The movement of plants around the world and their virtues are recorded in manuscripts. Within this area, João Curvo Semedo impacted society with "Memorial de Vários Simplícios Que da Índia Oriental, da América, e de Outras Partes do Mundo Vem ao Nosso Reino para Remédios de Muitas Doenças" (s/d, but before 1716), which describes around eight dozen drugs and medicines based on information from people who made observations in India and other parts of the world. In Amazonia, João Daniel was considered affectionate (*"afeiçoado"*) and tropicalized (*"tropicalizado"*) for coexisting with nature in the tropics in both dimension and depth. Due to his observations about the most notable (*"notáveis"*) herbs of Amazonas, and because he saw so many medicinal plants, he soon thought about the possibility of multiplying the herbs (*"multiplicados herbulados"*) for commercialization because they were rare, came from America, and could boost the economy of the 18th century.

The advancement of religious and exploratory missions in Amazonia favored greater rapprochement and contact between different peoples and ethnicities. This period, comprising the 18th century, recorded significant documentary data about diseases of various natures that proliferated during these encounters (Veríssimo and Pereira 2014). Sexually transmitted diseases, such as syphilis, became worrying and resulted in strong social consequences that included a change in behavior in the population, for example, the use of condoms made of sheep intestines (Veloso 2001). Guilherme Piso, in *História Natural do Brasil Ilustrada* (1648), highlighted the venereal evil ("*mal venéreo*") in Africans, Indians, Dutch and Portuguese due to contagion from intercourse. He considered this disease as endemic to the region and noted the efficiency of treating it with indigenous remedies ("[...] *sara logo, só com os remédios indígenas*").

The increased incidence of diseases resulted in a search for treatments using tests and formulas with therapeutic plants, which took place at houses built by the Jesuits near forests. According to Father João Daniel (1722–1776), the traditional pharmacies of the Jesuits were supplied with precious woods, miraculous barks, large forests of copaiba, *umeri* balsams, and the leaves and bark of cinnamon and quina, among other plants. This allowed them to take advantage of these natural resources and, with the help of the American Indians that were deeply knowledgeable about the forest and extremely skilled at extracting the resources, to use medicines in the missions and give them to white settlers so they would be sent to Europe where they were needed (Daniel 2004).

The commercial potential of Amazonian natural products was frequently valorized by João Daniel who, according to Pinto (2005), had a very peculiar intellectual and scientific vision for the formation of the missionaries that made up the religious order due to the importance of this knowledge as a mechanism of exploitation and power.

The mururé plant

The plant cited in *Tesouro Descoberto* with the common name mururé (Fig. 1.4) is *Brosimum acutifolium* Huber (Moraceae). This is a terrestrial tree native to Brazil that occurs in the Amazonia and Pantanal phytogeographic domains, known to occur in the North (Acre, Amazonas, Pará, Rondônia, and Roraima), Northeast (Maranhão) and Central-West (Mato Grosso) regions, and might also occur in the Southeast Region (São Paulo). The Flora of Brazil lists three subspecies for this taxon, including *Brosimum acutifolium* subsp. *interjectum* C.C. Berg that is endemic to Brazil and is known to occur in the states of Amazonas, Pará and Maranhão, in the Amazonian region (Romaniuc Neto et al. 2015).

The species is described as a large tree that can reach 35 m height, which has abundant white sap, is elongated, and has lanceolate leaves (Berg and Dewolf 1975, Rodrigues 1989). It occurs in tropical and subtropical climates, where the

Fig. 1.4. Specimen of mururé, donated by the "Rio de Janeiro Botanical Garden" to "The New York Botanical Garden", collected in the state of Pará in 1927 (Source: www.nybg.org).

average temperature is between 22°C and 30°C, and prefers sandy and clayey sand soils with an average amount of organic matter (Revilla 2001, *apud* Pinagé 2011).

Present time: phytochemical studies and therapeutic applications

The presence of secondary metabolites in plants (these compounds are often produced in a phase subsequent to growth) can be influenced by various environmental, climatic or temporal factors, and can be associated with handing methods related to collection and storage (Gobbo-Neto and Lopes 2007). Discovering useable therapeutic plant species is still an extremely complex process, especially because of the diversity of plants. In this context, the present traditional and popular knowledge, as well as information from communities that have produced this type of knowledge in the past, remain valuable resources for identifying medicinally useful plants (Pereira and Cardoso 2012, Medeiros 2009). Ratifying this, the WHO (2011) reported that the amount of modern medicines from plants is significant and could reach 60% for antitumor and antibacterial drugs.

Mururé (*Brosimum acutifolium* Huber), despite being used for other purposes (e.g., in cosmetics and food), is mostly employed medicinally because its latex and bark efficiently act against many illnesses and its sap has antiarthritic and antisyphilitic properties (Rodrigues 1989, Pinagé 2011). For treating syphilis, a daily dose of the sap should be used that does not exceed 8 g since an excess amount produces side effects, such as joint pain, polyuria (increased amount of urine), nausea and dizziness; however, literature also mentions the use of the bark prepared as a tea (Matta 2003, *apud* Pinagé 2011), contrary to João Daniel who only mentioned the sap.

Works reported the use of the species in Amazonian communities. According to Coelho-Ferreira et al. (2005), *B. acutifolium* is one of the 228 species that comprise the phytopharmacopoeia of the Marudá fishing community in the interior of Pará. In addition, Monteles and Pinheiro (2007) documented its use as a blood purifier (by preparing bottles with its bark) by a Quilombola community in the municipality of Presidente Juscelino which, according to IBGE (2016), is part of Maranhão in Amazonia.

In pharmacological studies, the flavonoid BAS1 (4'-hydroxy,7,8-(2",2"-dimethyl-pyran)-flavan) was isolated from *Brosimum acutifolium* (Moraes 2011). In this work, the mechanism of anti-inflammatory action of BAS1 in stimulated murine macrophages was characterized. The results demonstrated that BAS1, at high concentrations, had a cytotoxic effect. In addition, there was a reduction in the factors leading to inflammation and the flavonoid exhibited anti-inflammatory activity.

The antiproliferative and antineoplastic activities of four flavonoids from *B. acutifolium* were investigated by Maués (2013), who isolated two flavans, 4'-hydroxy-7,8-(2",2"-dimethylpyran)-flavan (BAS-1) and 7,4'-dihydroxy-8,(3,3-dimethylallil)-flavan (BAS-4), and two chalcones, 4,2'-dihydroxy-3',4'-(2",2"-dimethylpyran)-chalcone (BAS-6) and 4,2',4'-trihydroxy-3'-(3,3-dimethylallil)-chalcone (BAS-7), and verify the effects on rat glioblastomas *in vitro*. The study concluded that the BAS-1, BAS-4 and BAS-7 flavonoids had antineoplastic potential as an agent in therapy, and that the BAS-4 flavonoid was the most promising because its action on tumor cells was more efficient and its cytotoxic activity was the least harmful to healthy cells.

Other studies indicated the presence of various flavonoids in the bark of mururé (Torres et al. 2000, Takashima and Ohsaki 2002, Takashima et al. 2005). The latter study evaluated the activity of four flavonoids against leukemia cells, which had a cytotoxic effect on cells resistant to vincristine (the drug used to treat the illness).

Thus, some chemicals' structures of the isolated compounds from mururé sap/bark, e.g., flavans (BAS-1, and BAS-4), and chalcone (BAS-7) can be seen.

Baptista (2007), did ethnobotanical studies in the interior of Amazonia, as "Selected Concentrated *Brosimum acutifolium* (SCBA)", and used the sap for phytochemical experiments testing the antinociceptive (analgesia) and antiedematogenic actions in mice. This work reported that at certain doses SCBA prevented the development of edema in mices. The study also investigated the phytochemical constituents of *B. acutifolium* latex and found the presence of saponins, proteins and amino acids, phenols, and alkaloids. The results demonstrated that the analgesia promoted by SCBA is due to one or more secondary metabolites, such as saponin, which exhibit anti-inflammatory activity. The anti-oedematous activity was verified and explained by the action of the sap on the chemical mediators that promote an inflammatory state. According to Cruvinel et al. (2010), edema is a clinical sign of inflammation.

Baptista (2007) also studied the acute and sub-acute toxicity and the metabolites present in mururé bark. The work identified saponins that damage cells by altering the permeability of membranes, which destroys them. This type of alteration can be explained by the amphiphilic behavior of saponins, which can form complexes with steroids, proteins and membrane phospholipids that have a variable number of biological properties, including actions on cell membranes (Schenkel et al. 2004).

In a study about takini, a hallucinogen used by shamans in Amazonia that is prepared from the sap of *B. acutifolium*, Moretti et al. (2006) identified the presence of the alkaloid bufotenin (5-hydroxy-dimethyltryptamine) that is a compound with psychotrophic properties. This alkaloid is present in other species of *Brosimum as B. utile* (Kunth) Pittier and has been studied for its antifungal and antibacterial activities due to the toxic effect on the DNA of microorganisms (Haro 2015, Manotoa 2015).

Overall, saponins are important because in addition to their action on membranes, they help in the absorption of other compounds, which increases the immune response (Schenkel et al. 2004), and the alkaloid bufotenine promotes bactericidal activity.

Considerations about Syphilis and Present Treatment

The Ministry of Health, through the National STD and Aids Program, published that syphilis is one of the most common causes of genital ulcers (Brasil 2006) and, through the Department of Epidemiological Surveillance, declared that the profile of diseases leading to the formation of these ulcers (such as syphilis) is associated with an increase in risk of HIV contamination (Brasil 2010). Another major concern is the number of congenital syphilis cases in 25 countries in Latin America and the Caribbean, which includes 250 thousand new cases per year. Generally, this type of syphilis is neglected in epidemiological statistics, and is usually related to poverty and underdeveloped conditions (Organização Pan-Americana da Saúde 2009).

In Brazil, the epidemiological bulletin about syphilis (published by the Ministry of Health) noted that due to an increase in coverage of prenatal services, all regions in Brazil reported an increase in pregnant women with syphilis compared to the previous year (Brasil 2015a). The publication also states that the North Region of the country has the highest rates of late diagnosis (the last trimester of gestation), hindering treatment and increasing infant mortality rates, which continue to grow; the Northeast and North regions have the highest rates of infant mortality caused by congenital syphilis, which were above the national average of 2.2 in 2004 and rose to 5.5 per 100,000 live births in 2013.

Although penicillin is the most effective and inexpensive form of treatment for syphilis, there is a current national shortage of the medication due to the lack of raw material for its production in the global market (Brasil 2015b). Further, there are problems with the distribution of this medication to states and municipalities, mainly because, according to Sforsin et al. (2012), there is a high demand for this drug to treat patients and not all receive the treatment (Brasil 2015a). Thus, many people are left waiting for a solution.

As explained above, syphilis is currently treated mainly with penicillin, which inhibits enzymes that catalyze the formation of cell wall precursors and, thus, interferes with the integrity of the wall. This induces *T. pallidum* death due to the entrance of water into the organism (Guinsburg and Santos 2010). Except for rare cases where the patient is allergic to the medicine, it is advised to use penicillin first; as a last resource, alternative medication should be used, although many are not as effective as penicillin that can resolve primary and secondary cases with only one dose (Avelleira and Botino 2006).

According to Anvisa (2016d), penicillin has a nucleus with a ß-lactam ring, which is responsible for the antibacterial action in the medicine. There are four penicillin groups, including benzylpenicillins that are natural penicillins. The group of natural penicillins includes those used to treat syphilis: penicillin G benzathine, which is used to treat most cases; crystalline penicillin, which is a variety used for cases of neurosyphilis and congenital syphilis; penicillin G procaine, which is used for congenital syphilis; and penicillin V, which is not directly used to treat syphilis but to desensitize patients who are allergic to other types of penicillin (Avelleira and Botino 2006).

Anvisa (Brazilian Health Regulatory Agency, Ministry of Health) (2016a) notes that benzylpenicillins (benzathine and procaine penicillins) are administered intramuscularly, crystalline penicillin is administered intravenously, and only penicillin V is administered orally. Digestive activity interferes with the properties of penicillin and, for this reason, penicillin generally needs to be administered using something else (Fariña and Poletto 2010). This differs with the method mentioned by the Father, who indicated oral ingestion for the latex. The effects of mururé can be explained based on a study by Maués (2013), who found saponins in the latex that, like penicillin, destroy cell membranes.

Treatment with penicillin can produce side effects, including hypersensitivity, cutaneous manifestations, renal toxicity, haematological toxicity, and neurotoxicity (Anvisa 2016b). The Jarisch-Herxheimer reaction is common when treating diseases caused by spirochetes due to the release of antigens from the bacterial cell wall, which can occur soon after the treatment with the antibiotics, and includes symptoms characterized by a sudden onset of fever, chills, headaches, myalgia, rashes and, sometimes, refractory shock (Neves and Araújo 2013, Brasil 2016).

The adverse effects described by the Father, including shivering, tremors, and convulsions ("[...] *arrepiamentos, tremores, e convulsões*"), correspond to symptoms of the Jarisch-Herxheimer reaction. This shows that mururé destroys treponemas that cause syphilis by acting on the cell walls of the bacteria, which is evidence of metabolomic similarities between the action of mururé and penicillin. Another corresponding aspect is how fast the penicillin and sap act, where results are seen after one dose. A urine analysis is often conducted as a diagnostic parameter to look for harmful substances eliminated by the body. This factor has been observed in many studies, such as Han et al. (2016), who evaluated the effect of ketamine using urine as a study tool. In addition, saponins are important components for the action of many plant drugs, especially due to their diuretic effect, and are used as an adjuvant because they aid absorption mechanisms and increase the immune response (Schenkel et al. 2004).

Finally, it was observed that despite the different modes of ingestion, both penicillin and the exudate of mururé have a correlated reaction that results in the desired medicinal effect (i.e., fighting syphilis). Thus, clinical reports documented in past and present bring to light similarities and alternatives for proposing new syphilis drugs.

Conclusions

Current treatment methods for syphilis mostly involve the use of penicillin with no reference of using *Brosimum acutifolium*, although there are historical medical records that note mururé is effective. The search for new medicines and the value of Father João Daniel's documents are evidence of the great naturalistic ability of a plant that, due to the success of its medicinal effects, elevated mururé in the *Tesouro Descoberto do Máximo Rio Amazonas* to the potential drug category. Countries that currently lead the world in syphilis cases and suffer from elementary social problems (e.g., related to the production and distribution of medicines) should note this fact.

The search for medicinal properties in plants as a solution to the unavailability of medicine, mainly using past evidence, strengthens the relationship between people and plants, and is important to science today. The analysis made in this work show that the study of historical documents can be an important tool to develop medicine because these works are potential resources to find new drugs.

References

Almeida, L.F. 1976. Aclimatação de plantas do Oriente no Brasil durante o século XVII e XVIII. Revista portuguesa de História. 15: 339–431.

Almeida, M.A.P. 2012. O Porto e as epidemias: saúde e higiene na imprensa diária em períodos de crise sanitária, 1854–56, 1899 e 1918. Revista de História da Sociedade e da Cultura 12: 371–391.

Avelleira, J.C.R. and Botino, G. 2006. Sífilis: Diagnóstico, tratamento e controle. An. Bras. Dermatol. 81: 111–126.

Anvisa. 2016a. III Antimicrobianos - principais grupos disponíveis para uso clínico: ß-Lactâmicos – Penicilinas: Propriedades Farmacológicas. http://www.anvisa.gov.br/servicosaude/controle/rede_rm/cursos/rm_controle/opas_web/modulo1/penicilinas2.htm Consultado em: 11 de outubro de 2016.

Anvisa. 2016b. III Antimicrobianos - principais grupos disponíveis para uso clínico: ß-Lactâmicos – Penicilinas: Efeitos colaterais. http://www.anvisa.gov.br/servicosaude/controle/rede_rm/cursos/rm_controle/opas_web/modulo1/penicilinas10.htm Consultado em: 11 de outubro de 2016.

Anvisa. 2016c. III Antimicrobianos - principais grupos disponíveis para uso clínico: ß-Lactâmicos – Penicilinas: Indicações clínicas. http://www.anvisa.gov.br/servicosaude/controle/rede_rm/cursos/rm_controle/opas_web/modulo1/penicilinas6.htm Consultado em: 09 de outubro de 2016.

Anvisa. 2016d. Anvisa. III Antimicrobianos - principais grupos disponíveis para uso clínico: ß. http://www.anvisa.gov.br/servicosaude/controle/rede_rm/cursos/rm_controle/opas_web/modulo1/penicilinas10.htm Consultado em: 11 de outubro de 2016.

Baptista, E.R. 2007. Conhecimentos e práticas de cura em comunidades rurais amazônicas: recursos terapêuticos vegetais. Tese de Doutorado. Universidade Federal do Pará, Brasil.

Barros, A.M., Cunha, A.P., Lisboa, C., Sá, M.J. and Resende, C. 2005. Neurossífilis: Revisão Clínica e Laboratorial. ArquiMed. 19: 121–129.

Bessa, N.F.G., Borges, J.C.M., Beserra, E.P., Carvalho, R.H.A., Pereira, M.A.B., Fagundes, R. Campos, S.L., Ribeiro, L.U., Quirino, M.S., Chagas Junior, A.F. and Alves, A. 2013. Prospecção fitoquímica preliminar de plantas nativas do cerrado de uso popular medicinal pela comunidade rural do assentamento vale verde – Tocantins. Rev. Bras. Plantas Med. 15: 692–707.

Berg, C.C. and DeWolf, G.P. 1975. Moraceae. pp. 173–299. *In*: Lanjouw, J. and Stoffers, A.L. (eds.). Flora of Suriname. Vol. 5. Part 1. E.J. Brill, Leiden, Holanda.

Brasil. Ministério da Saúde. Secretaria de Vigilância em Saúde. Programa Nacional de DST e Aids. 2006. Manual de Bolso das Doenças Sexualmente Transmissíveis. Brasília, Brasil.

Brasil. Ministério da Saúde. 2010. Caderno 6: Aids, Hepatites Virais, Sífilis Congênita e Sífilis em Gestantes. pp. 1–60. *In*: Brasil. Ministério da Saúde. Secretaria de Vigilância em Saúde. Departamento de Vigilância Epidemológica. Guia de Vigilância Epidemiológica. Brasília.

Brasil. Ministério da Saúde. 2012. Práticas Integrativas e Complementares: Plantas Medicinais e Fitoterapia na Atenção Básica. Ministério da Saúde, Brasília, Brasil.

Brasil. Ministério da Saúde. Secretaria de Vigilância em Saúde. 2015a. Departamento de DST, Aids e Hepatites Virais. Boletim Epidemiológico: Sífilis 2015. Ministério da Saúde, Brasília, Brasil.

Brasil. Ministério da Saúde. Secretaria de Vigilância em Saúde. 2015b. Nota Informativa Conjunta nº 109/105/GAB/SVS/MS, GAB/SCTIE/MS. Ministério da Saúde, Brasília, Brasil.

Brasil. Ministério da saúde. 2016. Portal da Saúde, Ministério da Saúde - Tratamento. http://portalsaude.saude.gov.br/index.php/tratamento Consultado em: 11 de outubro de 2016.

Capone, D., Mogami, R., Lopes, A.J., Tessarollo, B., Cunha, D.L., Capone, R.B., Siqueira, H.R. and Jansen, J.M. 2006. Tuberculose Extrapulmonar. Revista Hope 5: 54–67.

Caixeta, L., Soares, V.L.D., Reis, G.D., Costa, J.N.L. and Vilela, A.C.M. 2014. Neurosífilis: Uma Breve Revisão. Rev. Soc. Bras. Med. Trop. 43: 121–129.

Calainho, D.B. 2005. Jesuítas e Medicina no Brasil Colonial. Tempo 10: 61–75.

Carrara, S. 1996. Tributo a vênus: a luta contra a sífilis no Brasil, da passagem do século aos anos 40. Editora Fiocruz, Rio de Janeiro, Brasil.

Castro, I. and Ramos, M.A. 1986. Estratégia e Táctica da Transcrição. pp. 99–122. *In*: Critique Textuelle Portugaise, Actes du Coloque. Fondation Calouste Gulbenkian, Paris, França.

Costa, K.S. 2007. Natureza, colonização e utopia na obra de João Daniel. Hist. Cienc. Saúde Manguinhos 14: 95–12.

Coelho-Ferreira, M.R.C. and Silva, M.F.F. 2005. A Fitofarmacopéia da Comunidade Pesqueira de Marudá, Litoral Paraense. Bol. Mus. Para. Emilío Goeldi, sér. Ciências Naturais 1: 31–43.

Cruvinel, W.M., Mesquita Júnior, D., Araújo, J.A.P., Catelan, T.T.T., Souza, A.W.S., Silva, N.P. and Andrade, L.E.C. 2010. Fundamentos da imunidade inata com ênfase nos mecanismos moleculares e celulares da resposta inflamatória. Rev. Bras. Reumatol. 50: 434–61.

Daniel, J. 2004. Tesouro Descoberto no Máximo Rio Amazonas. Vol. 1–2. Ed. I Contraponto, Rio de Janeiro, Brasil.

Dantas, C.F.N. and Ferreira, R.S. 2013. Os conhecimentos tradicionais dos (as) erveiros (as) da Feira do Ver-o-Peso (Belém, Pará, Brasil): um olhar sob a ótica da Ciência da Informação. Perspect. ciênc. inf. 18: 105–125.

Domingues, A. 1992. As remessas das expedições científicas no Norte Brasileiro na segunda metade do século XVIII. In Brasil, Nas vésperas do mundo moderno. Lisboa: Comissão Nacional para as Comemorações dos Descobrimentos Portugueses.

Fariña, L.O. and Poletto, G. 2010. Interações entre antibióticos e nutrientes: uma revisão com enfoque na atenção à saúde. Visão Acadêmic 11: 91–99.

Geraldes Neto, B., Soler, Z.A.S.G., Braile, D.M. and Daher, W. 2009. A sífilis no século XVI: o impacto de uma nova doença. Arq Ciênc Saúde 3(16): 127–129.

Gobbo-Neto, L. and Lopes, N.P. 2007. Plantas medicinais: fatores de influência no conteúdo de metabólitos secundários. Quím. Nova 30(2): 374–371.

Godoy, A.S. 1995. Pesquisa qualitativa: tipos fundamentais. Rev. adm. Empres. São Paulo, 35: 20–29.

Guinsburg, R. and Santos, A.M.N. 2010. Critérios diagnósticos e tratamento da sífilis congênita. Sociedade Brasileira de Pediatria. São Paulo.

Han, E., Know, N.J., Feng, L., Li, J. and Chung, H. 2016. Illegal use patterns, side effects, and analytical methods of ketamine. Forensic Sci. Int. 268: 25–34.

Haro, A.L.J. 2015. Estudio Fitoquímico y Evaluación de la Actividad Antidermatófica *in vitro* del Látex de *Brosimum utile* (Kunth c.s.) (Leche de Sandi). Tese de Doutorado. Escuela Superior Politécnica de Chimborazo. Ecuador.

IBGE—Instituto Brasileiro de Geografia Estatística: Presidente Juscelino. http://cidades.ibge.gov.br/painel/painel.php?codmun=210920 Consultado em 30 de Janeiro de 2017.

IPHAN—Instituto do Patrimônio Histórico e Artístico Nacional: Ver-o-Peso (PA). http://portal.iphan.gov.br/pagina/detalhes/828 Consultado em 26 de Março de 2017.

Laws, B. 2013. 50 plantas que mudaram o rumo da História. Sextante, Rio de Janeiro, Brasil.

Manotoa, A.A.M. 2015. Evaluación de la Actividad Antifúngica em *Moniliophthora roreri* de Frutos de Cacao (*Theobroma cacao* L.) de Extractos de Látex de sande de *Brosimum utile* Kunth. Tese de Doutorado. Escuela Superior Politécnica de Chimborazo. Ecuador.

Moraes, W. 2011. Caracterização do mecanismo de ação antiinflamatória do flavonóide BAS1 isolado da planta *Brosimum acutifolium*. Tese de Doutorado, Universidade Federal do Pará, Brasil.

Maués, L.A.L. 2013. Avaliação da atividade antiproliferativa e antineoplásica de flavonoides da espécie *Brosimum acutifolium* em modelos de glioblastoma *in vitro*. Tese de Doutorado, Universidade Federal do Pará, Brasil.

Medeiros, M.F.T. 2009. Etnobotânica Histórica: Princípios e Procedimentos. Ed I, NUPEEA Recife, Brasil.

Medeiros, M.F.T. 2010. Identificação de termos oitocentistas relacionados às plantas medicinais usadas no Mosteiro de São Bento do Rio de Janeiro, Acta bot. bras. 24: 780–789.

Monteles, R. and Pinheiro, C.U.B. 2007. Plantas medicinais em um quilombo maranhense: uma perspectiva etnobotânica. Rev. Biol. Ciênc. Terra 7: 38–48.

Moretti, C., Gaillard, Y., Grenand, P., Bévalot, F. and Prevosto, J.M. 2006. Identification of 5-hydroxytryptamine (bufotenine) in *takini* (*Brosimum acutifolium* Huber subsp. *acutifolium* C.C. Berg, Moraceae), a shamanic potion used in the Guiana Plateau. J. Ethnopharmacol. 106: 198–202.

Neves, C.D. and Araújo, E.C. 2013. Sífilis. pp. 1075–1083. *In*: Bichara C.N.C., Fahira Neto, H. and Vasconcelos, P.F.C. (eds.). Medicina Tropical e Infectologia na Amazônia. Ed. II Samauma Editorial, Belém, Brasil.

NYBG, Specimen Details: *Brosimum acutifolium* subsp. *interjectum* C.C.Berg. http://sweetgum.nybg.org/science/vh/specimen_details.php?irn=331150 Consultado em 11 de outubro de 2016.

Organização Pan-Americana da Saúde. Organização Mundial da Saúde. 2009. Eliminação de doenças negligenciadas e outras infecções relacionadas à pobreza. 49° Concelho Diretor, 61ª Sessão do Comitê Regional, Resolução CD49.R19. Washington, Estados Unidos.

Pinagé, G.R. 2011. Moraceae. pp. 2391–2453. *In*: Viana, C.A.S., Paiva, A.O., Jardim, C.V., Rios, M.N.F., Rocha, N.M.S., Pinagé, G.R., Arimoro, O.A.S., Suganuma, E., Guerra, C.D., Alvez, M.M. and Pastore, J.F. (eds.). Plantas da Amazônia: 450 espécies de uso geral. Universidade de Brasília, Brasília, Brasil.

Pinto, R.F. 2005. A viagem das ideias. Estudos avançados 19: 97–114.

Piso, G. 1648. História Natural do Brasil Ilustrada.

Pereira, R.J. and Cardoso, M.G. 2012. Metabólitos secundários vegetais e benefícios antioxidantes. Journal of Biotechnology and Biodiversity. 3: 146–152.

Prance, G.T. 2000. Ethnobotany and the future of conservation. Biologist 47: 65–68.

Rodrigues, R.M. 1989. A Flora da Amazônia. Cejup, Belém, Brasil.

Romaniuc Neto, S., Carauta, J.P.P., Vianna Filho, M.D.M., Pereira, R.A.S., Ribeiro, J.E.L.S., Machado, A.F.P., Santos, A., Pelissari, G. and Pederneiras, L.C. 2015. Moraceae. In Lista de Espécies da Flora do Brasil. Jardim Botânico do Rio de Janeiro. Disponível em: http://floradobrasil.jbrj.gov. br/jabot/floradobrasil/FB19770 Consultado em 16 de outubro de 2016.

Sá-Silva, J.R., Almeida, C.D. and Guindane, J.F. 2009. Pesquisa documental: Pistas teóricas e metodológicas. Rev. bras. Ci. Soc. 1: 1–15.

Salles, V. 2004. Apresentação: Rapisódia Amazônica de João Daniel. *In*: Daniel. J. (ed.). Tesouro Descoberto no Máximo Rio Amazonas. Contraponto, Rio de Janeiro, Brasil.

Santos, F.S. 2013. Indígenas, Jesuítas e a Farmacopeia verde das Terras Brasileiras: Os Segredos da Triaga Brasilica. Prometeica: Revista de Filosofia y Ciencias 4: 5–22.

São bento, V.M.C. and Santos, N.P. 2015. Jesuítas e ciência: a produção de medicamentos através da Colecção de Varias Receitas de 1766. Revista Maracanan 13: 146–157.

Sayahi, L. and Thomas, J.A. 2005. Morphophonological and semantic adaptation of Arabisms in Galician. Cahiers Linguistiques d'Ottawa 33: 23–42.

Semedo, J.C. (s/d, mas anterior a 1716). Memorial de Vários Simplícios Que da Índia Oriental, da América, e de Outras Partes do Mundo Vem ao Nosso Reino para Remédios de Muitas Doenças.

Sforsin, A.C.P., Souza, F.S., Sousa, M.B., Torreão, M.K.A.M., Galemback, P.F. and Ferreira, R. 2012. Gestão de Compras em Farmácia Hospitalar. Farmácia Hospitalar 16: 1–32.

Siewierski, H. 2014. O Tesouro da Alteridade Amazônica na Obra do Padre João Daniel. Revista Sentidos da Cultura 1: 81–92.

Schenkel, E.P., Gosman, G. and Athayde, M.L. 2004. Saponinas. pp. 711–740. *In*: Farmacognosia: da planta ao medicamento. 5 ed. Editora da UFRGS/ Editora da UFSC, Porto Alegre/ Florianópolis, Brasil.

Sousa, C.R. 2013. As práticas curativas na Amazônia colonial: Da cura da alma À cura do corpo (1707–1750). Amazônica: Revista de Antropologia 5: 362–384.

Sousa, J.G. 1996. Impacto Social da Sífilis: Alguns aspectos históricos. Medicina Interna 3: 184–192.

Takashima, J. and Ohsaki, A. 2002. Brosimacutins A-I, nine new flavonoids from *Brosimum acutifolium*. Nat. Prod. Res. 12: 1843–1847.

Takashima, J., Komiyama, K., Ishiyama, H., Kobayashi, J. and Ohsaki, A. 2005. Brosimacutins J-M, four new flavonoids from *Brosimum acutifolium* and their cytotoxic activity. Planta Med. 7: 654–658.

Tocantins, L. 1976. A Bíblia Ecológica do Padre João Daniel. pp. 7–24. *In*: Daniel, J. (ed.). Tesouro Descoberto no Máximo rio amazonas. Biblioteca Nacional, Rio de Janeiro, Brasil.

Torres, S.L., Arruda, M.S.P., Arruda, A.C., Muller, A.H. and Silva, S.C. 2000. Flavonoids from *Brosimum acutifolium*. Phytochemistry 8: 1047–1050.

Val, A.L. 2014. Amazônia um bioma multinacional. Ciênc. Cult. 66: 20–24.

Vargas, A.P., Carod-Artal, J.F., Negro, M.C.D. and Rodrigues, M.P.C. 2000. Demência por Neurossífilis: Evolução clínica e neuropsicológica de um paciente. Arq. Neuropsiquiatr. 58: 578–582.

Veloso, B. 2001. Da sífilis à sida. Med. Interna. 8: 56–61.

Vigier, J. 1718. Historia das plantas da Europa, e das mais usadas que vem da Ásia, de África, e da América, onde se vê suas figuras, seus nomes, em que tempo florescem e o lugar onde nascem.

Veríssimo, T.C. and Pereira, J. 2014. Floresta Habitada: História da ocupação humana na Amazônia. Imazon, Belém, Brasil.

Viegas Júnior, C., Bolzani, V.S. and Barreiro, E.J. 2006. Os produtos naturais e a química medicinal moderna. Quím. Nova 29: 326–337.

World Health Organization. 2011. The world medicines situation 2011: traditional medicines: global situation issues and challenges. Who, Geneva, Suíça.

World Health Organization. 2016. Who guidelines for the treatment of *Treponema pallidum* (syphilis). Who, Geneva, Suíça.

WRB. 1918. Careta. Ano IX. Edição 501. Rio de Janeiro, Brasil.

WRB. 1918. Careta. Ano IX. Edição 552. Rio de Janeiro, Brasil.

CHAPTER 2

Agroecology, Local Knowledge and Participatory Research

Articulation of Knowledge for Sustainable Use of Plant Resources in Agroecosystems

Santiago Peredo Parada[1,2,*] *and Claudia Barrera*[2]

Introduction

Sustainable management and conservation of natural resources fulfill the requirements of the rural communities involved and the necessities of the joint work among the different actors of a territory. For that, it is imperative to use an integrated approach, with established strategies allowing the deployment of all the ecological and social-cultural potential that a community has in the environment in which it is developed.

In this context, the local knowledge of peasants is fundamental because it represents a collective learning process mediated by the environment in which they live. This experiential knowledge has allowed, many times, to develop survival strategies based on the multifunctionality of their farms as a result of their capacity for cognitive adaptation.

Hence, it is necessary to promote space for the active participation of people through dynamic processes in which individuals are part of the decisions that affect them. Participatory research constitutes a methodology that seeks the participation of the community for its own benefit establishing a new relationship between theory and practice.

[1] Department of Agricultural Management, Universidad de Santiago de Chile, Chile.
[2] Agroecology and Environment Group, Universidad de Santiago de Chile, Chile.
* Corresponding author: santiago.peredo@usach.cl

Agroecology, as a scientific approach, constitutes an interaction between the knowledge of the peasant groups and that of technologists, which through participatory dynamics, allows, among other things, an alternative construction for the ecological management of the biodiversity at different territorial levels.

The purpose of this chapter is to present the local knowledge as an epistemic source for the agroecology. For its pursuit, in the first part, the main elements of peasant knowledge are addressed as a fundamental component of their ecosystems with the strategy of harmonious management of nature. Secondly, in a schematic way, the main features of participatory research and its application/adaptation to the peasant work on biodiversity management are presented. Finally, the experiences with rural communities in Central and Southern Chile are discussed.

Local Knowledge as an Epistemic Source for Agroecology

Norgaard (1985) was one of the first authors who laid the epistemological bases of agroecology. He described the following elements as important traditional farm systems: (1) ecological and social systems have an agricultural potential; (2) this potential has been explored/absorbed by traditional peasants by means of trials, errors, natural selection and cultural learning processes; (3) the knowledge is incorporated into traditional cultures through cultural learning, which stimulates and regulates the feedback of social systems to ecosystems, and (4) the nature of the potential of the socioecological system may be better understood by studying how traditional agricultural cultures have explored/absorbed that potential.

Based on these elements, the most important part of the agroecological work, related to the study, understanding, re-valorization and visibility of the great potential contained in the traditional agrarian systems and culture, is sustained. According to Altieri and Nichols (2000), four aspects of these systems of traditional knowledge are important for the agroecologists:

1. The knowledge of the environment concerns, among others, to the climatic seasonality, by using indicators such as the phenology of local vegetation, soil types, edaphic fertility grades and categories of use.
2. The native biological taxonomies that are well-known, e.g., the insects and arthropods playing preponderant roles because they serve as food, medicine, and also, those that are important for folklore and the mystic, in which the ethnobotanical aspects have been most and appropriately documented.
3. The knowledge of agricultural practices oriented to the understanding of adaptation mechanisms in under-stress and changing situations in which have been observed a horizontal as well as vertical crop organization, exploitation of microenvironment variety, maintenance of materials and waste closed cycles through practices of effective recycling, interdependence of ecological complexes, usage of local resources, and predominance of human and animal energy.

4. The experimental nature of traditional knowledge based not only on the acute observation but also on the experimental learning, such as the selection of seed varieties for specific environments, new agricultural methods to overcome the biological or socioeconomic particular limitations, among many others.

The characteristic of local knowledge pointed out by Gómez-Benito (2001) shows the close relation between the local culture and its environment that, through the appropriation processes of nature, has been configuring new landscapes evident in highly complex agroecosystems. However, this knowledge is not the result of a merely empirical exercise, but rather as a multipurpose strategy that seeks to maximize the benefit of the social group and thereby guarantee the conditions, allowing its social and material reproduction. As stated by van der Ploeg (1990), this knowledge has an internal organization. This local knowledge is transferred by way of language (Toledo and Barrera-Nassols 2008) and therefore, it is transmitted along the spatio-temporal structure using a different logic, that is, the orality.

Participatory Research and its Agroecological Application

Methodological strategies derived from the above argument are defined regarding the incorporation of diverse actors, who together complement the knowledge in a way such that it serves to achieve the objectives established in the research. An imperious element in this type of research is the participation because it provides the different actors (peasants and researchers) with equal positions (the same status) when defining the most suitable strategy. In general terms, it could be said that participatory research is a methodological proposal, incorporated into a strategy in action defined, which involves its own beneficiaries in the production of knowledge. It is a combination of action, learning, education and research (López de Ceballos 1992).

The object of this type of research (Ardón 2001) is not the advance for science, but the people themselves know and analyze a reality in three constitutive moments: (1) the objective processes; (2) the perception of peasants (level of consciousness) of these processes (in particular human being), and (3) the experience of peasants assimilated into their definite structures. Meanwhile, the role of the researcher is to contribute to the formulation of techniques explaining the social reality from its historical perspective, which are translated into these theories in the particular processes of the groups with which they interact.

Participatory research can be characterized as basically qualitative research, in which quantitative elements can be included, but always within the context of a qualitative problem. In addition, it is the production of knowledge about the dialectical relationships, which manifest themselves in the social reality, that is, between the objective structures and the way in which they are perceived in the historical relationship with these structures.

Another fundamental characteristic in this type of research is the importance of working together with various institutions shaping the social fabric of a territory.

The professional researcher, as a fellow of the research, is embarked, together with the others, in the process of learning, action and research. This implies invalidating its neutrality in the approach (if it has ever existed), which does not mean that the scientific criteria of precision in the observation and objectivity should not be maintained, but rather be combined with the interests of the community. Participation is not only visualized in the design and execution of the research, but also in the use of the results for the subjects' actions (Guzmán and Alonso 2007).

In a systematic and selective way, it could be said that the main characteristics of participatory research are the following:

1. The community must be immediately and directly benefited. It cannot be justified only as a basis for intellectual exercises. It is important that the community or population take advantage, not only of the results of the investigation, but also of the process itself.
2. The research should involve the community/population throughout the whole process, from the problem formulation to the interpretation of the discoveries and the discussion of the solutions.
3. The process must be considered as part of the educational experience serving to determine the needs of the community.
4. The effective participation of professional researchers and promoters is defined based on their theoretical contribution (contributions that facilitate a logic of analysis) and practice (background information, training capacity and committed participation).
5. Research is a dialectical process, a dialogue through time and it is not the static image of a point in time.
6. Its purpose is the liberation of the creative potential and the mobilization of human resources for the solution of the problem.
7. This is a permanent research process: if the needs change, the actions generate new research.
8. Process is done from the inside and under the optics.
9. The study of the problematic, in its historical structural interrelations and in the context of the global society, is encouraged.

Guzmán and Alonso (2007) reported that in the case of work with farmers, the holistic understanding of reality by researcher is due to the overlap of the set of techniques with popular knowledge, and the systemic perception that the farmer has of his agroecosystem. This type of research has proven to be effective for promoting social changes in rural/countryside areas. The processes of agroecological transition through technological innovation/incorporation, the acquisition and deployment of skills of various actors, and the organization by the groups involved so that they can continue the process by themselves is an example of this.

One of the first stages carried out in a participatory research process is based on the needs and motivations expressed by the community; the research is designed and planned according to the endogenous potential, both in its ecological and social dimension. This potential constitutes the starting point of a process that necessarily implies establishing symmetrical relationships among the different actors involved in such process. Therefore, the knowledge forms of each actor are in a position of equality, which is crucial to generate the necessary conditions allowing this dialogue between knowledge.

Once these conditions are generated (through a series of meetings and workshops), the methodological strategy used in the research, whose purpose is to determine the potential of bio-cultural diversity (Peredo 2009), is carried out in two phases (Fig. 2.1) in which the first two practical techniques are simultaneously applied. The first one consists of a survey based on a semi-open questionnaire, whose objective is to collect the information related to biodiversity (e.g., natural, planned, cultivated, associated) present in the community, its use or property, the way in which the families obtained it, its availability throughout an annual cycle, the preparation form and how this knowledge is transmitted/exchanged/socialized into the community. In summary, it is to have an approach to local knowledge tied to the resources and the socialization agents.

Through this technique—typical of the social sciences—it is intended to establish a register of the species present in the community and their respective complementary information relevant to the objective of participatory process.

The second of the mentioned techniques consists of plotting/making transect, census, or collection (depending on the agroecosystems and the important species for the investigation) with the objective of verifying and elaborating in the field, an inventory of those species (biodiversity) distributed in the community. In this way, it is possible to determine the "local" and "foreign" species found in the agroecosystems.

INTERVIEWS

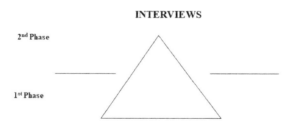

2nd Phase

1st Phase

SURVEYS   **TRANSECT**

Fig. 2.1. Technique articulation on the methodological strategy.

In the second stage, the interviews are performed with specific members of the community, chosen for the detailed and precise knowledge they have about the attributes of the species of interest for the community. The application of this technique (in-depth interviews) is recommended among older adults (men/women) as it allows the establishment of symmetrical relationships enriching the horizontal dialogue, thereby expanding the research scope on the "secrets" of local wisdom.

Using the aforementioned methodological strategy allows an approach to work with agricultural communities in order to diagnose its biocultural potential;

the manner in which these techniques are articulated enables the expansion of the known and unknown information by the different members of the community. The simultaneity in the application of the techniques, in addition, allows to connect and complement the quantitative antecedents with those of qualitative nature. On the other hand, the complementarity of the techniques (in quali- and quantitative terms) permits to extend the previously established attributes (characteristics, variables) of the study. The combination of the techniques, consequently, is indispensable insofar as the agroecosystem is an agrarian unit where both sociocultural and biophysical elements converge.

It is convenient to reinforce that in the proposed methodology the main actors are the community members, which acquire a decisive role for the achievement of the final objectives through the execution of concrete actions. The methodological design, therefore, requires that the strategies be established, adjusted, and modified as a result of this interaction and as an emerging product during the development of the research (Peredo and Barrera 2002a). Pursuant to Guber (2004), the research process is flexible, creative, and heterodox because it is joined to the constant and parallel relationship between observation and elaboration, obtaining information and data analysis. This process generates new concepts and explanatory connections based on initial assumptions, which are reformulated and enriched by the actors' categories and contextualized in the social life. For this, times and spaces are made more flexible by obeying the reason (rationality) of life in the countryside (Toledo 1993), where the different tasks and the time distribution are managed on the basis of the daily activity of each one.

Agroecological Research: The Potential of Endemic Species and the Importance of the Knowledge Socialization

In reference to the methodological application previously described, two concrete cases are presented. The first one referred to the action developed with the species *Dasyphyllum diacanthoides* as a result of the participatory research carried out with a group of peasants from the Andean area of Huechelepún (Araucanía Region, Chile). After having identified the potential of this community, it was agreed to investigate two scientific aspects unknown for the species: the nutritional value and the reproductive strategy, considering the potential of the species as forage and its inclusion in the design of resilient agroforestry systems.

The second case refers to a work carried out with the Andean community of Armerillo (Maule Region, Chile) in which, as a result of the methodological application, the knowledge level and socialization of the floristic resources of its environment was determined.

The Trevo: Endemic Species of "Unknown" Use as Forage

The trevo (*Dasyphyllum diacanthoides*) is an evergreen tree that can reach up to 1 m in diameter and 10 m in height. It resists extreme weather conditions such as the snow season in sub-mountainous areas inhabited by peasants and their livestock. A very interesting phenotypic characteristic of the species is the presence of a pair of thorns of 0.8–1.5 cm long in the young branches, which disappear at the height of 1.5 m. The old branches are glabrous and without thorns, located in the upper part of the tree (Abarzúa et al. 2007), which makes it difficult to access for the animals. Among all its properties, the best known is medicinal (Hoffmann et al. 2003); it is used as a febrifuge, diuretic, antiseptic, antitussive, purgative and astringent, to eliminate warts and heal wounds, against rheumatism and bruises, and liver attacks (Muñoz et al. 1981, Montenegro et al. 1994, Estomba et al. 2005, Zampini et al. 2007, Valencia 2013). In addition, trevo has the characteristic of being palatable for the animals used by peasants in the extreme mountainous areas, where the climatic conditions are adverse for the establishment of pastures.

In relation to the possibility of reproducing this species, after having developed a research based on local knowledge and the endogenous resources of the community (Peredo et al. 2015), the results indicated that it is possible to propagate *D. diacanthoides* by means of semi-lignified cuttings (Table 2.1) and the values, in agreement with the variables defined by the farmers, can be relevant for the establishment of massive plantations. The propagation prospect of *D. diacanthoides* coincides with research carried out on other native species in Chile, such as those from *Nothofagus* genus (Santelices and García 2003) as well as for endemic species such as *Berberidopis corallina* Hook.f (Latsague-Vidal et al. 2008), *Eucryphia glutinosa* Poepp & Endl. (Latsague-Vidal et al. 2009) and *Guindilia trinervis* Gillies ex Hook. et Arn. (Jordan et al. 2010). Therefore, the potential established by the peasants from *D. diacanthoides*, considering its native-endemic nature and economic importance for both medicinal and forage uses, is worth noting.

On the other hand, an important aspect related to the reliability of propagation of plants is the callus formation and rooting: the percentage values between the formation and rooting in each treatment suggested that the callus formation was different in the rooted stakes. This is relevant since the formation of calluses does not always lead to rooting in the stake. Non-formation of roots in all the calluses was described by Priestley and Swingle (1929), and by Latsague-Vidal et al.

Table 2.1. Survival, callus formation, rooting and radicular growth in *D. diacanthoides* on different treatments.

	SEwt	SEwht	NEwht	NEwt
Survival (%)*	8	44	92	88
Callus formation (%)*	8	56	96	84
Rooting (%)*	50	81,8	83	81,8
No of roots (x ± de)	2 ± 0	2,7 ± 0,4	2,6 ± 0,6	2,9 ± 0,7
Root length (cm) (x ± sd)	4,4 ± 0,8a	5,2 ± 4,7a	7,2 ± 4,7 b	5,3 ± 1,5a
Root length max (cm)	4,6	15,7	17,7	5,7

a, b: means with different letters—significantly different (p ≤ 0.05) U Mann-Whitney's test. * Logistic regression. SE: south exposition; NE: north exposition; wt: with thorns; wht: without thorns. X: mean; sd: standard deviation.

(2009). The high percentage of callus formation in *D. diacanthoides*, especially in treatments with stems obtained from mother plants with northern exposure, would indicate that the maintenance conditions of stalks are adequate for the rhizogenesis process (Santelices and García 2003, Latsague-Vidal et al. 2009).

De Vastey (1962) was a pioneer in demonstrating the importance of the branch location in the upper part of the mother tree from where the stakes will be obtained, as well as the position of the stakes in the same branch, which has a localized rooting due to the unequal distribution of phytohormones and nutritional reserves in the different parts of the tree. Leakey and Coutts (1989) attributed rooting only to changes and redistribution of carbohydrates in the cuttings; Santelices and Bobadilla (1997) identified that for *Quillaja saponaria,* the greatest rooting is achieved in stakes obtained from the lower parts of the plant.

This study showed that rooting is greater in spineless cuttings from mother plants with northern exposure, that is, those obtained from branches located more than 2 m high. Conforming to Abarzúa et al. (2007), the spineless branches of *D. diacanthoides* would be the oldest; however, since these tissues are ontogenetically younger, they are more easily differentiated (Taiz and Zeiger 1991, Rodríguez et al. 2005). This would explain the greater rooting of spineless stakes from mother plants with northern exposure. In the literature reviewed, reports were not found on the importance of environmental conditions where the mother plant is located and from which the stakes that will be propagated are obtained. The study also allowed to verify the greatest success of propagation, measured through survival and root growth, which occurred in stakes from mother plants located in the northern exposure. According to Taíz and Zeiger (1991), this condition would favor the photosynthetic metabolism of the plant and, therefore, it would have more reserves of carbohydrates as they are exposed to the sun for a longer time.

The low survival presented in the stakes with thorns from mother plants with southern exposure could indicate their lower suitability for vegetative propagation. The most probable cause would be their diminished physiological condition (Taíz and Zeiger 1991) or their genotypic condition different from mother plants exposed to the north. This is suggested by Santelices (2005), who attributed differential rhizogenesis, in stakes of *Nothofagus alessandri* Espinosa, from different mother plants.

Regarding the nutritional value indicated by the farmers, of the different edible parts (sprout, leaf and stem) of *D. diacanthoides*, the results of the proximal analysis performed on them are presented in Table 2.2. The values of crude protein (CP) and crude fiber (CF) for the stem are highlighted, since they are the lowest and the highest, respectively, possibly due to the high lignin content of the species. The lower PC, DM and P values presented by this part, together with the higher fiber values, would indicate that due to the nutritional contribution of the younger parts, it would be the most suitable to be used as fodder. However, since it is a potential feed based on the browsing, the field observations indicated that there was no selection by the livestock.

Table 2.2. Chemical composition of the parts of *D. diacanthoides* eaten by cattle.

	DM M	DM SD	CP M	CP SD	ME M	ME SD	EE M	EE SD	P M	P SD	TA M	TA SD	CF M	CF SD	ADF M	ADF SD	NDF M	NDF SD
Shoot	39.95 a	1.52	7.76 a	0.15	2.49 a	0.02	2.36 a	0.07	0.13 a	0.004	11.11 a	0.45	23.53 a	0.56	32.91 a	0.92	39.42 a	0.82
Leaf	39.49 a	1.52	8.72 b	0.25	2.68 b	0.03	2.78 a	0.25	0.15 b	0.004	13.34 a	0.68	19.15 b	0.68	32.38 a	0.82	36.68 a	0.18
Steam	46.53 b	1.08	5.03 c	0.10	1.94 c	0.05	2.28 a	0.27	0.10 c	0.01	5.33 b	0.22	35.97 c	1.06	46.38 c	0.25	57.15 c	0.21
Mean Value	41.78	1.06	7.23	0.36	2.38	0.07	2.46	0.12	0.13	0.01	10.04	0.78	25.95	1.62	35.09	1.64	42.15	2.32

DM = dry matter; CP = crude protein; ME = metabolizable energy; EE = ether extract; P = phosphorus; TA = total ash; CF = fiber crude; ADF = acid detergent fiber; NDF = neutral detergent fiber. a, b, c: significantly different (p ≤ 0.05). M = mean; SD = standard deviation.

Otherwise, when comparing the average values of the metabolizable energy and the percentage of digestible dry matter determined for *D. diacanthoides* with the studies carried out in other species of trees used as forage in different parts of the world, this species stands out as an important source of energy and its relative potential value as food (Ayala et al. 2006). It is a relevant fact considering that *D. diacanthoides* is used as a nutritional source in sub-mountainous areas where very-low temperatures (< 0°C) are recorded for more than half a year (March to September), and it can be determinant for the maintenance of the animal body temperature in such conditions (Bondi 1989) and in the compensation of the growth rate (Zea and Díaz 1990).

Based on the report by Anrique et al. (2014), the most important factor to optimize the absorption of nutrients in ruminant animals is the energy-protein ratio. Depending on this relationship and considering the main food sources for the central-southern region of Chile, it was observed that the CP value (7.2) for *D. diacanthoides* was lower than for forage species (in green) such as *M. sativa* (18.9), *T. pratense* (14.4) and *A. nuda* (12.9), but higher than *T. aestivum* (3.5) and similar than *Z. mays* (7.5). Despite the fact that some species mentioned above presented CP values higher than "trevo", a species widely known as forage (hay) showed an average value (7.3) closer and comparable to that of *D. diacanthoides*. Nevertheless, it should be clarified that hay has improved/increased its value due to the use of an appropriate management and better technology, whereas the value of "trevo" is intrinsic (not manipulated/modified through improvements). With respect to the energy values, *D. diacanthoides* (2.38) was only surpassed by *A. nuda* (3.34) and *Z. mays* (2.62); in turn, it was slightly higher than *M. sativa* (2.35) and even considered in the range of good (2.3–2.6) when it was compared with energetic contribution value of prairie silage.

The balanced energy-protein relationship of *D. diacanthoides*, according to the proximal analysis, positions it as a potential supplementary nutritional alternative, also allowing for its availability in winter as an energy source. If it is considered for its palatability (Abarzúa et al. 2007), endemic condition (hence, adapted to the ecosystem), uses and knowledge on the part of the peasants, and reproductive strategy (Peredo et al. 2015), *D. diacanthoides* would constitute a suitable endogenous resource for the establishment of agroforestry systems in conditions of the Andean sub-mountains of the central-southern zone of Chile.

The Socialization of Knowledge: Do We All Know What You Know? Do You Know What We All Know?

Regarding the second case that will be treated for the purpose of this chapter (Peredo and Barrera 2002b), it was highlighted that the community identified more than 70 species with medicinal attributes, which differed from what was stated by San Martín in 1983, who reported only 26 species for the same sector of San Clemente. Besides, the same author mentioned some species which were unknown to the people who were surveyed, such as quillay (*Quillaja saponaria*) and maitén

(*Maytenus boaria*), etc. Moreover, the results of the survey showed the presence and/or use of species that had not been reported for this area but for other places within the region studied by San Martín, such as linden (*Tilia* cf. *europeae* L.), medlar (*Eriobotrya japonica*), culén (*Psoralea glandulosa*), plantain (*Plantago major*), paico (*Chenopodium ambrosioides*), among others.

Other plants registered in the surveys, which San Martín (1983) mentioned for medicinal uses but the surveyed population did not tie-in, were concli (*Xanthium spinosum*) for stomach pain, the diuretic character of quinchamalí (*Quinchamalium majus*), and the aphrodisiac properties of the herb of nail (*Geum chiloense* Bald), inter alia. In a like manner, the surveys identified uses that were not reported by San Martín, such as the properties attributed to cachanlahue (*Centaurium cachanlahuen*), plantain (*Plantago major*) and avocado (*Persea americana*), to name a few.

Of the species used by the locality, only 50% are available all through the year. The remaining 50% are characterized by being seasonal products. Regarding this particularity and based on the background obtained by the surveys, the community established three categories: (a) those with permanent availability, (b) those with marked seasonality and sporadic availability, and (c) those of seasonality and permanent availability. From the first category (a), some herbs can be mentioned such as pennyroyal (*Mentha pulegium*), mint (*Mentha piperita*), plantain (*Plantago major*), and horizon (*Tetraglochin alatum*). With regard to those species defined as marked seasonality and sporadic availability [second category (b)], other plants can be named, e.g., cohile (*Lardizabala bitermata*), avocado (*Persea americana*), nalca (*Gunnera chilensis*), medlar (*Eriobotrya japonica*), and prickly pear (*Opuntia ficus-indica*), etc. Finally, for those species defined as seasonal and having a permanent availability (category c), some plants can be included such as walnut (*Junglans regia*), linden (*Tilia* cf. *europaea* L.), pile-pile (*Modiola caroliniana*), chamomile (*Matricaria chamomilla*) and blackberry (*Rubus ulmifolius*). The condition of permanent availability of this last group is due to how different peasant families have developed multiple strategies that have allowed them to prolong the availability of those resources that have a marked seasonality. One of these strategies has been to apply/use various conservation techniques such as drying or the preparation of preserves in the cases of leaves, and fruits, respectively. Other technique consists, for example, in the application of some forced cultivation practices in the backyard of the houses, which represents the prolongation of the forest ecosystem, also contributing to the conservation of biodiversity. In this way, the products are obtained out of season, which in no way implies the intensification of production, since most of them are obtained from the collection and natural reproduction.

On the subject of the socialization of traditional knowledge, Fig. 2.2 clearly shows that there was little exchange between families regarding the uses of non-traditional products, since none of the reported species was known by all of the families surveyed. In general, knowledge about the attributes and properties of these species is exchanged (socialized) among family groups in which there is some link or degree of friendship.

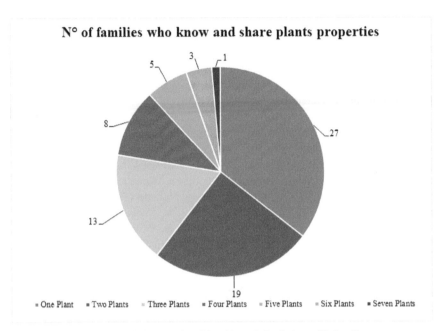

N° of families who know and share plants properties

■ One Plant   ■ Two Plants   ■ Three Plants   ■ Four Plants   ■ Five Plants   ■ Six Plants   ■ Seven Plants

Fig. 2.2. Socialization of traditional knowledge in Armerillo locality.

Grouping the species according to the number of families that knew its use, only one of them—rue (*Ruta graveolens*)—turned out to be the species whose attributes were known by the greatest number of families. On the other hand, a total of 26 species were only known individually by the families, but they were not necessarily the same species for each family. Some examples of these species were pile-pile (*Modiola caroliniana*), bailahuén (*Haplopappus baylahuen*), donkey tea (*Viviania marifolia*), culén (*Psoralea glandulosa*), perlilla (*Margyricarpus pinnatus*), etc. From the results, it was observed that of the all the families surveyed, there was none who knew the use or properties of all the species registered.

The mechanism of assimilation and transfer of this knowledge is focused on what is transmitted orally from generation to generation. This clearly shows that the main socialization agency is the family, where the person from childhood is able to know the forms of culture in which he/she was born.

Conclusions and Future Perspectives

The purpose of this chapter was to show empirical evidence that it is possible to obtain concrete quantitative information from an agroecosystem based on joint work between peasants and technicians. It is noteworthy that the reviewed works were the result of processes in which the participation and collaboration between peasants and technicians were performed at the different stages of research, beginning with the formulation of the problem question from which the research was designed, including the selection of the relevant indicators, up to the evaluation of the impact that such projects had on the quality of life and the sustainability of their territories, among others.

In this way, it was possible to observe, from the developed experiences, the relevance of the selection of indicators in a participatory mode (Peredo and Barrera 2016) because, fundamentally, these must exactly correspond (or fit) to concrete (or specific) needs in a precise historical context, that is, the indicators are the "reflection of the certain interests" of an evaluation in a "particular historical moment". This condition acquires greater relevance when the categories and ranges of valuation have also been established in a participatory manner (Peredo et al. 2016) due to the valuation exercises which achieve a sense of belonging to the community through processes of permanent iteration in which the elements of subjectivity are considered (Peredo and Barrera 2005).

As a final point, a special mention is made of the collective learning process that was generated during these processes amongst the different actors, promoting a real dialogue of knowledge. From the academic point of view, the multidisciplinary integration carried out by professionals and undergraduates of diverse curricular formations, who participated in research with rural communities, confirmed the importance of dialogue between different perspectives, the cost of the reductionist vision when dealing with complex phenomena (in terms of loss of diversity to address the phenomenon), the inseparable relationship between theory and practice, and the importance of a critical look at the modern perspective of development (Peredo and Aedo 2016, Aedo et al. 2017).

Acknowledgements

The authors thank Dr. Amner Muñoz (Universidad del Norte, Colombia) for his comments and reviews that improved the text, and to Vice Rectory of Research, Development and Innovation (Universidad de Santiago de Chile) to support a stay in the Agroecosystems History Lab (Universidad Pablo de Olavide, Sevilla-Spain).

References

Abarzúa, A., Donoso, P. and Donoso, C. 2007. *Dasyphylum diacanthoides* (Asteráceas). Trevo, tayo, tevo, palo santo, palo blanco, Familia Asteraceae (Compositae). *In*: Donoso, C. (ed.). Las Especies Arbóreas de los Bosques Templados de Chile y Argentina. Autoecología. Marisa Cueno Ediciones, Valdivia, Chile.

Aedo, M.P., Peredo, S. and Schaeffer, C. 2017. From an essential being to an actor's becoming: political ecology transformational learning experiences in adult education. Environm. Educ. Res.

Altieri, M. and Nichols, C. 2000. Agroecología: teoría y práctica para una agricultura sustentable. Serie Formación Ambiental Programa de la Naciones Unidas para el Medio Ambiente. PNUMA México.

Anrique, R., Molina, X., Alfaro, M. and Saldaña, R. 2014. Composición de alimentos para el ganado bovino. Cuarta Edición. Universidad Austral de Chile-Inia, Valdivia, Chile.

Ardón, M. 2001. Métodos e instrumentos para la etnoecología participativa. Etnoecología 6: 129–143.

Ayala, A., Capetillo, C., Cetina, R., Sandoval, C. and Zapata, C. 2006. Composición química-nutricional de árboles forrajeros. 1ªEd. Facultad de Medicina Veterinaria y Zootecnia, Universidad Autónoma de Yucatán, Mérida, México.

Bondi, A. 1989. Nutrición Animal. 1ªEd. Acribia S.A., Madrid, España.

De Vastey, J. 1962. Estudios sobre propagación de especies forestales por estacas. Tesis Magister Agronomía. IICA. Turrialba, Costa Rica.

Estomba, D., Ladio, A. and Lozada, M. 2006. Medicinal wild plant knowledge and gathering patterns in a Mapuche community from North-western Patagonia. J. Ethnopharmacol. 103: 109–119.

Gómez-Benito, C. 2001. Conocimiento local, diversidad biológica y desarrollo. *In*: Altieri M. y Labrador J. (eds.). Agroecología y Desarrollo. MundiPrensa, Madrid.

Guber, R. 2004. El salvaje metropolitano. Reconstrucción del conocimiento social en el trabajo de campo. Paidós, Bs. Aires.

Guzmán, G. and Alonso, A. 2007. La investigación participativa en agroecología: una herramienta para el desarrollo sustentable. Ecosistemas 16: 24–36.

Hoffmann, A., Farga, C., Lastra, J. and Vegahzi, E. 2003. Plantas medicinales de uso común en Chile. Editorial Fundación Claudio Gay, Santiago de Chile.

Jordan, M., Prehn, D., Gebahuer, M., Neumann, J., Parada, G., Veloso, J. and San Martín, R. 2010. Iniciación adventicia de raíces en estacas adultas y juveniles de *Guindilia trinervis*, una planta endémica de Chile, apta para producción de biodiesel. Bosque 31: 195–201.

Latsague-Vidal, M., Saéz-Delgado, P. and Hauenstein, E. 2008. Inducción de enraizamiento en estacas de *Berberidopsis corallina* con ácido indolbutírico. Bosque 29: 227–230.

Latsague-Vidal, M., Saéz-Delgado, P. and Yáñez-Delgado, J. 2009. Efecto del ácido indolbutírico en la capacidad rizogénica de estacas de *Eucryphia glutinosa*. Bosque 30: 102–05.

Leakey, R.B. and Coutts, M.P. 1989. The dynamics of rooting in *Triplochiton scleroxylon* cutting: their relation to leaf area, node position, dry weight accumulation leaf water potential and carbohydrate composition. Tree Physiol. 5: 135–146.

López de Ceballos, L. 1992. Un método para la investigación acción. Editorial Popular, Madrid.

Montenegro, S., Gómez, M., Iturriaga, L. and Timmermann, B. 1994. Potencialidad de la flora nativa chilena como fuente de productos naturales de uso medicinal. Rojasiana 2: 49–66.

Muñoz, M., Barrera, E. and Meza, I. 1981. El uso medicinal y alimenticio de plantas nativas y naturalizadas en Chile. Publicación Ocasional MNHN 33.

Norgaard, R. 1985. Las bases científicas de la Agroecología. *In*: Agroecología: bases científicas de la agricultura alternativa. Cetal Ediciones, Valparaíso.

Peredo, S. and Barrera, C. 2002a. Importancia de la socialización del conocimiento local para la conservación de la diversidad bio-cultural: el caso de Armerillo, (Chile). *In*: Tomo, I., Dapena, E. and Porcuna, J. (eds.). La agricultura y ganadería ecológicas en un marco de diversificación y desarrollo solidario. Serida-Seae, Asturias, España.

Peredo, S. and Barrera, C. 2002b. Desarrollo rural endógeno: condiciones para una transición agroecológica desde una experiencia de producción orgánica. Cuhso 6: 71–90.

Peredo, S. and Barrera, C. 2005. El impacto de programas de desarrollo en la calidad de vida de una comunidad rural en la Región de la Araucanía (Chile). Un análisis agroecológico. Rev. Antropol. Experim. (RAE) No. 5. Universidad de Jaén (España).

Peredo, S. 2009. Una propuesta metodológica para potenciar la biodiversidad y el conocimiento local en procesos de investigación participativa. Rev. Bras. Agroecología 4: 2340–2343.

Peredo, S. and Aedo, M.P. 2016. Complejidad y multidisciplinariedad en el aprendizaje de la sustentabilidad: la experiencia del Diplomado en Educación para el Desarrollo Sustentable de la Universidad de Santiago de Chile. Rev. Sustentabil. 7: 76–87.

Peredo, S. and Barrera, C. 2016. Definición participativa de indicadores para la evaluación de la sustentabilidad de predios campesinos del sector Boyeco, Región de la Araucanía. Idesia 34.

Peredo, S., Vela, M. and Jiménez, A. 2016. Determinación de los niveles de resiliencia/vulnerabilidad en iniciativas de Agroecología urbana en el suroeste andaluz. Idesia 34: 5–13.

Peredo, S.F., Parada, E., Álvarez, R. and Barrera, C. 2015. Propagación vegetativa por estacas de *Dasyphyllum diacanthoides* mediante recursos endógenos. Una aproximación agroecológica. Bol. Latinoam. Caribe Plant. Med. Aromat. 14: 301–307.

Priestley, J. and Swingle, F. 1929. Vegetative propagation from the standpoint of the plant anatomy. US Department of Agriculture. Technical Bull. 151.

Rodríguez, P., Fernández, M., Pacheco, J. and Cañal, M.J. 2005. Envejecimiento vegetal, una barrera a la propagación. Alternativas. *In*: Sánchez-Olate, M.E. and Ríos-Leal, D.G. (eds.). Biotecnología Vegetal en Especies Leñosas de Interés Forestal. Concepción, Chile.

San Martín, J. 1983. Medicinal plants in Central Chile. Econ. Bot. 37: 216–227.

Santelices, R. and Bobadilla, C. 1997. Arraigamiento de estacas de *Quillaja saponaria* Mol y *Peumus boldus* Mol. Bosque 18: 77–85.

Santelices, R. and García, C. 2003. Efecto del ácido indolbutírico y la ubicación de la estaca en el rebrote de tocón sobre la rizogénesis de *Nothofagus alessandri* Espinoza. Bosque 24: 53–61.

Santelices, R. 2005. Efecto del árbol madre sobre la rizogénesis de *Nothofagus alessandri*. Bosque 26: 133–136.

Taiz, L. and Zeiger, E. 1991. Plant physiology. The Benjamin/Cummings Publishing Company Inc., Los Angeles, California.

Toledo, V. 1993. La racionalidad ecológica de la producción campesina. *In*: Sevilla y González de Molina (eds.). Ecología, campesinado e historia. La Piqueta, Madrid.

Toledo, V. and Barrera-Bassols, N. 2008. La memoria biocultural. La importancia ecológica de las sabidurías tradicionales. Serie Perspectivas agroecológicas. Editorial Icaria, Barcelona, España.

Valencia, E. 2013. Validación y actualización del uso de plantas medicinales presentes en la selva valdiviana. Tesis Químico Farmacéutico, Universidad Austral de Chile.

Van der Ploeg, J.D. 1990. Sistemas de conocimiento, metáfora y campo de interacción: el caso del cultivo de la patata en el altiplano peruano. Agricultura y Sociedad 56: 143–166.

Zampini, I., Cudmani, N. and Isla, M.I. 2007. Actividad antimicrobiana de plantas medicinales argentinas sobre bacterias antibiótico-resistentes. Acta Bioquim. Clin. Latinoam. 41: 385393.

Zea, J. and Díaz, M. 1990. Producción de carne con pastos forrajes. Centro de Investigaciones Agrarias de Mabegondo. Xunta de Galicia, Santiago de Compostela, España.

CHAPTER 3

The Path of Ethnopharmacobotany

From Economic Botany to Ethnobotany

Marcelo L. Wagner, * *Leonardo M. Anconatani, Rafael A. Ricco,*
Beatriz G. Varela and *Gustavo C. Giberti*[†]

Introduction

A transformation of knowledge is happening in the scientific field. There is a fresh look where western science is not the only theoretical-practical model of assuming reality. Other ways of knowing and perceiving nature are being raised. We started studying how knowledge is passed on and has been passed on from generation to generation, millenium to millenium, town to town, person to person. Therefore, other kind of realities and other truths come together, prevail, are kept, are reproduced and give rise to new knowledge, new perceptions, and cosmovisions about the world, life, and nature.

This knowledge or "folk wisdoms" or "traditional knowledge" are the subject of study of almost all scientific disciplines. Natural science is the most developed area; fields like Ethnography, Ethnology, Antrophology, and Sociology have been focusing on the study of the peoples and their associated knowledge. It is thus that the prefix *ethnos* makes reference to human aspects and specific knowledge of the peoples and ethnic groups, i.e., to the knowledge of groups of individuals of the same culture (Arenas and Martínez 2013).

Cátedra de Framacobotánica, Museo de Farmacobotánica "Juan Aníbal Domínguez", Facultad de Farmacia y Bioquímica, Universidad de Buenos Aires, Junín 956, 1113 Ciudad Autónoma de Buenos Aires, Argentina.
[†] In memorial
* Corresponding author: mlwagner@ffyb.uba.ar

Nevertheless, we must bear in mind that human groups neither parcel nor sectorize their knowledge, but have a general worldview, i.e., they do not have a unidirectional view "the whole makes the part of the whole". Therefore, they have a holistic view of the world, they join past and future, lifeless with life, material with spiritual, they link norms, values, nature, traditions, health, sickness, and many other concepts with magic, legends, and myths that are an indissoluble part of their knowledge and their daily practices (Arenas and Martínez 2013, Rengifo-Salgado et al. 2017).

Let's consider as premise that "traditional knowledge" is the practical knowledge of ethnic groups or local communities which is based upon accumulated and selected experience throughout thousands of years to obtain the best results in the utilization of natural resources and their survival (Arenas and Martínez 2013).

As mentioned above, this knowledge must be grasped not only in a utilitarian view, for example, when Botany investigates these sources of knowledge it gives shape and sense to many of these concepts in Economic Botany, Medicine, Agriculture, and Ecology, but we should take into account the worldview of the community, who provided that knowledge. Thus, new fields of research are created, such as Ethnobotany, Ethnomedicine, Agroecology, and Ethnoecology (Reyes García and Martí Sanz 2007, Nolan and Turner 2011, Johnson and Davison-Hunt 2011).

At an early stage, information about medicinal flora, its usefulness, and ways of usage were collected without considering the cultural setting in which that knowledge was produced. Then, there was a progress towards a multidisciplinary study, where humankind and the plants that they use to heal and the different ways of usage were taken into account according to their worldview. We suggest a journey through the history of different fields to disentangle the path towards Ethnopharmacobotany and the importance of herbaria so that we can document this ride.

Ethnoscience

Ethnoscience (or cognitive anthropology) is relatively young and it is placed on the border of natural and social sciences. The term ethnoscience first appeared in 1950, in the book *Outline of Cultural Materials*, where the author, George Murdock, intended to make a thorough list of all the elements constituting human culture. He subdivided this field of science in Ethnometeorology, Ethnophysics, Ethnogeography, Ethnobotany, Ethnozoology, Ethnoanatomy, Ethnophysiology, Ethnopsychology, and Ethnosociology (Murdock 1950). In this first stage, western science is seen as superior with regard to cultures that were called "primitive" back then. In such manner, he calls "exact sciences" to western knowledge while he defines traditional knowledge as "non-systematized ideas about nature and humankind" or "speculative and popular notions" (Beaucage 2000).

Stephen Tyler (1969) considered the ethnosciences as the study of the native speech, not placing it within a western scientific order but to reinforce the ability that communities have to produce knowledge and observations about the environment where they live. In other words, ethnoscience is the cultural understanding that the community has of the world, its conceptual models and worldviews.

Most recently, Darrel Posey highlighted the importance of stories (tales, myths) as being responsible for the transmission of knowledge about the ecosystem where they live. Therefore, knowledge is an integration of wisdoms and practices (López Garcés and de Robert 2012).

However, the third and main approach is the utilitarian one, which is the most predominant. This approach presents the recovery and scientific reassessment of these systems of knowledge.

> "... The rescue and reassessment of traditional practices does not imply
> to "scientificize" traditional knowledge to incorporate it to new packets
> of knowledge, but to redirect the efforts of research, articulating the
> popular knowledge with science, in a process of inclusive investigation
> that recreates the knowledge of the own communities and gives back an
> enriched knowledge, assimilable, reappropriated by the same communities
> to strengthen their ability of self-management of their productive
> resources..."   (Leff and Carabias 1993)

But it only seeks the possibility of benefiting from what can be used to raise innovation and solutions to specific problems, and above all, the possibility of patenting discoveries from the knowledge and biodiversity supported by local communities.

Ethnobotany

Ethnobotany is the scientific study of the relationship between human groups and their natural environment, whose name comes from the combination of two fields of study: Ethnology (the study of culture) and Botany (the study of plants). Researchers focus on the topic of this field from two perspectives. The first one pursues a practical or utilitarian ideology and the second one is of philosophical character. The most quoted definitions of this field are stressed in the research of the relationship or interaction between humankind and the plant world (Jones 1941, Ford and Jones 1978, Schultes and Von Reis 1995), in the impact that plants have on human culture (Balick and Cox 1997), or a full record of the uses and concepts of the plant kingdom in societies (Berlin 2014, Plotkin and Famolare 1992).

The term "Ethnobotany" was used for the first time in 1895 by John William Harshberger, during his classes in the University of Pennsylvania (Harshberger 1895). However, the history of this academic field starts long before, since the interest in Ethnobotany goes from the beginning of civilization, when human beings perceived plants as a source of survival. Thus, the first men are practically considered ethnobotanists, since they classified plants in different categories and were capable of distinguishing those species that are beneficial from those which are harmful (Choudhary et al. 2008). Teofrasto (c. 370–285 BC), the father of Botany, describes the uses of economically important plants and sets the generic names (for example: *Crataegus*, *Daucus*, and *Asparagus*) which are still being used. On the other hand, Caius Plinius Secundus, better known as Plinio el Viejo, collected information about the growth of medicinal plants in his work "Historia Natural" (Bennet 2013). In the year 77, the Greek philosopher Dioscorides publishes the work "De Materia Medica", a record of around 600 plants of the Mediterranean watershed that includes information related to the use (specially medicinal), collection, toxicity, and edibility of such plants. Dioscorides also specifies the economic potential of plants, anticipating in this way, the creation of Economic Botany, interested in the use of plants which are useful to human beings and their economic value (Wickens 2001).

The information of the classical age was repeated for over 1500 years. It was only in the sixteenth century that European herbalists recorded new observations about the use of plants. In 1542, a Renaissance artist, Leonhart Fuchs, listed 400 native plants from Germany and Austria in his book "De Historia Stirpium", followed by "Historia Plantarum" by John Ray, where the first definition of "species" was published and subsequently, "Species Plantarum" by Carl Linnee, which includes related information of about 5900 plants.

The collection of data about the use of plants was not only a European interest. In the sixteenth century, Martín de la Cruz wrote the Aztec herbarium. His book, known as "Manuscrito Badianus", contains a description about the therapeutical and psychoactive properties of 251 Mexican plants. This manuscript symbolizes the first herbarium of medicinal plants written in the New World (Pease et al. 2000). Hipólito Ruiz López and José Antonio Pavón y Jiménez, Spanish botanists, collected plant species in the viceroyalty of Peru and published them in "Flora Peruviana et chilensis" (1798–1802). It is important to know that Chinese, Arab, and Indian texts are produced in parallel, generally less known in the western world, with valuable descriptions of the plant kingdom. However, the study of this historical material is the subject of analysis of Historical Botany and not that of Ethnobotany (Bennet 2013).

Linnee, whose latinized name represents the synonym of modern taxonomy, was famous for the invention of the binomial method of nomenclature, where the scientific name assigned to a species is made up by the combination of two words (genus, species) (Loonen 2008). Linnee was also a pioneer of modern Ethnobotany studies because he published "A tour in Lapland" where he made detailed observations about the use of plants by the Sami peoples in Laponia (Wickens 2001). Linnee's successors did not limit their investigations to taxonomy, for example, Alphonse de Candolle, wrote "Origen of cultivated plants" in 1885, a classical work related to the origin of cultivated plants.

The Summit of botanical exploration took place in the eighteenth and nineteenth century. It is worth mentioning the trips of Alexander von Humboldt and Aimé Bonpland, and James Cook. On the other hand, the English botanist Richard Spruce, one of the great Victorian botanist explorers, took 15 years to explore the Amazon (mainly all Brazil). His collections, indexed in the Royal Botanical

Fig. 3.1. Cover of the Spanish version of the book "Natural History" of Cayo Plinio (1624).

Gardens, represent an important ethnobotanical resource. In the same way, botanic samples from North and Central America were collected by the American botanist and archaelogist Edward Palmer, a field assistant for the Office of American Ethnology (McVaugh 1956). Palmer provides information related to the life and use of plants by the natives of North America. Likewise, Henry Throreau spreads the same content in his essay "Walden" (Thoreau 1906). The field of study called

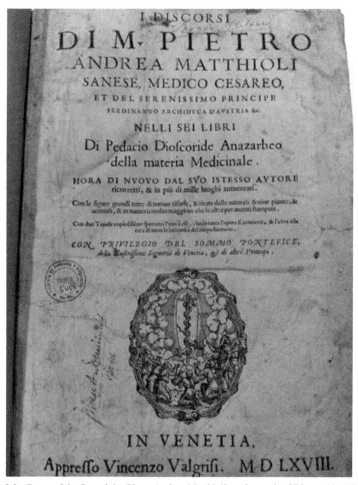

Fig. 3.2. Cover of the Speech by Pierre Andrea Matthioli on the work of Dioscoride (1568).

"native botany" (which is devoted to the research of those types of plants used by natives aimed, among others, to the obtention of food, medicine, and fabrics) (Fig. 3.4). The term "native botanical" was used for the first time in 1874 by Stephen Powers. An essential part of study in this field is called "folk" or popular taxonomy, referring to the method used to ease the recognition and designation of plants used by members of a certain linguistic community (Sánchez et al. 2007, Berlin 2014). Sometimes, native nomenclature tells a lot about the characteristics of the plant, vegetation, or its effects (if it is toxic or nutritious, or laxative, astringent, sedative, or without any active ingredient) (Powel 1877). A publication of Leopold Glueck, a German physician who worked in Sarajevo, which is about traditional medicine and the plants used by rural population in Bosnia from an emic point of view (created from phonemes), is considered to be the first modern ethnobotanical

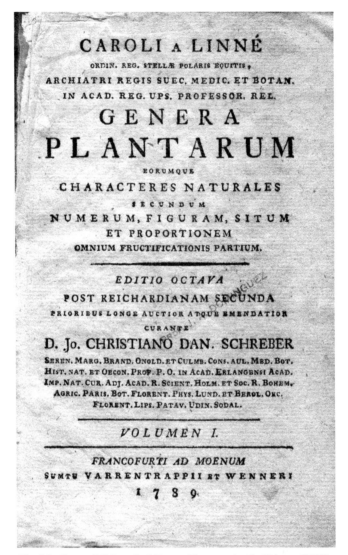

Fig. 3.3. Cover of Volume I of *Genera Plantarum* by Carl Linnee (1789).

work (Cunningham 2001). "Emic" and "ethic" are technical terms originally derived from the suffixes of the words 'phonemic' and 'phonetic' suggested by the linguist Kenneth Pike in 1967. The first term relates to any unit of sound in a particular language and the second, to the system of notations cross-culturally useful which those "vowel sounds" represent (McCutcheon 1999).

At the beginning of the twentieth century, the neologism by Harsberger "Ethnobotany" is approved, although it is only about a semantic replacement. The change of paradigm that gave rise to a more conceptual and methodological

Fig. 3.4. Aimé Bonpland (1773–1858).

approach evolved progressively. The beginning of modern Ethnobotany as an academic discipline is deeply linked to its founder, the biologist Richard Evans Schultes (Sequeira 2006). At first, the concept of Ethnobotany is more ecological, centered specifically on the relationship and interaction between humankind and plants. Such researchers begin to study plants as an integral part of the ecosystem

Fig. 3.5. *Ilex paraguariensis* St. Hill. (Aquifoliaceae) "yerba mate".
Yerba mate was utilized by the Guaraní native and in some Tupí communities in southern Brazil, prior to European colonization. Yerba mate is widely known as the source of the beverage called mate in both Spanish and Portuguese, also called by various other names in Portuguese like chimarrão, and tereré. It is traditionally consumed in central and southern regions of South America, primarily in Paraguay, as well as in Argentina, Uruguay, southern and central-western Brazil, the Chaco region of Bolivia and southern Chile.
Yerba mate can also be found in various energy drinks in the market today.

in which they are. Later, the Ethnobotany becomes a cultural topic and currently, scientists are trying to understand it from this perspective. Lastly, the scope of this field redefines "humankind" and creates a new concept. At present, the term "traditional" is used since it is less derogatory than the one used before, that is, "primitive" (Ford and Jones 1978, Cotton 1996).

An important point is to know if there is a key difference between the traditional and modern use and handling of plants. This distinction can be artificial, since etymologically, there is no reason to restrict Ethnobotany from traditional societies. The prefix "ethno" refers to any people or cultural group, not only traditional societies (Rodríguez-Echeverry 2010, Bennet 2013).

The current frame of Ethnobotany emphasizes the different skills needed by the scientists of this field: a necessary botanical background to identify and preserve plant samples, an anthropological training that helps the researcher understand cultural concepts, and a linguistic background that allows the researcher to be able to transcribe terms and understand the morphology, syntax, and native semantics (Choudhary et al. 2008). The investigation of the practical functions of plants dominates the research agenda. The Ethnobotany, as discipline, is oriented to the exploration of new plant resources, the collection of genetic material, the discovery of drugs or medicines resulting from plants, and the development of new natural products (Plotkin et al. 1992, Schultes and Von Reis 1995, Balick and Cox 1997).

Ethnopharmacology

The term Ethnopharmacology was used for the first time by Efron et al. (1967). It was implemented in the title of the book *Ethnopharmacological Search for Psychoactive Drugs,* which deals with the ethnopharmacological search of psychoactive drugs (Efron et al. 1970, Holmstedt 1967).

Ethnopharmacology is, by definition, a scientific approach to the study of the biological activities of any preparation used by human beings traditionally, which have in a broad sense, beneficial, toxic or other pharmacological direct effects. As such, we do not attempt to describe uses (generally local or traditional) but some characteristics about the large antrophological and pharmaco-toxicological study of these preparations.

Furthermore, studies that describe the medicinal use of plants are generally included in this definition, but these studies are performed with the aim of leading to an experimental investigation of medicinal plants, in other words, validating the use that one human group gives to plants (Heinrich et al. 2009).

The study of traditional drugs by ethnopharmacologists does not have the purpose to advocate for a return to the use of these remedies as used by communities, nor to promote traditional medicine. The objective is, thus, to rescue and document a cultural heritage before it gets lost to be able to investigate and evaluate it.

Therefore, Ethnopharmacology involves a multidisciplinary work in which botanists, phytochemists, and pharmacologists are found, and where anthropologists have an active participation. The first investigation with this multidisciplinary profile was performed by the French naturalist Leschenault de la Tour in 1803 by investigating the poison of arrows in the island of Java. The botanical part was performed by Jussie who identified the *Strychnos*; later, Magendie and Raffenau de Lille studied the effects of the poison on animals and determined the target organ in which it acts (spinal marrow). A decade later, Pelletier and Caventou isolated the responsible alcaloid, strychnine (Holmstedt and Bruhn 1983).

The final objective of Ethnopharmacology is the validation (or non-validation) of remedies used by traditional communities for the treatment of health. This process is performed by isolating active substances and characterizing the pharmacological action of those substances. These achievements allow isolating new targets for the development of new medicines.

Hence, Ethnopharmacology is not only a science of the past that uses an outdated approach, but it also constitutes the backbone of the development of active therapeutics based on traditional medicine of different ethnic groups. On the other hand, Ethnopharmacology contributes to a better and safer use of such resource (Heinrich 2014).

Herbaria in the Study of Ethnosciences

An interesting topic to include in this chapter is that of herbaria. Currently, in the age of genomics, proteomics, and metabolomics, it may seem odd that herborization is still made and herbaria are still kept as in previous centuries. However, herborization is an important practice not only for botanical works but it is key to Ethnopharmacobotany. This is due to the fact that the samples deposited in herbaria certify the identity of a plant and they act as a document for researchers (Bebber et al. 2010, Fertig 2016).

Even though the history of plant sciences begins in Ancient Greece with Teofrasto (3rd century B.C.), the term herbarium is relatively recent. It was employed by Joseph Pitton of Tournefort and Carl von Linneo in the seventeenth and eighteenth centuries to design a collection of dead plants (*hortus mortus*), dissected and flattened (*hortus siccus*), more available in any season of the year than in an outdoor botanical garden (*hortus hiemalis*), for study purposes (Giberti 1996).

The practice of saving dissected plants started in the times of Luca Ghini (1490–1556), Italian physician and botanist, founder of the Botanical Gardens of Florence and Pisa, professor of Botany in Pisa and Pharmaceutical Botany in Bologna (Staufleu and Cowan 1976–1988, Cristofolini et al. 1993). Apparently, no samples of Ghini have survived, but there are some of his disciples, such as Ulises Aldrovani (1522–1605). Aldrovani also taught Botany in Bologna, and his herbarium, the greatest of the ones preserved from the sixteenth century, is bound in 17 volumes in foil and it features almost 5,000 plants. The method of binding the sheets of herbarium was abandoned due to its impracticality, but it is often seen in some classical collections, such as the herbarium collected in the Middle East by Leonhard Rauwolf (1535–1596), in Leiden, and the one by Andrea Cesalpino

(1519–1603), that contains almost 800 samples and it is preserved in Florence (Staufleu and Cowan 1976–1988, Cristofolini et al. 1993, Giberti 1996).

It is considered that Aldrovani's collection has been precisely the origin of the first institutional herbarium that is known: the one of the *Museum rerum naturalium* of Bologna. A second modern meaning of herbarium is thus noticed, since not only it is applied to the collections but also to the institutions that contain them (for example: botanical gardens, agricultural or forest research facilities, germplasm banks, museum offices, universities, scientific societies, among others) (Giberti 1998).

A third meaning of the word herbarium refers to ancient botanical texts, pioneers of the modern Flora and Herbal Pharmacopoeia, some preceding the press, which Anglo-Saxon authors know as herbals. Examples of herbals, herbaria or Kraüterbüchen are: the Herbarium by Pseudo-Apuleius (fourth century); the New Kreüterbuch. Basel: Michael Isingrin (1543) by Leonhard Fuchs; the A New Herball (1551) by William Turner; the Cruydeboeck (1554) by Rembert Dodoens, and the Materia Médica Misionera (circa 1700) attributed to the Jesuit brother Montenegro of the Jesuit Missions in South America (Giberti 1998). These and many other similar works, while they deal with strictly botanical issues with more or less rigor, they often share a great interest in the ethnohistory of medical and pharmaceutical sciences, since most of ancient botanists have been herbalists and physicians, even up to the beginning of the nineteenth century, for example, Aimé Bonpland (1773–1858) (Giberti 2008, Bell 2010).

On the other hand, the accumulation of 500 years of plant collections by botanists and travellers is a rich source and yet, there is not enough explored data about useful plants (useful in a broad sense) nor of the communities that use them (Nesbitt 2014). This is because the collectionists used to herborize those plants that had certain utility. Therefore, it is probable that more than half of the samples

Fig. 3.6. Herbarium of Pharmacobotany Museum "Juan Aníbal Domínguez", Buenos Aires, Argentina.

housed in great herbaria are plants of useful or symbolic value to human beings. An interesting work would be an ethnopharmaceutical investigation of the botanical collections of herbaria placed in museums or academic institutions (Bebber et al. 2010, Nesbitt 2014).

During the twentieth century, herbaria were driven away from their roots since they went from collections of useful plants to be the material used by taxonomists for alpha taxonomy, detection, description, and classification of taxa. Researchers lost their interest in human interaction with plants (Anderson 1952). Perhaps, this was not by chance. Towards the end of the twentieth century, a crisis in the botanical collections was recorded which gave rise to a decrease of funding and a lack of confidence in the value of the collections of the museums (Clifford et al. 1990).

At present, opinion has changed since there is, as mentioned before, an important source of information which revalues the importance of herbaria (Funk 2003, Suarez and Tsutsui 2004).

It is important to keep in mind that herbarium samples fulfil three vital functions: they allow the identification of the specimen botanically, they allow identifications to be controlled by other researchers, and they allow updating the work in the light of new taxonomical concepts. Why are these functions so important?

Most of the plants that are found in the ethnobiological field of work, certainly, already have an identification in the form of a local vernacular name. As far as preservation of traditional knowledge and record of ethnotaxa goes, vernacular names are essential. However, the identification of a species by botanical name (latin) is also essential since it allows to have a comprehensive knowledge of the plant, for example, the uses that it might have in other cultures, phytochemical, pharmacological, ecological studies, and plant growing (Bennett and Balick 2013). Botanical names are also, in theory, less bound to ambiguity than vernacular names; they can often be written in different ways when they are registered and the same name can be used for more than one species (Bennet and Balick 2013). On the other hand, the identification of a plant requires that the material contains flowers or fruits to use the standard botanical tools such as the flora or the comparison with samples of reference herbaria, that are often missing in the ethnopharmacological specimens, such as the plant drug formed by some of the vegetative organs of the plant or exudates (Eisenman et al. 2012). In many cases, the identification is made in the field, for which handbooks or the botanical expertise of the researcher are used, but there is a risk of confusing the well-known species with new species that have a similar aspect. Field identifications are uncertain and it is necessary to confirm them with morphological studies in the laboratory, and many times it is necessary to resort to chemical studies or to DNA molecular biology to confirm them (Eisenman et al. 2012).

In genetics, herbarium samples have allowed the reinterpretation of results of hybridization experiments decades after they had occurred, in light of new taxonomical arrangements (Sauer 1953, Barkworth and Jacobs 2001). Likewise, the new taxonomical work of Schmidt-Lebuhn (2008) about the aromatic gender

Minthostachys allowed the reinterpretation of many inconsistent records of plant chemistry, their use, and vernacular naming.

When a sample is herborized, apart from the plant material, a label is incorporated with standard data, such as the date and place or georeference where the collection was carried out. Data, such as the ethnobotanical, is included which enables the use in future investigations (Giberti 1998).

Historical bonds between herbaria and Economic Botany (Ethnobotany, Ethnopharmacology, and Ethnopharmacobotany) that had weakened in the twentieth century, have been reinforced at present. Herbarium samples are currently widely acknowledged as an essential document to validate the ethnopharmacobotanical investigations. It is important to note that collection techniques of herbarium samples continue being simple and easy to learn as they were 400 years ago. On the other hand, herbarium samples are also subject of new questions and research techniques: they act as records of use through their labels and associated data, and as biological samples that can be measured and sampled.

It is interesting that while herbaria are being incorporated to databases, their usefulness increases. This will allow to easily find herbarium samples, independently of the changes in the botanical name or the herbarium where the sample is placed and it will enrich ethnopharmacological studies.

Conclusion

Traditional medicine, just like folk medicine, is a system of practices and own remedies proper of native societies, even if in certain cases and for geographical reasons it is possible to isolate their components. Traditional medicine as well as folk medicine contain an indeterminate number of followers and mostly non-explicitly written rules. In general, the historical information known comes from secondary testimonies.

The limits among the three medical systems (scientific, traditional, and folk) are not very clear, though. Clarity is blurred inasmuch as their study is deepened, contacts and conceptual confluences appearing as a consequence. Adoption and use of different remedies and medical practices encourage to reflect on the similarities and not only on differences and contrasts.

It is necessary to keep in mind that the three great medical groups–scientific medicine, folk medicine, and traditional medicine—share a common characteristic: rational thinking. This is not a privilege of western societies but it is also observed in different contexts. Lévi-Strauss expressed that conceptual proliferation is not an exclusive heritage of the "civilized" but that the "primitive" thinking also introduces abstract and complex elements, led by an interest in classifying and organizing the environment. In other words, there is a systematically developed knowledge among the people we call "primitive", even without a practical utility, i.e., it is not related to vital needs, such as food or medicine, but with previous intellectual requirements. The will to introduce an order, a taxonomy, is a common point with modern science and it is found in the basis of all human thought (Di Liscia 2003).

Lastly, the Ethnosciences allow a more detailed understanding about the underlying action modes and they provide a better basis for the rescue and evaluation of knowledge. Therefore, the Ethnosciences contribute as a link between the different paradigms and create a harmonious relationship among diverse cultures.

It is then, once reviewed the common origins of these Ethnosciences, we can define Ethnopharmacobotany as the branch of Ethnobotany whose subject of study is the prevailing relationship between human beings and the plants they use to heal themselves, their different ways of use, as well as the habits and traditions related to them under their worldview.

References

Anderson, E. 1952. Plants, Man and Life. Little, Brown and Company, Boston.
Arenas, P. and Martínez, G.J. 2013. Estudio etnobotánico en Regiones Áridas y Semiáridas de Argentina y zonas limítrofes. Experiencias y reflexiones metodológicas de un grupo de investigación. pp. 11–43. *In*: Arenas, P. (ed.). Etnobotánica en Zonas Áridas y Semiáridas del Conosur de Sudamérica., CONICET, Buenos Aires.
Balick, M.J. and Cox, P.A. 1997. Plants, People, and Culture: The Science of Ethnobotany. New York, EEUU: Scientific American Library.
Beaucage, P. 2000. La etnociencia, su desarrollo y sus problemas actuales. Cronos 3(1): 47–92.
Bebber, D.P., Carine, M.A., Wood, J.R.I., Wortley, A.H., Harris, D.J., Prance, G.T., Davidse, G., Paige, J., Pennington, T.D., Robson, N.K.B. and Scotland, R.W. 2010. Herbaria are a major frontier for species discovery. PNAS 107(51): 22169–22171.
Bell, S. 2010. A Life in Shadow. Aimé Bonpland in Southern South America, 1817–1858. Stanford University Press.
Bennet, B.C. 2013. Economic botany—Ethnobotany and economic botany: Subjects in search of definitions. In: Encyclopedia of Life Support Systems. (EOLSS). Paris, Francia: EOLSS Publishers.
Bennett, B.C. and Balick, M.J. 2013. Does the name really matter? The importance of botanical nomenclature and plant taxonomy in biomedical research. J. Ethnopharmacol. 152(3): 387–392.
Berlin, B. 2014. Ethnobiological classification: principles of categorization of plants and animals in traditional societies. Princeton, New Jersey: Princeton University Press.
Clifford, H.T., Rogers, R.W. and Dettmann, M.E. 1990. Where now for taxonomy? Nature 346: 602.
Cotton, C.M. 1996. Ethnobotany: Principles and applications. Chichester, West Sussex, UK: Wiley.
Cristofolini, G., Mossetti, U. and Bonfiglioli, C. 1993. Pre-Linnean herbaria in Bologna: some newly discovered collectors from time of Ulisses Aldavani. Webbia 48: 555–565.
Cunningham, A.B. 2001. Applied ethnobotany: People, wild plant use and conservation. London, UK: Earthscan Publishing Limited.
de Candolle, A. 1885. Origin of cultivated plants. Nueva York, EEUU: D. Appleton.
Choudhary, K., Singh, M. and Pillai, U. 2008. Ethnobotanical survey of Rajasthan—An update. AEJB 1(2): 38–45.
EFSA. 2011. Scientific opinion on pyrrolizidine alkaloids in food and feed. EFSA Journal 9(11): 1–134.
Eichele, K. 2010. Phytotherapy—An introduction. The Journal of the European Medical Writers Association 19(1): 67.
Eisenman, S.W., Tucker, A.O. and Struwe, L. 2012. Voucher specimens are essential for documenting source material used in medicinal plant investigations. Journal of Medicinally Active Plants 1: 8.
Fertig, W. 2016. Are Herbaria Still relevant in the 21st Century? https://biokic.asu.edu/blog/are-herbaria-still-relevant-21st-century (September 2017).
Ford, R.I. and Jones, V.H. 1978. The nature and status of ethnobotany. Ann Arbor, MI: Museum of Anthropology, University of Michigan.
Fresquet Febrer, J.L. 1994. Guía para la realización de trabajos de folkmedicina y otros sistemas médicos. Cuadernos Valencianos de Historia de la Medicina y de la Ciencia. Seria A (monografías).

Valencia, España: Univer sitat de València - C. S. I. C., Instituto de estudios documentales e históricos sobre la ciencia.

Funk, V. 2003. 100 uses for a herbarium (well at least 72). American Society of Plant Taxonomists Newsletter 17: 17–19.

Giberti, G.C. 1998. Herborización y herbarios como referencia en estudios técnicos-científicos. Herbarios de Argentina. Dominguezia 14(1): 19–39.

Giberti, G.C. 2008. El archivo Bonpland en el Museo de Farmacobotánica "Juan Aníbal Domínguez". Dominguezia 24(1): 5–9.

Harshberger, J.W. 1895. The plants cultivated by aboriginal people and used in primitive commerce. The Evening Telegraph (daily), 64(134): 2.

Johnson, N.M. and Davison-Hunt, I. 2011. Ethnoecology and landscapes. pp. 267–284. *In*: Anderson, E.N., Pearsall, D.M., Hunn, E.S. and Turner, N.J. (eds.). Ethnobiology. John Willey & Sons. Publ., Hoboken, New Jersey.

Jones, V.H. 1941. The nature and scope of ethnobotany. Chronica Botanica 6: 219–221.

Kadiri, A.B., Adekunle, A.A. and Ayodele, A.E. 2010. An appraisal of the contributions of herbalism to primary health care delivery in South West Nigeria. Ethnobotanical Leaflets 14: 435–444.

Leff, E. and Carabia, J. 1993. Cultura y Manejo Sustentable de los Recursos Naturales, 2 volúmenes, CIICH-UNAM/Miguel Ángel Porrúa, México.

Loonen, M.J.J.E. 2008. Linnaeus as biologist. The importance and limitations of Linnaean systematics in biology. TijdSchrift voor Skandinavistiek 29(1&2): 145–152.

López Garcés, C.L. and de Robert, P. 2012. El legado de Darrell Posey: de las investigaciones etnobiológicas entre los Kayapó a la protección de los conocimientos indígenas. Boletim do Museu Paraense Emílio Goeldi. Ciências Humanas 7(2): 565–580.

McCutcheon, R.T. 1999. The insider/outsider problem in the study of religion: A reader. London, UK: Cassell.

McVaugh, R. 1956. Edward Palmer: Plant explorer of the American West. Norman, OK: University of Oklahoma Press.

Mirtha Parada, V. 2012. Legislación en Chile sobre fitofármacos y plantas medicinales. Rev. Farmacol. Chile 5(2): 7–11.

Murdock, G. 1950. Outline of Cultural Materials. New Haven, Human Relations Area Files.

Nesbit, M. 2014. Use of Herbarium specimens in ethnobotany. Chapter 22 *In*: Jan Salick, Katie Konchar, and Mark Nesbit (eds.). Curating Biocultural Collections. Kew: Royal Botanic Gardens, Kew.

Nolan, J.M. and Turner, N.J. 2011. Ethnobotany: The study of people-plant relationships. pp. 133–147. *In*: Anderson, E.N., Pearsall, D.M., Hunnand, E.S. and Turner, N.J. (eds.). Ethnobiology. New York Botanical Garden Bronx, New York.

Pardal, R. 1937. Medicina aborigen americana. Buenos Aires, Argentina: Facsímiles Renacimiento.

Pease, F., Rabiela, T.R., Pons, F.M. and Damas, G.C. 2000. Historia general de América Latina: El primer contacto y la formación de nuevas sociedades. París, Francia: UNESCO.

Plotkin, M. and Famolare, L. 1992. Sustainable harvest and marketing of rain forest products. Washington D. C., EEUU: Island Press.

Powel, J.W. 1877. Aboriginal botany. In: Contributions to North American Ethnology. Washington D. C., EEUU: Govt. Print. Off. 3: 419–431.

Ray, J. 1686. Historia plantarum: Species hactenus editas aliasque insuper multas noviter inventas & descriptas complectens: in qua agitur primo de plantis in genere, earumque partibus, accidentibus & differentiis; deinde genera omnia tum summa tum subalterna ad species. Londres, UK: Faithorne.

Rengifo-Salgado, E., Rios-Torres, S., Fachín Malaverri, L. and Vargas Arana, G. 2017. Saberes ancestrales sobre el uso de flora y fauna en la comunidad indígena Tikuna de Cushillo Cocha, zona fronteriza Perú-Colombia-Brazil. Rev. Peru Biol. 24(1): 67–68.

Reyes García, V. and Martí Sanz, N. 2007. Etnoecología: punto de encuentro entre naturaleza y cultura. Ecosistemas 16: 45–54.

Rodríguez-Echeverry, J.J. 2010. Uso y manejo tradicional de plantas medicinales y mágicas en el Valle de Sibundoy, Alto Putumayo, y su relación con procesos locales de construcción ambiental. Rev. Acad. Colomb. Cienc. 34(132): 309–326.

Sequeira, L. 2006. Richard Evans Schultes: January 12, 1915–April 10, 2001 (A Biographical memoir). Washington, D. C., USA: National Academy Press.

Shirwaikar, A., Verma, R. and Lobo, R. 2009. Phytotherapy—Safety aspects. Nat. Prod. Rad. 8(1): 55–63.

Schultes, R.E. and Von Reis, S. 1995. Ethnobotany: Evolution of a discipline. New York, USA: Dioscorides Press.

Stafleu, F.A. and Cowan, R.S. 1976–1988. Taxonomic Literature (2º edition) Vol. 1-7. Regnum Vegetabile 94, 98, 105, 110, 112, 115, 116. Bohn, Scheltema & Holkema, Utrecht.

Suarez, A.V. and Tsutsui, N.D. 2004. The value of museum collections for research and society. BioScience 54: 66–74.

Thoreau, H.D. 1906. Walden. La vida en los bosques. http://consumoetico.webs.uvigo.es/textos/walden.pdf.

Thoreau, H.D. 1906. The writings of Henry David Thoreau (B. Torrey, Ed.). Boston, New York: Houghton, Mifflin and Company.

Tyler, S. 1969. Cognitive Anthropology. New York, Holt, Rinehart, and Winston.

Wickens, G.E. 2001. Economic Botany. Principles and Practices. Kluwer Academic Publishers.

CHAPTER 4

Patagonian Berries

An Ethnobotanical Approach to Exploration of their Nutraceutical Potential

Melina F. Chamorro,[1] *Ana Ladio*[1,]* and *Soledad Molares*[2]

Introduction

Inhabitants of rural communities in Argentine-Chilean Patagonia are mainly of Mapuche-Tehuelche origin, while others have Creole ancestry and/or are descendants of different waves of settlers who arrived in the región more than a century ago (Ladio and Molares 2014). The vast Patagonian territory is characterised by its great environmental diversity, including mountainous regions, steppe, forests and ecotones, which have provided inhabitants with subsistence resources since ancient times (Vignati 1941, Nacuzzi and Perez de Micou 1984, Prattes 2009, Llano and Barberena 2013). Even today, despite the marked increase experienced in urbanisation and markets, inhabitants of rural Patagonia continue to make use of the different ecological environments, and have therefore accumulated extensive knowledge of their plant surroundings (Ladio and Lozada 2001, 2008, Richeri et al. 2013). The close relationship they have established with plants is particularly evident when considering the species used by local families to treat illness and provide food (Ladio 2001, Ladio and Lozada 2004, Ladio 2005, Molares and Ladio 2009b, Ochoa and Ladio 2015).

Native fruits have been much used by local populations as part of their diet, according to archaeological and recent records (Martínez-Crovetto 1982, Dillehay

[1] INIBIOMA.Quintral 1250-S.C. de Bariloche, Río Negro, Argentina.
[2] CIEMEP, CONICET-Universidad Nacional de la Patagonia San Juan Bosco. Roca 780, Esquel (9200), Chubut, Argentina.
* Corresponding author: ahladio@gmail.com

1988, Rapoport and Ladio 1999). In this work, we are particularly interested in the berries which are native to this región. Pochettino (2015) proposed that berries correspond principally to the fleshy fruits of trees and shrubs of temperate areas, generally red to violet in colour, and are also called "soft fruits". In their natural state they normally contain more than 50% water, which they contribute to the people's diet, and are low in calorific value. The sugars they contain are easily digested and absorbed by the body, and they also enrich the diet with fibre, minerals and vitamins (Hurrell et al. 2010). Berries can be classified generally into simple fruits, with pips (pome fruit), stones (drupes and polydrupes) and aggregate fruits (thickened receptacles containing partial fruits).

Most of the best known species throughout the world were domesticated a long time ago, and have become valued resources in the international and the local market for Argentina and Chile: strawberry (*Fragaria* x *ananassa*), raspberry (*Rubus idaeus*), highbush blueberry (*Vaccinium corymbosum*) and cherry (*Prunus avium),* among others (Hurrell et al. 2010). In Northwestern Patagonia, there is a sizeable market for food products based on exotic berries, which provides work for the local population and also migrant workers who come from other localities to work on the cultivation and harvesting of the soft fruit (Ladio et al. 2013). However, there are also numerous wild or semi domesticated "berries" which have been little studied, as in the case of the native Patagonian species.

In addition to their attraction as edible fruits, berries have become recognised on a global level as conferring health benefits, which has led to research on their bioactive compounds (Szajdek and Borowska 2008). Epidemiological evidence demonstrates that certain plant compounds are related to effects that protect against the development of cancer, cardiovascular disease, diabetes, osteoporosis and neurodegenerative illnesses (Pandey and Rizvi 2009). New concepts arise from the perspective that considers there to be a strong link between diet and health, such as functional foods and nutraceuticals. Functional foods are those which, in addition to their nutritional properties, also contribute to the prevention of disease. The name "nutraceutical" is generally given to a product that takes on a pharmaceutical form (e.g., capsules, pills, granules), whose production is based on a food substance, or part of one (Mahabir 2014). In this text, we use "nutraceutical potential" to refer to any foodstuff that may provide the dual benefit of nutrition and health.

On studying the reasons behind many of the beneficial effects of diets rich in fruit and vegetables, it was found that their antioxidant properties were mainly responsible for providing these benefits (Szajdek and Borowska 2008). An antioxidant substance has a high capacity for absorption of free radicals, even at a low concentration. These are radical oxygen and nitrogen species, highly reactive molecules which can damage different cell components (Shahidi and Naczk 2004). In our diet we can find natural antioxidants made up mainly of phenolic and polyphenolic compounds, such as anthocyanins, phenolic acids, flavonoids, stilbenes and tannins, all from plants (Shahidi and Naczk 2004).

In general, uptil now, the most studied berry species have been from the northern hemisphere, due to their high levels of natural antioxidant, specifically phenolic compounds (Szajdek and Borowska 2008). However, interest is increasing in exploration of the potential of South American fruits (Schreckinger et al. 2010). New inroads are therefore being made into this subject in Argentine-Chilean Patagonia, with very promising results (Arena and Curvetto 2008, Simirgotis and Schmeda-Hirschmann 2010, Schmeda-Hirschmann and Simirgiotis 2011, Damascos 2011, Arena et al. 2014, Jiménez Aspee et al. 2015, 2016).

An interesting first step in the development of this topic is the articulation of ethnobotanical studies that describe the history and use patterns of plants in the different cultures that inhabit Patagonia, and make effective use of the native berries. These case studies reveal the cultural, symbolic and material value of the berries, and the use practices that sustain their gathering, supply, processing and consumption (Ladio 2000, Ladio and Lozada 2009, Ochoa and Ladio 2015).

The INIBIOMA Ethnobiology group has been studying the use of wild and cultivated plants in the Patagonian región for many years, conducting field work in different enclaves there, and also drawing up bibliographical databases of sources relating to different social and biological disciplines. The ethnobotanical approach seeks to integrate sociocultural, biophysical, and phytochemical criteria, and also bioactivity, which together explain the selection of species and enable us to reflect on the plants' cultural importance.

According to a large number of authors (Davidson-Hunt and Berkes 2003, Toledo and Barreras Bassols 2008, Capparelli et al. 2011), the study of traditional management practices of useful plants reveals the local solutions found for their conservation and sustainable use, and so their inclusion takes on significance for modern science. In this chapter, therefore, based on an exhaustive bibliographic review and our own results, we offer a preliminary overview of the 10 native species of Patagonian berries which hold the most cultural significance, identifying local practices and their history. In addition, we present phytochemical or pharmacological information, which give us evidence of the nutraceutical potentiality of these fruits.

Archaeobotanical, Ethnohistorical and Ethnobotanical Bibliographic Review

Our bibliographic analysis was based on the interpretation of different kinds of available documents (Medeiros 2009). We worked on information referring to publications from 1940 up to the present, which includes the first ethnographical/ ethnobotanical studies on native fruits in the region (Molares and Ladio 2009b). Existing scientific publications were considered, as well as a compilation of documents and records found in the library and databases of the Ethnobiology Group. Literature was also examined through search engines such as Scopus, Science Direct and Scielo, using the following key words in both Spanish and English: medicinal plants of Patagonia, Patagonian flora and its uses, ethnobotany in Patagonia, medicinal and alimentary uses, Patagonia, native fruits, berries, and Patagonia, Mapuche. Publications prior to the year 1940, and those which are ethnohistorical, i.e., studies revealing posthispanic dietary information (from the

16th century up to 1940), were also reviewed. Existing archaeobotanical sources were also analysed (prehispanic). This research was performed following the same procedures as mentioned above, and enriched with the historical texts and documents of travellers, explorers and colonists. The guidelines and sources of Ochoa (2015) were followed in carrying out the survey. In this work, the fragmentary nature of available information on the region was highlighted, but what was also emphasized was the importance of its systemisation and interpretation in an integral, coordinated fashion. We also include botanical information to briefly describe species and their fruits.

Criteria in the Selection of Literature

Analysis of the information began with construction of a database. The following criteria were followed in the selection of articles and documents: (1) the study should have been conducted in an Argentine-Chilean Patagonian community or cite species found in Argentine Patagonia, (2) species lists should ideally be obtained from ethnobotanical/ethnographic field work, that is, preferably not from a compliation or review of literature from miscellaneous sources, (3) work carried out by chroniclers, naturalists and/or expedition members was included, but those which used doubtful or difficult taxonomy to apply were excluded, (4) scientific archaeobotanical work on the use of Patagonian berry species was included, (5) summaries and congress, symposium and conference proceedings, were not included, (6) ethnopharmacological studies related to antioxidants and other relevant biological properties were included, if found, for the cited species. A base of 710 entries was constructed, which included 63 studies published between 1917 and 2016. In order to choose the principal 10 species, the use consensus index was used (CU), which considers the number of publications that cited each species, in relation to the total number of sources [(no. of sources that cited species i *100)/total no. of sources] (Molares and Ladio 2009a). This index, according to various authors, gives a direct indication of the cultural importance of the species (Molares and Ladio 2009a, Silva et al. 2014). We therefore consider that the current corpus of existing literature on the subject of Patagonian berries, together with our fieldwork, indirectly reflect the relative importance of these species. In addition, the ethnographic information examined and the interest shown by ethnopharmacological research in certain species, highlight the resources which display the most potential. Nevertheless, it should be pointed out that published sources remain very scarce for certain species and phytogeographical environments; consequently, our conclusions are tentative in nature.

Included in this analysis were the species that can be categorised as Patagonian berries: fleshy fruits, round or oval in shape, either simple or aggregate. These include berries, drupes, polydrupes, polyachenes, and even false fruits such as the bracts of the genus *Ephedra* which are referred to as edible fruits by inhabitants of Patagonia. We did not include references to the *Fabaceae* family, although their pulses are sometimes fleshy or any other type of fruit which does not fit into the definition of "berry" in the wide sense. The botanical nomenclature and geographical distribution of the species were corroborated using *Flora del cono sur* (Zuloaga and Morrone 2009), and only species occurring in Argentine-Chilean Patagonia were included.

Species with records of only edible and/or edible and medicinal uses in their sources of origin were selected, excluding the records which mentioned only medicinal use. This criterion was based on the likelihood that the overlapping of alimentary and medicinal uses would reveal those species with high nutraceutical potential (Ladio 2011). We recorded a species as edible when the source referred to the use as food, drink with or without alcohol, sweets, jams, or the consumption of fresh or dried fruit (Ladio 2005). The use was recorded as medicinal when the literature indicated that it was considered capable of curing an illness, relieving pain, or treating or counteracting a symptom of any kind (Estomba et al. 2006, Molares and Ladio 2009a). Unfortunately, in the records the medicinal property is often associated with a species, and details as to which part of the plant is used do not always figure.

Berries Most Cited in the Literature of Patagonia

Table 4.1 and Fig. 4.1 show the 10 most frequently cited species in the literature examined, in order of use consensus. They are all sub-Antarctic forest species, with the exception of *B. microphylla,* whose distribution extends as far as the Patagonian steppe and monte of Argentina. These species belong mainly to the *Myrtaceae* and *Berberidaceae* families, which are highly representative of the patagonian flora (Correa 1969–1999).

Berberis microphylla (Calafate, Michay, Box-leaved Barberry, Magellan Barberry)

Berberis microphylla is a spiny evergreen shrub which can reach 1.5 m in height and is native to the southwest of South America (Argentina and Chile) (Fig. 4.1.1). Its berries are subglobose, 7–11 mm in diameter and blackish in colour, as described by Damascos (2011). The fruit matures in January and can still be found till March. Berries have a tangy, sweet flavour and when eaten will dye the mouth violet, they weigh 0.3 g on average, and the seeds account for between 17 and 36% of the total weight (Rapoport et al. 2005).

According to various sources (Ragonese and Martínez-Crovetto 1947, Muñoz et al. 1981, Mösbach 1992), the Mapuche people consumed these berries fresh or used them to make jams and syrups by mixing the fruit with sugar. In parallel with this, there are records of their consumption by the Alacaluf, Selk-nam and Yaganes indigenous people (Martínez-Crovetto 1982). Also mentioned in the literature is ingestion of the fruit to quench thirst or in the preparation of alcoholic drinks (Ragonese and Martínez-Crovetto 1947, Martínez-Crovetto 1980). One of these drinks is "chicha", prepared with the fermented berries and commonly used in rituals or festivities. According to Martínez-Crovetto (1982), the Tehuelche people

Table 4.1. The ten most cited Patagonian berries in the bibliography which has both alimentary and medicinal uses. Key to gathering environments: f: forest; sc: scrubland; pb: peat bog; s: steppe; CU: consensus of use (N = 63).

Species	Common Names	Family	Distribution	Environment	Life Form	Fruit	CU (%)
Berberis microphylla G. Forst.	calafate, michay, box-leaved, barberry, magellan barberry	Berberidaceae	ARG (Chubut, Neuquén, Río Negro, Santa Cruz, Tierra del Fuego)/CHL (VI, VII, VIII, IX, X, XI, XII)	f, sc	shrub	berry	62
Aristotelia chilensis (Mol.) Stuntz	maqui, clon, queldrón, chilean wineberry	Elaeocarpaceae	ARG (Catamarca, Chubut, La Pampa, La Rioja, Mendoza, Neuquén, Río Negro, San Juan, San Luis)/CHL (IV, V, VI, VII, VIII, IX, X, XI, metropolitan, Juan Fernández archipelago)	f	shrub or tree	berry	43
Potentilla chiloensis (L.) Mabb.	frutilla silvestre, llahuén, wild strawberry	Rosaceae	ARG (Neuquén, Río Negro)/CHL (VI, VII, VIII, IX, X, XI, IJF)	f	perennial plant	conocarp	33
Ribes magellanicum Poir.	parrilla, zarzaparrilla, mulul, magellanic currant	Grossulariaceae	Variety magellanicum ARG (Río Negro, Santa Cruz, Tierra Del Fuego)/CHL (VI, VII, VIII, IX, X, XI, XII, metropolitan)	f, sc	shrub	berry	33
Berberis darwinii Hook.	michay, calafate, darwin's barberry	Berberidaceae	ARG (Chubut, Neuquén, Río Negro, Tierra del Fuego)/CHL (VII, VIII, IX, X, XI)	f, sc	shrub	berry	29
Gaultheria mucronata (L.f.) Hook. & Arn.	chaura, prickly heath	Ericaceae	ARG (Chubut, Neuquén, Río Negro, Santa Cruz, Tierra del Fuego)/CHL (VIII, IX, X, XI, XII)	f	shrub	berry	25
Empetrum rubrum Vahl ex Willd.	mutilla, red crowberry	Empetraceae	ARG (Chubut, Mendoza, Neuquén, Río Negro, Santa Cruz, Tierra del Fuego)/CHL (V, VII, VIII, IX, X, XI, XII, metropolitan, Juan Fernández archipelago)	f, sc, pb, s	shrub	drupe	24
Amomyrtus luma (Molina) Legrand et Kausel	luma, palo madroño	Myrtaceae	ARG (Chubut, Neuquén, Río Negro)/CHL (VII, VIII, IX, X, XI)	f	shrub or tree	berry	22
Ugni molinae Turcz.	murta, ugni, chilean guava, strawberry myrtle	Myrtaceae	ARG (Chubut, Neuquén, Río Negro)/CHL (V, VI, VII, VIII, IX, X, XI, XII)	f	shrub	berry	22
Luma apiculata (DC.) Burret	arrayán, palo colorado, chilean myrtle	Myrtaceae	ARG (Chubut, Neuquén, Río Negro)/CHL (IV, V, VI, VII, VIII, IX, X, XI, metropolitan)	f	shrub or tree	berry	19

Fig. 4.1. General appearance of the 10 most frequently cited berries' species in order of use consensus. *1. Berberis microphylla, 2. Aristotelia chilensis, 3. Potentilla chiloensis, 4. Ribes magellanicum var. magellanicum, 5. Berberis darwinii, 6. Gaultheria mucronata, 7. Empetrum rubrum, 8. Amomyrtus luma, 9. Ugni molinae, 10. Luma apiculata.*

Photographic Credits
1. Melina F. Chamorro; 2. Melina F. Chamorro; 3. Melina F. Chamorro; 4. Ana H. Ladio; 5. Melina F. Chamorro; 6. Ana H. Ladio; 7. Ana H. Ladio; 8. Melina F. Chamorro; 9. Melina F. Chamorro; 10. Ana H. Ladio

also prepared a drink with the fruit of this species, but in contrast to chicha, it is not fermented. The use of *B. microphylla* for medicinal purposes has been described in the Mapuche medical system as being refreshing for feverishness and useful for treating indigestion (Gusinde 1917, Mösbach 1992). Mösbach (1992) also refers to use of the fruit for treating diarrhea.

In modern times, gathering practices still persist in the Mapuche communities of Neuquén, Río Negro and Chubut (Ladio and Rapoport 1999, Ladio 2001, Ladio and Lozada 2004, Ladio 2006, Lozada et al. 2006). The fruit is collected despite the great distances locals have to cover to find the plants, as in the case of the Rams Mapuche community, from Paraje de la Media Luna (Neuquén province). The gathering method these communities use is of particular interest, as it helps them avoid being pricked by the plant's spines. In their walks into the countryside, the women and children carry a recipient and a stick or branch with which to hit the bush so that the mature fruit falls to the ground. Although the small children prefer to eat the fruit raw, in general it is used to make jellies, jams, and juices (Ladio 2001, Rapoport et al. 2005). For medicinal purposes, the fruit of *B. microphylla* is used as a febrifuge, for example, in the Curruhuinca Mapuche community in San Martin de los Andes (Conticello 1997).

It was also found that they possess a higher content of anthocyanins and a higher antioxidant capacity than 10 other Chilean berry species (Ruiz et al. 2013). When comparing calafate berries with five berries (including *Ugni molinae* and *Luma apiculata*), they not only had higher levels of phenolic compounds, but also had the highest antioxidant capacity (Ramirez et al. 2015). In the same study, calafate fruits, were also highlighted in one trial in which the inhibition of peroxidation of human erythrocytes was evaluated (Ramirez et al. 2015).

Aristotelia chilensis (Maqui, Chilean Wineberry)

Aristotelia chilensis is a medium dioecious tree that can reach a height of up to 4 m (Fig. 4.1.2). The leaves are petiolate, with a toothed edge and positioned in the shape of a cross with respect to the other leaves on the stem (Rapoport et al. 2005). The fruit is a shiny black berry 4–6 mm in diameter, which has 4–8 seeds inside. The flavour is tangy-sweet and the berry dyes the mouth and hands a deep violet colour. The tree flowers in November and December and produces fruit in summer.

Bibliographic records describe its direct consumption when fresh, or in the form of a drink similar to chicha, called "tecu" by the Mapuche people (Ragonese and Martínez-Crovetto 1947, Muñoz et al. 1981, Mösbach 1992). According to Houghton and Manby (1985), the Mapuche have used the leaves and fruit in infusions to cure diverse ailments of the throat. In addition, it has been used ground and in paste form as a febrifuge, and the leaves have been used in powder form to treat wounds (Meza and Villagrán 1991).

More recent studies carried out in Mapuche communities in Neuquén province have reported its use as a snack, and also in preparations such as jams (Ladio 2006). *A. chilensis* was and is still recognised for its gastrointestinal properties as a treatment for diahrrea (Molares and Ladio 2009b), its fruit is ingested directly (Muñoz et al. 1981) or used to make infusions, for both adults and children (San Martin 1983). This fruit has become popular in the world market as a superfruit, mainly due to its antioxidant properties. Chile, the principal exporter of the fruit, saw a 168% increase in its exports in 2015 compared to the previous year. The Chilean government's Agricultural Research and Policies Office (ODEPA) reported that the main destinations are Japan, South Korea, Italy and the United States, amongst others. From maqui, powders, extracts, capsules, nectar and antioxidant drinks are produced commercially, among other nutraceutical products. Similarly, mainly in Chile but also in Argentina, the number of family businesses dedicated to the production of jams, liqueurs, and even maqui coffee are increasing. These products are all prepared with fruit that comes directly from the gathering practices of wild populations, and therefore sustainable forms of management and cultivation are being sought.

The antioxidant capacity of *A. chiloensis* is principally attributed to its high content of anthocyanins, and these have been studied in Chile for some time now (Escribano-Bailón et al. 2006). In order to identify genotypes that will prove suitable for cultivation, in recent years studies have focused on finding the fruit with the best anthocyanin profile (Fredes et al. 2014). On the other hand, many pharmacological properties are being investigated, among which its antidibetogenic and analgesic effects seems to be outstanding (Romanucci et al. 2016).

Potentilla chiloensis (Frutilla silvestre, Wild strawberry)

Potentilla chiloensis is a perennial plant with silk-like stolons (Fig. 4.1.3). The leaves are dark green to reddish, and from a distance can be confused with the fruit. The flowers are white with yellow stamens. The fruit is compound and formed by a fleshy receptacle which is oval in shape and 2 cm in length, on which the achenes are positioned. These highly perfumed wild strawberries are an attractive red colour (although there is also a white variety) with a pleasant, sweet flavour, all of which make them very appetising. Due to their morphological and organoleptic qualities, they were taken to Europe to be hybridised in order to obtain the variety now sold around the world, *Fragaria* x *annanasa* (Rapoport et al. 2005). Their fruiting season is summer.

Several authors (e.g., Martínez-Crovetto 1980, Muñoz et al. 1981) have recorded the use of *P. chiloensis* fruit in jams, drinks and syrups in rural Creole, Chiloen and Mapuche populations. The strawberries were also dried, like raisins, to be conserved for winter, or ground for production of chicha for the Mapuche festivals (Ragonese and Martínez-Crovetto 1947). According to Martínez-Crovetto (1982), in the past the strawberries were collected and eaten raw by most of the Tehuelche, Gununakene, Alacaluf, Selk-nam, Yaganes and Mapuche groups. Records describe their medicinal use in the Argentine and Chilean Mapuche groups, mentioning all parts of the plant. It was used in cases of indigestión, haemorrhage or diahrrea, due to its emollient and astringent properties, and was prepared as herbal tea (Mösbach 1992). A decoction of the leaves and roots was also prepared and was drunk by women following a birth, or to clean the genital tract (Martínez-Crovetto 1980).

This fruit continues to form a part of the diet of Patagonian inhabitants, whether through direct ingestion of the fresh fruit or prepared as jams, liqueurs or included in desserts (Ladio and Rapoport 1999, Ladio 2001, Ladio and Lozada 2004, Lozada et al. 2006). With regard to its medicinal use, our ethnobotanical work revealed its persistant use in obstetrics, administered in the same ways as mentioned above (Lozada et al. 2006, Molares and Ladio 2009b, 2014).

The wild strawberry was studied in terms of its phenolic compounds and antioxidant activity of the fruit, leaf and rhizomes (Simirgotis and Schmeda-Hirschmann 2010, Schmeda-Hirschmannn and Simirgiotis 2011), as well as the different parts of the fruit (Cheel et al. 2007). In addition, Avila et al. (2017) have verified the protective activity of extracts of *Potentilla chiloensis f chiloensis* in gastric cells exposed to free radicals and proposed the prevalent mechanism of cytoprotection.

Ribes magellanicum (Parrilla, Magellanic Currant)

Ribes magellanicum is a deciduous bush of up to 2 m in height, and is very common in undergrowth and secondary shrublands (Damascos et al. 2011). The fruit is a blackish globose berry, 5–8 mm in diameter, which grows in racemes (Fig. 4.1.4). The plants bear fruit in December and January. This fruit was used in the past in various ways by the Mapuche people, whether ingested directly, in the form of jams and also fermented drinks (Ragonese and Martínez Crovetto 1947, Martínez-Crovetto 1980). According to the literature, these people also used a decoction of the root to heal the blood (Martínez-Crovetto 1980).

Many of these uses persist nowadays (Ladio et al. 2007). In the rural population of Arroyo Las Minas (Río Negro) and in the Curruhuinca Mapuche community in San Martin de los Andes, this species is still used to treat circulatory and blood complaints (Estomba et al. 2006, Ochoa et al. 2010). Tacón et al. (2006) reported that the fruit perishes quite rapidly, so it is eaten immediately by the gatherers themselves or used in traditional fermented drinks.

This species was studied in Argentina by Arena and Coronel (2011) and they have determined the moment of highest anthocyanin content during the fruiting period. *R. magellanicum* also presented a strong antioxidant profile in comparison with other fruits (Ruiz et al. 2013). The University of Talca, Chemistry of Natural Products Laboratory and the Ethnobiology group of the INIBIOMA Ecotono Laboratory, have recently published a joint study which compares the antioxidant profile of *Ribes* fruit on both sides of the Cordillera. First results show that the Argentine fruit contains a higher content of anthocyanins and has a more complex chemical profile (Jiménez Aspee et al. 2015).

Berberis darwinii (Michay, Darwin's Barberry)

This is a shrub that can reach 2.5 m in height, with typical coriaceous, dark green leaves which are rhomboidal in shape, with spines on the edge (Fig. 4.1.5). The flowers, borne on attractive racemes, are yellow-orange in colour. The fruit of this species is a bluish globose berry, 6 mm in diameter, on average, with 4 to 6 seeds. Flowering occurs from October to February and the fruiting period is summer, between December and January.

Martínez-Crovetto (1982) and Mösbach (1992) reported that *B. darwinii* fruit was consumed by communities of the Tehuelche, Gununakene, Mapuche and Selk-nam peoples. Its use is well documented in the preparation of drinks, including chicha, in the Mapuche communities that lived in Argentine-Chilean Patagonia (Mösbach 1992). Gusinde (1917) recorded the refreshing and febrifuge qualities of these berries, and the same properties have been cited for the leaves (Houghton and Manby 1985).

Several more recent studies (Lozada et al. 2006, Eyssartier 2011a,b) reveal that the fruit is consumed in different rural populations of Argentina. This use is also referred to on the island of Chiloé by Contreras Vega (2007), who also mentions the febrifuge properties of its fruit, leaves and roots. *B. darwinii* has been studied for its alkaloid content (Srivastava et al. 2015). This species has a similar antioxidant profile to *B. microphylla* flavonols, a fact that provides some suggestions for future research.

Gaultheria mucronata (Chaura, Prickly Heath)

Gaultheria mucronata is a small dioecious shrub with an average height of 1.5 m, and small coriaceous leaves with a thickened edge and 4 or 5 scallops ending in a mucro (Correa 1969–1999). The berries, 6–8 mm in diameter, are globose and their colour varies from white, through pink, to purple (Fig. 4.1.6). The plants flower between December and March, fruiting from February to July (Rapoport et al. 2005).

The literature shows that this fruit has been consumed in the past, either fresh or prepared as chicha by different societies: Tehuelche, Gununakene, Selk-nam, Yaganes and Mapuche (Ragonese and Martínez-Crovetto 1947, Martínez-Crovetto 1982). Although there are not many sites mentioning its medicinal properties, Gusinde (1917) documented its use for treating sores and ulcers.

The small *G. mucronata* berries are still consumed today by the Curruhinca Mapuche community of San Martin de los Andes, Neuquén province (Conticello et al. 1997), and knowledge of this has been registered among teachers of different cities in Chubut province (Ladio et al. 2007). Tacón et al. (2006) reported that the fruit loses its flavour and texture rapidly, so it is eaten immediately by the gatherers themselves, or in traditional fermented drinks.

On examining anthocyanin content, qualitative and quantitative differences were found between these and other, better known berries. In particular, *G. mucronata* berries have a low concentration of anthocyanins, and these are found in a higher proportion on the skin of the fruit (Ruiz et al. 2013).

Empetrum rubrum (Murtilla, Red Crowberry)

Empetrum rubrum is a dwarf bush, mainly dioecious, and has small, ericoid leaves with a longitudinal groove (Fig. 4.1.7). The berries are globose drupes, 4–7 mm in diameter, of different shades of red, including wine colours (Correa 1969–1999). Of the three species defined by the genus, *E. rubrum* is exclusive to South America. It can be found in Argentine-Chilean Patagonia in the high-Andean forests (Correa

1969–1999). The plants flower at the end of the winter and bear fruit in spring and early summer.

According to various sources (Ragonese and Martínez-Crovetto 1947, Martínez-Crovetto 1982, Mösbach 1992), the fruit has an acid flavour and has been eaten raw in the past by Mapuche, Alacalauf, Selk-nam and Yaganes communities. The berries were considered refreshing and good for the stomach (Mösbach 1992).

Data obtained from current studies, for example in the Paineo Mapuche community and the rural Creole community of Cuyín Manzano in Neuquén province, have shown persistence in its use as a food resource, but not in its medicinal use (Ladio and Lozada 2004, Lozada et al. 2006).

In this review, phytochemical studies on *E. rubrum* with regard to antioxidant substances have not been found.

Amomyrtus luma (Luma)

Amomyrtus luma is a shrub, or sometimes a tree with light brown bark (Fig. 4.1.8). The leaves are simple, coriaceous and very shiny and aromatic. The fruit is a globose berry 0.8 cm in diameter. It is violet-black in colour with few seeds, aromatic, and it maintains the sepals of the flowers (Rapoport et al. 2005). The plants bear fruit from the end of November till the beginning of April.

According to several authors (Muñoz et al. 1981, Mösbach 1992), in the past the Mapuche communities prepared fermented drinks, such as chicha, with the fruit. In some cases, the process consisted in grinding the berries in water and mixing them with previously fermented fruit (Martínez-Crovetto 1982). The literature cites the plant as being medicinal, with stimulant and astringent properties (Muñoz et al. 1981). It is still used today (although rarely) as a food resource in the region of Bariloche, either eaten as fresh fruit or in non-alcoholic beverages (Ladio and Rappoport 1999).

So far, there are no records of analyses having been carried out on this species in relation to its antioxidant properties. Nevertheless, fruits of *Amomyrtus meli* were studied, and six anthocyanins were identified (Ramirez et al. 2005). On the other hand, extracts made from dry material from *A. luma*, have showed platelet antiaggregant activity (Falkenberg et al. 2012). Emerging evidence shows the increasing interest to study the nutraceutical potential of this plant.

Ugni molinae (Murta, Chilean Guava, Strawberry Myrtle)

Ugni molinae is a shrub that can reach 2 m in height, with pubescent branches and leaves which are dark green on the upper side and lighter green on the underside, with petioles (Fig. 4.1.9). The fruit is a dark red, globose berry, up to 1.5 cm in diameter. It is characterised by maintaining the flower calyx as five scales on the apical part of the berry. The berries contain a large number of seeds which go unnoticed when eaten. The average weight of each fruit is 0.27 g, but they can reach 0.5 g, so that a large bush can produce 1 kg of these berries (Rapoport et al. 2003). The fruit is produced mid summer, and is sweet and aromatic.

According to Gusinde (1917), the Mapuche made wine with the fruit, as well as chicha and jams, or the berries were eaten fresh or dried (Muñoz et al. 1981, Mösbach 1992). Tonic, stimulant and astringent properties were attributed to the beverages prepared (Muñoz et al. 1981, Mösbach 1992).

At present, *U. molinae* stands out as one of the non-woody forest product, mostly commercialised as a food resource in the IX region of Chile (3278 kg/year in the Valdivian region) (Tacón et al. 2006). From a medicinal point of view, although still mentioned in current bibliography (Pardo and Pizarro 2005), its use has not been confirmed. In craft fairs and/or street markets in the south of Chile, jams, liqueurs and murta in syrup can be obtained. In Argentina, its commercialisation as a food resource has not been recorded.

The anthocyanin and antioxidant profile of this species has been studied, and it was found to have a simple pattern with only six anthocyanins. Nevertheless, this species has a strong activity in inhibition of lipid peroxidation in human erythrocytes and scavenging of superoxide anion tests (Ramírez et al. 2016).

Luma apiculata (Arrayán, Chilean Myrtle)

This tree is easily identified due to its brown-orange coloured bark, and the species is characteristic of Patagonia. Its leaves are dark green, shiny and aromatic, and the flowers are white with numerous stamens of the same colour (Fig. 4.1.10). The fruit is a violet, almost black, berry, and is round, approximately 1.5 cm in diameter and has three seeds. It tends to form groups, and even forests.

Several sources (Ragonese and Martínez-Crovetto 1947, Martínez-Crovetto 1982, Pardo and Pizarro 2005) have reported that this fruit was ingested fresh or in the form of chicha by the Mapuches of Chile and Argentina. The Mapuche communities used its stalks, leaves and bark for medicinal purposes, as an antibiotic, stimulant, astringent, and in the treatment of sores and herpes (Houghton and Manby 1985, Muñoz et al. 1981).

The persistence of its use at present has been registered for both alimentary and medicinal purposes in the city and surrounding rural zones of San Carlos de Bariloche (Funes 1999, Ladio and Rapoport 1999). In the case of medicinal use, in both Chile and Argentina the bark and leaves are used to treat several ailments, but mainly gastrointestinal complaints (Conticello 1997).

It is also known that the fruit of *L. apiculta* has very good antioxidant properties, together with high polyphenol content (Simirgiotis et al. 2013, Ramírez et al. 2016). Recent studies have also shown that mature arrayán fruit extracts could exert a protective effect on the vascular endothelium after glucose overload, that is, could prevent the development of some cardiovascular diseases (Fuentes et al. 2016).

Conclusions and Future Perspectives

The ten species described here constitute part of the cultural heritage of the region. They have been used as food and medicine from ancient times up to the present, forming part of both the material and symbolic lives of the different Patagonian inhabitants (Rapoport et al. 2005, Ladio and Lozada 2000, 2001). Their importance is reflected in the oral history of both countries, through legends (e.g., the legend of the calafate), the regional toponymy, and in the everyday lives of the people. Their biocultural value in the zone can be attributed to diverse factors that are multifactorial–sociocultural and biological-environmental, each one enriching the others, which are briefly detailed below:

These species have a broad geographical distribution, mainly in the Argentine-Chilean Andean-Patagonian forests and in areas of the Andean pre-cordillera (Zuloaga and Morrone 2009). Various studies have shown that the more widespread species have tended to be the most used as a food or medicinal resource, since different cultures are more likely to have experience of them and transmit this knowledge to other communities living in areas where the same species grow (Lozada et al. 2006, Molares and Ladio 2014).

Therefore, they are apparent resources for local people. The life forms of the 10 species are mostly perennial shrubs, followed by low bushes and herbs, which are relatively abundant and visible, can be monitored throughout the whole year, and the fruit is easily accessible, growing at an easy height for gathering by hand. Various authors have highlighted the importance of the availability, abundance and accessibility of the plants in space and time, so that the different societies can make use of them and experiment with them (Stepp and Moerman 2001, Molares and Ladio 2012).

In the rural diet, these 10 species basically fulfil the function of buffer and/or emergency resources. The ethnobotanical studies of this group reveal that the use of native berries diversifies the diet, offering a variety of minerals and micronutrients, and are alternative resources in times of scarcity, when the vegetable garden has not yet produced a crop (in autumn, for example, *L. apiculata* offers fruit from March to July, and *R. magellanicum* in spring, from October to February), when working on the land means spending many hours away from home, and when the children go on horseback to the rural schools, among others (Ladio and Lozada 2000, 2001, 2003).

Perhaps for these reasons, the plants are significant elements in the processes of transmission of plant knowledge. Diverse studies in rural communities have shown that the acquisition of plant knowledge occurs at an early age (Lozada et al. 2006). Nevertheless, in many communities the practice of gathering is being lost, due to numerous factors, both social and ecological (Ladio and Lozada 2004, Ochoa and Ladio 2011, Molares and Ladio 2015). In these scenarios, for example, michay and calafate berries seem to fulfil a substantial role in childhood learning about plants. Rural children learn to gather them when they are small, especially if their parents continue to follow the pastoral lifestyle. This is possibly due, in part, to their visibility compared to the other eight berries highlighted in this review, an aspect which should be studied in greater detail.

In general, we could say that these berries could be considered prototypical: they are globose, red-bordeaux, shiny, spineless, turgid, etc., they have everything that would be expected from an edible fruit—*"they are tasty and they are not bad for you"*—according to local inhabitants. In this sense, these species are easy to identify, to harvest, and they are not confused with toxic plants, which means that children can easily gather them. In general, their sweet, delicate flavours are considered inoffensive, suitable for consumption by children according to the criteria of the rural inhabitants we have worked with (Molares and Ladio 2009b). *"It's sweet, so we can eat it"*, mentioned various informants. If, on the other hand, the berries were tasteless, or bitter, the people would be in doubt.

Our interpretation is that they are elements with a certain duality, being used in both alimentary and health contexts. Various case studies show that from a local perspective, health and diet are intimitely linked (Ladio 2011). For example, ethnobotanical studies carried out in Mapuche communities reveal a marked overlap between edible and medicinal species (Ladio 2006, Molares and Ladio 2015). This duality forms part of a traditional health system based on varied cultural contributions, the best known and the most widespread in the region being the Mapuche traditional health system, which has become hybridised with elements from Creole and other cultural heritages (Citarela et al. 1995, Molares and Ladio 2009b). Significantly, this view of integrality in traditional systems now has its equivalent in scientific environment within the subject of functional foods.

The long period of time over which sources cite their use reveals the cultural importance of these ten species that persists in the societies, even though there have been slight differences and changes over time. Although the original communities in the region have given new meanings to their alimentary and health systems over time, mainly due to the imposition of strong measures following the conquest and national politics, these species continue to be used and appreciated in the lives of rural workers.

Their use does not appear to have been affected by the drastic introduction of new foods or the notorious change in landscapes these people underwent (Eyssartier et al. 2011b, Torrejón and Cisternas 2002). Many traditional practices have withstood the hybridisation processes of traditional wisdom with new knowledge that came from European immigrants or from the societies that emerged after the formation of the Argentine and Chilean states. Furthermore, some species, such as maqui, are taking on even more importance on a commercial and global scale.

Finally, this list is intended to give an overview of the principal fruits consumed by the first inhabitants of this part of South America, and up to the present time. From an ethnobotanical point of view, further research is necessary, in particular on the subject of traditional managment practices these species are subject to, the social rules relating to extraction and conservation, and the conservation guidelines followed by local inhabitants in order to prevent or recover from their overexploitation.

Preliminary studies carried out in rural communities of Chubut province show that rural dwellers are selective in the gathering process used for *B. microphylla,* they prefer certain individual plants to others, and prefer to pick the larger, darker berries. The most productive calafate plants are protected in their natural environments by homemade fences, constructed with sticks and wire. This has also been observed for maqui and parilla plants. Most of the berry species analysed here already have propagation protocols (Rovere 2006) and are even sold in plant nurseries in the region, although principally for ornamental purposes. Therefore, the potential for encouraging their cultivation and use on a regional scale as multipurpose resources for ornamental, alimentary and medicinal use should not be underestimated. In addition, this work may constitute a useful starting point from which to consider the Patagonian species as a focus of deeper study, in particular their application in nutrition as functional foods, in phytomedicine, phytotherapies or phytocosmetics (creams and emulsions). This would provide an answer for the sizeable sectors of society which seek to improve their health and wellbeing through incorporation of new plant products.

Acknowledgements

We would like to thank the men and women from the different Argentine Patagonian communities (especially Cuyín Manzano, and the Mapuche communities of Catán Lil, Lago Rosario and Nahuelpan) for their friendly conversation and the affection shown to us during our work. Our thanks also go to CONICET. This investigation has been financed by PIP 2013-0466 held by Phd. Ana Ladio.

References

Arena, M.E. and Curvetto, N. 2008. Berberis buxifolia fruiting: Kinetic growth behavior and evolution of chemical properties during the fruiting period and different growing seasons. Sci. Hortic. 118: 120–127.

Arena, M.E. and Coronel, L.J. 2011. Fruit growth and chemical properties of Ribes magellanicum "parrilla." Sci. Hortic. 127: 325–329.

Arena, M.E., Giordani, E. and Radice, S. 2014. Flowering, fruiting and leaf and seed variability in berberis buxifolia, a native patagonian fruit species. pp. 1–20. *In*: Marin, L. and Kovac, D. (eds.). Native Species: Identification, Conservation and Restoration. Nova Science, New York, USA.

Ávila, F., Theoduloz, C., López-Alarcón, C., Dorta, E. and Schmeda-Hirschmann, G. 2017. Cytoprotective mechanisms mediated by polyphenols from chilean native berries against free radical-induced damage on AGS cells. Oxid. Med. Cell. Longev. 2017: 1–13.

Capparelli, A., Hilgert, N. and Ladio, A. 2011. Paisajes culturales de Argentina: Pasado y presente desde las perspectivas etnobotánica y paleoetnobotánica. Argent. Ecol. Paisajes. 2(2): 67–79.

Cheel, J., Theoduloz, C., Rodríguez, J.A., Caligari, P.D.S. and Schmeda-Hirschmann, G. 2007. Free radical scavenging activity and phenolic content in achenes and thalamus from *Fragaria chiloensis* ssp. chiloensis, *F. vesca* and *F.* x *ananassa* cv. Chandler. Food Chem. 102: 36–44.

Citarella, L., Conejeros, A.M., Espinosa, B., Jelves, I., Oyarce, A.M. and Vidal, A. 1995. Medicinas y culturas en La Araucanía. Ed. Sudamericana, Santiago, Chile.

Conticello, L., Gandulo, R., Bustamante, A. and Tartarglia, C. 1997. El uso de plantas medicinales por la comunidad mapuche de San Martín de los Andes. Provincia de Neuquén (Argentina). Parodiana. 10(1-2): 165–180.

Contreras Vega, M. 2006. Plantas medicinales y alimenticias de Chiloe. Colección Cultura Insular. Castro, Chile.

Correa, M.N. (ed.). 1969–1999. Flora Patagónica 8, Partes 1–7. Col. Ci. INTA, Buenos Aires, Argentina.

Damascos, M.A. 2011. Arbustos silvestres con frutos carnosos de Patagonia. Fondo Editorial Rionegrino, Viedma, Argentina.

Davidson-Hunt, I.J. and Berkes, F. 2003. Nature and society through the lens of resilience: toward a human-in-ecosystem perspective. *In*: Berkes, F., Colding, J. and Folke, C. (eds.). Navigating Social-Ecological Systems: Building Resilience for Complexity and Change. Cambridge University Press, Cambridge, UK.

Dillehay, T.D. 1988. Monte Verde: a Pleistocene settlement in Chile. Vol I. Paleoenvironment and site context, Smithsonian Institution Press, Washington DC, USA.

Escribano-Bailón, M.T., Alcalde-Eon, C., Muñoz, O., Rivas-Gonzalo, J.C. and Santos-Buelga, C. 2006. Anthocyanins in berries of Maqui (Aristotelia chilensis (Mol.) Stuntz). Phytochem Anal. 17(1): 8–14.

Estomba, D., Ladio, A. and Lozada, M. 2006. Medicinal wild plant knowledge and gathering patterns in a Mapuche community from North-western Patagonia. J. Ethnopharmacol. 103(1): 109–119.

Eyssartier, C., Ladio, A.H. and Lozada, M. 2011a. Traditional horticultural knowledge change in a rural population of the Patagonian steppe. J. Arid. Environ. 75(1): 78–86.

Eyssartier, C., Ladio, A.H. and Lozada, M. 2011b. Horticultural and gathering practices complement each other: a case study in a rural population of northwestern Patagonia. Ecol. Food Nutr. 50: 429–451.

Falkenberg, S.S., Tarnow, I., Guzman, A., Mølgaard, P. and Simonsen, H.T. 2012. Mapuche herbal medicine inhibits blood platelet aggregation. Evidence-based Complement. Altern. Med. 1–9.

Fredes C., Yousef, G.G., Robert, P., Grace, M.H., Lila, M.A., Gómez, M. and Montenegro, G. 2014. Anthocyanin profiling of wild maqui berries (Aristotelia chilensis [Mol.] Stuntz) from different geographical regions in Chile. J. Sci. Food Agric. 94(13): 2639–2648.

Fuentes, L., Valdenegro, M., Gómez, M.G., Ayala-Raso, A., Quiroga, E., Martínez, J.P., Vinet, R., Caballero, E. and Figueroa, C.R. 2016. Characterization of fruit development and potential health benefits of arrayan (Luma apiculata), a native berry of South America. Food Chem. 196: 1239–1247.

Funes, F. 1999. Estudio etnobotánico del Valle del Río Manso Inferior, provincia de Río Negro. UNComa, CRUB, Bariloche, Argentina.

Guisinde, M. 1917. Medicina e higiene de los antiguos araucanos. Imprenta Universitaria. Chile.

Houghton, P.J. and Manby, J. 1985. Medicinal plants of the Mapuche. J. Ethnopharmacol. 13(1): 89–103.

Hurrell, J.A., Ulibarri, E.A., Delucchi, G. and Pochettino, M.L. 2010. Frutas: frescas, secas y preservadas. Ed. LOLA, Buenos Aires, Argentina.

Jiménez-Aspee, F., Thomas-Valdés, S., Schulz, A., Ladio, A., Theoduloz, C. and Schmeda-Hirschmann, G. 2015. Antioxidant activity and phenolic profiles of the wild currant *Ribes magellanicum* from Chilean and Argentinean Patagonia. Food Sci. Nutr. 4(4): 595–610.

Jiménez-Aspee F., Theoduloz, C., Ávila, F., Thomas-Valdés, S., Mardones, C., von Baer, D. and Schmeda-Hirschmann, G. 2016. The Chilean wild raspberry (*Rubus geoides* Sm.) increases intracellular GSH content and protects against H_2O_2 and methylglyoxal-induced damage in AGS cells. Food Chem. 194: 908–919.

Ladio, A. and Rapoport, E.H. 1999. El uso de plantas silvestres comestibles en una población suburbana del noroeste de la Patagonia. Parodiana. 11(1-2): 49–62.

Ladio, A. and Lozada, M. 2000. Edible wild plant use in a Mapuche community of northwestern Patagonia. Hum. Ecol. 28(1): 53–71.

Ladio, A. 2001. The Maintenance of Wild Edible Plant Gathering in a Mapuche Community of Patagonia. Econ. Bot. 55: 243–254.

Ladio, A. and Lozada, M. 2001. Non-timber forest product use in two human populations from NW Patagonia: A quantitative approach. Hum. Ecol. 29(4): 367–380.

Ladio, A. and Lozada, M. 2003. Comparison of wild edible plant diversity and foraging strategies in two aboriginal communities of northwestern Patagonia. Biodivers Conserv. 12: 937–951.

Ladio, A. and Lozada, M. 2004. Patterns of use and knowledge of wild edible plants in distinct ecological environments: A case study of a Mapuche community from northwestern Patagonia. Biodivers Conserv. 13(6): 1153–1173.

Ladio, A. 2005. Malezas exóticas comestibles y medicinales utilizadas en poblaciones del NO patagónico: aspectos etnobotánicos y ecológicos. Bol. Latinoam Caribe Plant Med. Aromat. 4(4): 75–80.

Ladio, A. 2006. Gathering of wild plant foods with medicinal use in a Mapuche community of Northwest Patagonia. pp. 297–321. *In*: Pieroni, A. and Price, L. (eds.). Eating and Healing: Traditional Food as Medicine. The Haworth Press, New York, USA.

Ladio, A., Molares, S. and Rapoport, E. 2007. Conocimiento etnobotánico de plantas comestibles entre los maestros patagónicos: patrones de variación ambiental oeste-este. Kurtziana 33(1): 141–152.

Ladio, A. and Lozada, M. 2008. Medicinal plant knowledge in rural communities of North-western Patagonia, Argentina. A resilient practice beyond acculturation. Current Topics in Ethnobotany 661(2): 39–53.

Ladio, A. and Lozada, M. 2009. Human ecology, ethnobotany and traditional practices in rural populations inhabiting the Monte region: Resilience and ecological knowledge. J. Arid. Environ. 73: 222–227.

Ladio, A. 2011. Underexploited wild plant foods of North-Western Patagonia. pp. 297–321. *In*: Filip, R. (ed.). Multidisciplinary Approaches on Food Science and Nutrition for the XXI Century. Transworld Re-search Network, Kerala, India.

Ladio, A., Molares, S., Ochoa, J. and Cardoso, B. 2013. Etnobotánica aplicada en patagonia: La comercialización de malezas de uso comestible y medicinal en una feria urbana de San Carlos de Bariloche (Río Negro, Argentina). Bol. Latinoam Caribe Plant Med. Aromat. 12(1): 24–37.

Ladio, A. and Molares, S. 2014. The dynamics of use of nontraditional ethnobiological products: Some aspects of study. pp. 311–319. *In*: Albuquerque, U.P., Vital Fernandes Cruz da Cunha, L., Farias Paiva de Lucena, R. and Nobrega Alves, R.R. (eds.). Methods and Techniques in Ethnobiology and Ethnoecology. Springer Science.

Llano, C. and Barberena, R. 2013. Explotación de especies vegetales en la Patagonia septentrional: El registro arqueobotánico de Cueva Huemul 1 (Provincia de Neuquén, Argentina). Darwiniana 1: 5–19.

Lozada, M., Ladio, A. and Weigandt, M. 2006. Cultural transmission of ethnobotanical knowledge in a rural community of Northwestern Patagonia, Argentina. Econ. Bot. 60(4): 374–385.

Mahabir, S. 2014. Methodological challenges conducting epidemiological research on nutraceuticals in health and disease. Pharma Nutrition. 2: 120–125.

Martínez-Crovetto, R. 1980. Apuntes sobre la vegetación de los alrededores del lago Cholila (Noroeste de la Provincia de Chubut). Publicación Técnica, Universidad Nacional Del Nordeste Facultad de Ciencias Agrarias, Corrientes, Argentina.

Martínez-Crovetto, R. 1982. Breve panorama de las plantas utilizadas por los indios de Patagonia y Tierra del Fuego. Suplemento Antropológico Vol XVII. Nro. 1, Universidad Católica, Asunción 61–97.

Medeiros, M.F.T. 2009. Etnobotânica Histórica: Princípios e Procedimentos. NUPEEA/Sociedad Brasileira de Etnobiología e Etnoecología, Recife, Brasil.

Meza, I.P. and Villagran, C. 1991. Etnobotánica de la Isla Alao, Archipielago de Chiloe, Chile. Bol. del Mus. Nac. Chile. 42: 39–78.

Molares, S. and Ladio, A.H. 2009a. Ethnobotanical review of the Mapuche medicinal flora: Use patterns on a regional scale. J. Ethnopharmacol. 122(2): 251–260.

Molares, S. and Ladio, A.H. 2009b. Plantas Medicinales de los Andes Patagónicos: una revisión cuantitativa del conocimiento etnobotánico Mapuche. pp. 87–128. *In*: Vignale, N.D., Pochettino, M.L. (eds.). Avances sobre Plantas Medicinales Andinas, CYTED, S.S. de Jujuy, Argentina.

Molares, S. and Ladio, A. 2012. The usefulness of edible and medicinal Fabáceae in argentine and Chilean Patagonia: Environmental availability and other sources of supply. Evidence-based Complement Altern. Med. eCAM. 2012; 2012: 901918.

Molares, S. and Ladio, A. 2014. Medicinal plants in the cultural landscape of a Mapuche-Tehuelche community in arid Argentine Patagonia: an eco-sensorial approach. J. Ethnobiol. Ethnomed. 10: 61.

Molares, S. and Ladio, A.H. 2015. Complejos vegetales comestibles y medicinales en la Patagonia Argentina: sus componentes y posibles procesos asociados. Bol. Latinoam Caribe Plant Med. Aromat. 14(3): 237–250.

Mösbach, E.W. 1992. Botánica indígena de Chile. Ed. Andrés Bello, Santiago, Chile.

Muñoz, M.S., Barrera, E.M. and Meza, I.P. 1981. El uso medicinal y alimenticio de plantas nativas y naturalizadas en Chile. Pub. Ocac. Mus. Hist. Natural de Chile. 33: 3–89.

Nacuzzi, L.R. and Perez de Micou, C. 1984. Los recursos vegetales de los cazadores de la cuenca del Río Chubut. Cuad. del Inst. Nac. Antropol. 10: 407–423.

Ochoa, J.J., Ladio, A.H. and Lozada, M. 2010. Uso de recursos herbolarios entre Mapuche y criollos de la comunidad campesina de Arroyo Las Minas (Río Negro, Patagonia Argentina). Bol. Latinoam Caribe Plant Med. Aromat. 9(4): 269–276.

Ochoa, J.J. and Ladio, A.H. 2011. Pasado y presente del uso de plantas silvestres con órganos de almacenamiento subterráneos comestibles en la patagonia. Bonplandia 20(2): 265–284.

Ochoa, J.J. 2015. Uso de plantas silvestres con órganos de almacenamiento subterráneos comestibles en Patagonia: perspectivas etnoecológicas. Tesis doctoral, Universidad Nacional del Comahue, Bariloche, Argentina.

Ochoa, J.J. and Ladio, A.H. 2015. Plantas silvestres con órganos subterráneos comestibles: transmisión cultural sobre recursos subutilizados en la Patagonia (Argentina). Bol. Latinoam Caribe Plant Med. Aromat. 14(4): 287–300.

Pandey, K.B. and Rizvi, S.I. 2009. Plant polyphenols as dietary antioxidants in human health and disease. Oxid. Med. Cell. Longev. 2(5): 270–278.

Pardo, O. and Pizarro, J.L. 2005. Especies Botánicas consumidas por los Chilenos Prehispánicos. Colección Chile Precolombino, Santiago, Chile.

Pochettino, M.L. 2015. Botánica económica. Las plantas interpretadas según tiempo, espacio y cultura. Sociedad Argentina de Botánica, Corrientes, Argentina.

Prates, L. 2009. El uso de recursos por los cazadores-recolectores posthispánicos de Patagonia continental. Relac la Soc Argentina Antropol. 34: 201–229.

Ragonese, A.E. and Martínez-Crovetto, R. 1947. Plantas indígenas de la Argentina con Frutos o Semillas comestibles. Revista de Investigaciones Agrícola 1(3): 147–216.

Ramírez, J.E., Zambrano, R., Sepúlveda, B., Kennelly, E.J. and Simirgiotis, M.J. 2015. Anthocyanins and antioxidant capacities of six Chilean berries by HPLC-HR-ESI-ToF-MS. Food Chem. 176: 106–114.

Rapoport, E. and Ladio, A. 1999. Los bosques andino-patagónicos como fuentes de alimento. Bosque 20(2): 55–64.

Rapoport, E.H., Ladio, A.H. and Sanz, E. 2003. Plantas nativas comestibles de la Patagonia andina argentino-chilena. Parte II. Imaginaria, Bariloche, Argentina.

Rapoport, E.H., Ladio, A.H. and Sanz, E. 2005. Plantas nativas comestibles de la Patagonia andina argentino-chilena. Parte I. Imaginaria, Bariloche, Argentina.

Richeri, M., Ladio, A.H. and Beeskow, A.M. 2013. Conocimiento tradicional y autosuficiencia: La herbolaria rural en la meseta central del chubut (Argentina). Bol. Latinoam Caribe Plant Med. Aromat. 12(1): 44–58.

Romanucci, V., D'Alonzo, D., Guaragna, A., Di Marino, C., Davinelli, S., Scapagnini, G., Di Fabio, G. and Zarrelli, A. 2016. Bioactive compounds of Aristotelia chilensis Stuntz and their pharmacological effects. Curr. Pharm. Biotechnol. 17(6): 513–523.

Rovere, A.E. 2006. Cultivo de Plantas Nativas Patagónicas: árboles y arbustos. Ed Caleuche. Bariloche, Argentina.

Ruiz, A., Hermosín-Gutiérrez, I., Mardones, C., Vergara, C., Herlitz, E., Vega, M., Dorau, C., Winterhalter, P. and von Baer, D. 2010. Polyphenols and antioxidant activity of calafate (berberis microphylla) fruits and other native berries from Southern Chile. J. Agric. Food Chem. 58(10): 6081–6089.

Ruiz, A., Hermosín-Gutiérrez, I., Vergara, C., von Baer, D., Zapata, M., Hitschfeld, A. and Mardones, C. 2013. Anthocyanin profiles in south Patagonian wild berries by HPLC-DAD-ESI-MS/MS. Food Res. Int. 51(2): 706–713.

San Martín, J.A. 1983. Medicinal plants in central Chile. Econ. Bot. 37(2): 216–227.

Schmeda-Hirschmann, G. and Simirgiotis, M. 2011. Chemistry of the Chilean Strawberry (*Fragaria chiloensis* spp. chiloensis), Gene, Genomes and Genomics 5: 85–90.

Schreckinger, M.E., Lotton, J., Lila, M.A. and de Mejia, E.G. 2010. Berries from South America: a comprehensive review on chemistry, health potential, and commercialization. J. Med. Food. 13: 233–246.

Shahidi, F. and Naczk, M. 2004. Phenolic compounds in fruits and vegetables. pp. 131–156. *In*: Shahidi, F. and Naczk, M. (eds.). Phenolics in food and nutraceutical, CRC Press, Francis Taylor Group. Boca Raton, Florida, USA.

Srivastava, S., Srivastava, M., Misra, A. and Pandey, G. 2015. Review article: a review on biological and chemical diversity. EXCLI J. 14: 247–267.

Silva, V.A., Nascimento, V.T., Taboada Soldati, G., Medeiros, M.F.T. and Albuquerque, U.P. 2014. Techniques for Analysis of Quantitative Ethnobiological Data: Use of Indices. *In*: Albuquerque, U.P., Fernandes Cruz Cunha, L.V., Lucena, R.F.V., Nobrega Alves, R.R. (eds.). Methods and Techniques in Ethnobiology and Ethnoecology. Springer, New York, USA.

Simirgiotis, M.J. and Schmeda-Hirschmann, G. 2010. Determination of phenolic composition and antioxidant activity in fruits, rhizomes and leaves of the white strawberry (*Fragaria chiloensis* spp. chiloensis form chiloensis) using HPLC-DAD–ESI-MS and free radical quenching techniques. J. Food Compost. Anal. 23(6): 545–553.

Simirgiotis, M.J., Bórquez, J. and Schmeda-Hirschmann, G. 2013. Antioxidant capacity, polyphenolic content and tandem HPLC-DAD-ESI/MS profiling of phenolic compounds from the South American berries Luma apiculata and L. chequén. Food Chem. 139(1-4): 289–299.

Stepp, J.R. and Moerman, D.E. 2001. The importance of weeds in ethnopharmacology. J. Ethnopharmacol. 75: 19–23.

Szajdek, A. and Borowska, E.J. 2008. Bioactive compounds and health-promoting properties of Berry fruits: A review. Plant Foods Hum. Nutr. 63(4): 147–153.

Tacón, C.A., Palma, J.M., Fernández, U.V. and Ortega, F.B. 2006. El Mercado de los Productos forestales no madereros y la conservación de los bosques del Sur de Chile y Argentina. Red de Productos forestales No Madereros de Chile, WWF Chile.

Toledo, V.M. and Barreras-Bassols, N. 2008. La memoria biocultural. La importancia agroecológica de las sabidurías tradicionales. Ed. Icaria, Barcelona, España.

Torrejón, F. and Cisternas, M. 2002. Alteraciones del paisaje ecológico araucano por la asimilación mapuche de la agroganadería hispano-mediterránea (siglos XVI y XVII). Revista Chilena de Historia Natural. 75: 729–736.

Vignati, M.A. 1941. El pan de los patagones protohistóricos. Notas del Museo de La Plata. Antropología 23: 321–336.

Zuloaga, F. and Morrone, O. 2009. Flora del Cono Sur. Catálogo de las Plantas Vasculares, Instituto de Botánica Darwinion, Buenos Aires, Argentina.

CHAPTER 5

Status of Research on Medicinal Plants in the Cajamarca's Region, Peru

Juan F. Seminario Cunya,* Berardo Escalante Zumaeta and
Alejandro Seminario Cunya

Introduction

Medicinal plants are a subject of concern to researchers, traders, users and states and, in general terms, the problems facing these species in the world are similar. On one hand, a greater proportion is located in certain areas of the planet, in developing countries and its use and conservation is linked to traditional cultures, implying knowledge, beliefs and spirituality. On the other hand, the fact that most are wild (40.5%) or naturalized (33.3%) and only 3.3% are cultivated (Singh et al. 2012) means that their conservation depends mainly on collection pressure.

The most important risk factors for medicinal species in the world would be competition for land use, which entails the destruction of natural habitats, excessive collection due to the growth of market demand—a direct consequence of the growth of the human population—and the increase of industrial products derived from plants. This last factor implies permanent bio prospecting in searching new medicinal compounds (Roberson 2008, Castle et al. 2014). Bio prospecting leads to the risk of bio piracy. This means that large companies patent genetic resources, remedies and traditional knowledge of local or indigenous populations and commercialize them with great economic benefits, but with little or no benefit

Programa de Raíces y Tubérculos Andinos, Facultad de Ciencias Agrarias, Universidad Nacional de Cajamarca.
* Corresponding author: jfseminario@yahoo.es

to the countries or communities of origin (Singh et al. 2014, Barnett cited by Roberson 2008). The latter generates fear among actors and hinders cooperation for the sustainable management of wild populations and studies for their incorporation into agro-ecosystems. However, in each space or geographic area, the intensity of the effect of the mentioned factors is different and to them can be added others, whose local importance for the conservation and sustainable use of the species may have an even greater preponderance.

The Cajamarca Region, located in the north of Peru, is part of the *health axis* that includes the territory between Loja (South of Ecuador) and Lambayeque–North of Peru—covering the departments of Piura, Lambayeque, La Libertad, Cajamarca and San Martín. The term *health axis*, established by Camino (1992), is defined as a geographical space delimited by common concepts regarding health. This region is located between 4° 36"S and 7° 45'S and 77° 44'E and 79° 27'W, with a total area of 33317.5 km² and covers an altitudinal range that, for the purposes of this presentation and the species we are going to treat, we divide into two parts: high and lower zones.

The high zone ranges from 3000 masl to the highest peaks (4000 masl). This territorial strip is called Jalca or Páramo (Weberbauer 1945, Sánchez and Dillon 2006), which includes the geographical regions Quechua alta and Suni, according to Pulgar Vidal (1966). The Jalca is a biogeographic formation that houses a high diversity of plants, including an undetermined number of wild medicinal species, along with several grass species that form the grassland, as the predominant community (Sánchez 2014). In addition to the landscape value, the Jalca is a fundamental hydrological source due to the presence of abundant lagoons and wetlands and because raindrops infiltrate the mountains and provide watersheds throughout the year.

The lower zone is comprised between 500 masl and 3000 masl. It includes the geographical regions Yunga and Quechua baja (Pulgar Vidal 1966). Most of the human populations are located in this territory. Land use is intensive for agriculture, livestock and urbanization, while the medicinal plants of the Jalca are wild and numerous. The medicinal plants of the lower zone are few and can be wild, cultivated, or tolerated into the arable fields.

The Cajamarca's Region has an ancient tradition in the use of medicinal plants, whose antecedents are located in the Caxamarca culture itself—1500 BC–1532 AD—(Watanabe 2002), as well as in the different regional cultures which were extended to this territory like Chavín (1300 BC to 300 BC), Cupisnique (1500 BC to 200 BC) and Mochica (100 AD to 700 AD) (Towle 1961, Ayasta 2012). These cultures based their subsistence on the use of cultivated and wild plant species. Part of this tradition in the use of medicinal plants was documented by Martínez Compañón (1789, volume V) through 138 aquarelles of a similar number of species, of the jurisdiction of the bishopric of Trujillo, of which Cajamarca was part. Also, the evidence is documented by pre and post Colombian ethnobotany (Yacovleff and Herrera 1934, Towle 1961, Ugent and Ochoa 2006).

There are five protected natural areas (San Andrés de Cutervo, Tabaconas-Naballe, Pagaibamba, Udima and Sunchubamba) in the Cajamarca's Region, where 908 genera and 2699 plant species are housed (Sagástegui et al. 1999). Like the Huánuco's Region, it occupies the first place in endemism in Peru, with 948 species (17%). Of this total, 296 (31%) are rare species. Because of this rich flora, it is a supplier of various vegetable resources and especially of medicinal plants for the markets of the Peruvian coast. These medicinal species mainly come from the Jalca, as demonstrated by the studies of Bussmann and Sharon (2006), Bussmann et al. (2008), Bussmann and Sharon (2009), Ramírez et al. (2006), Vásquez et al. (2010) and Sánchez (2011).

Since the 1970s, an aggressive intervention process of the Jalca has been initiated through new human settlements, intensive grazing, burning of natural pastures to provoke its regrowth (traditional peasant practice) and agriculture of cereals and tubers. The disturbance of these ecosystems was accelerated in the 1990s, due to the intensification of open-pit mining (Ramírez et al. 2006, Seminario and Sánchez 2010, Sánchez 2014). These mountains, in addition to their wealth of flora and fauna, are rich in minerals such as gold, copper and silver. As a result, besides the mining Yanacocha, a company operating in one of the largest gold deposits of Latin America, there are another companies in exploration and operation, although there are no official data on the total concessions.

In this scenario, the study by Vásquez et al. (2010), which documents in detail the morphology, distribution and conservation status and uses of 130 medicinal species of northern Peru, emphasizes that several wild medicinal species that are harvested for the market are in the disappearing process. Among these, the authors cited the valeriana (*Valeriana pilosa*), huamanripa (*Senecio tephrosioides*), vira vira (*Senecio canescens*), azarcito (*Isertia krausei*), cascarilla (*Cinchona pubescens*), añasquero (*Dyssodia jelskii*) and uña de gato (*Uncaria tomentosa*). However, some of the mentioned species are not considered within the categorization of endangered species of wild flora of Peru (DS-043-2006-AG 2006), so it is necessary to review and update it.

The objective of this document is to summarize the advances of research in medicinal plants in the Cajamarca's Region. This state of knowledge should serve to know how much has been done and what remains to be done. For this purpose, formal publications on medicinal plants of the region were revised, taking as a starting point the studies of Luis Iberico at the beginning of the decade of 1980. Also taken into account is some gray literature that, by the importance of the data, allows a better vision of what has been done so far on the subject. Another special consideration was that, in all cases, the determination of the botanical species was directed or supported by the botanists A. Sagástegui Alva and I. Sánchez Vega and was carried out in the herbaria of the National University of Cajamarca (Isidoro Sánchez Vega), National Agrarian University La Molina (A. Weberbauer), National University of San Marcos (Museum of Natural History) and the Antenor Orrego Private University. From this review, a matrix with the 15 studies, the families and the species was generated. This served for the corresponding analysis.

The information is presented in three blocks. The first block exposes research on the diversity of medicinal plants in the region. We have included here two investigations that refer to floristic diversity: the first one was developed in the province of Contumazá (Sagástegui 1995) with its eight districts and the second (Marcelo et al. 2006) one was realized in a Páramo of the province of Jaén. In the first, a Chapter (8, p. 97–118) is dedicated to medicinal plants and in the second, in the list of species, several species are known as medicinal in the region. The second block summarizes research related to the market of medicinal species in Cajamarca and, in the third block, specific studies on the use of medicinal species in the region are presented.

Studies on the Diversity of Medicinal Species in Cajamarca

Scientific research on the diversity of medicinal plants in Cajamarca began around the beginning of the 1980s. Luis Iberico, with an anthropological focus on agrarian folklore (Iberico 1981), collected fables and stories about plants in the life of peasants, involving medicinal species such as corn, dandelion, orange, rue, valerian, chamomile, chicory, mallow, oregano, aloe and coca. In *medical folklore* (Iberico 1984), he mentions (through the vulgar names) about 130 plants used to treat or cure 90 psychophysical dysfunctions in the traditional rural environment of Cajamarca. Plant systematic studies were then initiated with Castañeda and Vargas (1991). Studies are then carried out for various purposes, but at the same time they constitute inventories of the diversity of medicinal species (Sagástegui 1995, La Torre 1998). To date, 15 studies are recorded (Table 5.1), in which a list of the species, determined according to international norms of botanical nomenclature, is presented. These studies cover 46 localities in 19 districts (16% of the total) and six provinces, with a notable predominance in the southern provinces and especially in the provinces of Cajamarca and Contumazá (Fig. 5.1). They cover all the ecological floors from the Yunga (500 masl) to the Jalca, 4496 masl, at its highest peak, Cerro Rumi Rumi–Cajabamba (Montoya and Figueroa 1991), and there was a clear tendency towards the jalca, above the 3000 masl. These studies mention between 12 to 155 species each, mostly different, so that the general inventory shows 457 species, including 105 families (Table 5.2) and 296 genera. The six best represented families were: Asteraceae, Lamiaceae, Fabaceae, Solanaceae, Rosaceae and Piperaceae. The number of species allows us to build up an idea about the richness of the Cajamarca region in medicinal plants, mainly, if we compare it with the report of Bussman and Sharon (2006). The authors worked between Trujillo and Chiclayo (Peruvian coast) and collected 510 medicinal species in the field, in the markets and in the homes of healers.

On the other hand, on an average, 78% of the reported species are wild; this is consistent with the fact that most studies included areas of this region. Likewise, the predominance of native species of the Andean region (80% approximately) is clear.

The most frequently mentioned species (up to 10 studies) were Ishpingo verde (*Achyrocline alata*) and manzanilla (*Matricaria chamomilla*). These species were

Table 5.1. Localities and districts where the studies of medicinal plants in the Cajamarca region were made, number of species and % of wild.

No	Locality	District/Province	Altitude (masl)	Utm Coordinates East	Utm Coordinates North	Species No	Species Wild (%)	Source
1	Contumazá y 7 localidades más	Ocho distritos/Contumazá	350–4333	742292.23	9185161.54	155	61	Sagástegui 1995
2	Yanacancha	Chumuch/Celendín	2800–4000	811019.61	9266816.44	82	83	La Torre 1998
3	Chilete	Chilete/Contumazá	500–1000	738647.29	9201202.43	57	77	Orozco 2003
	Contumazá	Contumazá/Contumazá	2300–2800	742127.70	9184837.03			
	Sorochuco	Sorochuco/Celendín	2600 4100	803615.79	9235100.42			
	Chetilla	Chetilla/Cajamarca	2800–4000	756704.60	9209411.55			
4	Espino y Palambe	Sallique/Jaén	3000–3560	693850.89	9382559.44	12	100	Marcelo et al. 2006
5	Encañada	Encañada/Cajamarca	2700–3400	793678.78	9216374.10	65	100	Alvitres et al. 2007
6	Chigden	San Juan/Cajamarca	2566–2700	774540.66	9196361.29	76	55	Bussmann et al. 2008
7	Higuerón	San Juan/Cajamarca	2240–2600	772346.90	9195085.86	40	57	Revene et al. 2008
8	Higuerón y 5 localidades más	San Juan/Cajamarca	2240–2600	772346.90	9195085.86	58	88	Seminario and Sánchez 2010
9	Combayo y 6 localidades más	Encañada/Cajamarca	3100–4200	9222334–9239636	781636 a 786023	58	90	Castañeda and Condori 2010
10	Llacanora	Llacanora/Cajamarca	2600–3000	784491.01	9204345.38	65	71	Sánchez 2014
11	Capulipampa y 7 localidades más	Chetilla y Magdalena/Cajamarca	3400–4000	773156.68–721502.41	9214612.53–9196425.97	39	72	Ramos 2015
12	Cutervillo y 4 comunidades más	Huambos/Chota	2100–2300	721847.76	9286330.26	17	100	Seminario and Escalante 2016
13	Quecherga	Encañada/Cajamarca	3600–4100	793195.84	9230296.73	63	72	Ayay 2017
14	Chilincaga	Cajamarca/Cajamarca	3200–3500	768646.78	9225563.12	85	61	Castillo-Vera et al. 2017
15	Cajabamba	Cajabamba/Cajabamba	1200–4496	826175.67	9156185.86			

Fig. 5.1. Map of the CajamarcaÂ´s Region, indicating the districts where the studies on medicinal plants were carried out.

followed in frequency of mention in nine studies: cola de caballo (*Equisetum bogotense*), pie de perro (*Desmodium mollicum*), chancua (*Minthostachys mollis*). In eight studies: hinojo (*Foeniculum vulgare*). In seven studies: ajenjo (*Artemisa absintium*), carqueja (*Baccharis genistelloides*), escorzonera (*Perezia multiflora*), cerraja (*Sonchus oleraceus*), orégano (*Origanum vulgare*) and llantén (*Plantago major*). The majority of species (54%) included in the 15 studies were mentioned only once.

Studies on the Market of Medicinal Plants in Cajamarca

Ethnobotany of medicinal plants in the Cajamarca's market

In the city of Cajamarca, between the years 2000 and 2003, Manuel Aldave carried out the inventory, recorded collection and gathering centers, uses and their forms, and exports to the coastal markets of medicinal plants. He found that the population of this city treated their diseases with 305 medicinal plant species, corresponding to 246 genera and 94 families. Asteraceae, Fabaceae, Lamiaceae and Solanaceae were the most representative families. 67% of inventoried species in the markets

Table 5.2. Medicinal plant families of the Cajamarca's Region, mentioned in 15 studies.

Family	No of Species	%	Family	No of Species	%
Asteraceae	71	15.54	Campanulaceae	2	0.44
Lamiaceae	31	6.78	Dioscoreaceae	2	0.44
Fabaceae	21	4.60	Ephedraceae	2	0.44
Solanaceae	19	4.16	Equisetaceae	2	0.44
Rosaceae	17	3.72	Hypericaceae	2	0.44
Piperaceae	16	3.50	Nyctaginaceae	2	0.44
Amaranthaceae	11	2.41	Proteaceae	2	0.44
Apiaceae	11	2.41	Santalaceae	2	0.44
Euphorbiaceae	11	2.41	Tropaeolaceae	2	0.44
Verbenaceae	9	1.97	Xanthorrhoeaceae	2	0.44
Caprifoliaceae	8	1.75	Acantaceae	1	0.22
Gentianaceae	8	1.75	Alstroemeriaceae	1	0.22
Adiantaceae	7	1.53	Anemiaceae	1	0.22
Brasicaceae	7	1.53	Annonaceae	1	0.22
Malvaceae	7	1.53	Arecaceae	1	0.22
Poaceae	7	1.53	Balanophoraceae	1	0.22
Ericaceae	6	1.31	Bixáceae	1	0.22
Geraniaceae	6	1.31	Burseraceae	1	0.22
Lycopodiaceae	6	1.31	Calceolariaceae	1	0.22
Polygonaceae	6	1.31	Cannaceae	1	0.22
Myrtaceae	5	1.09	Caricaceae	1	0.22
Polypodiaceae	5	1.09	Cletharaceae	1	0.22
Borraginaceae	4	0.88	Clusiaceae	1	0.22
Cucurbitaceae	4	0.88	Columeliaceae	1	0.22
Chenopodiaceae	4	0.88	Commelinaceae	1	0.22
Lauraceae	4	0.88	Convolvulaceae	1	0.22
Loasaceae	4	0.88	Coriaciaceae	1	0.22
Loranthaceae	4	0.88	Cyperaceae	1	0.22
Onagraceae	4	0.88	Dipsacaceae	1	0.22
Orchidaceae	4	0.88	Erytroxylaceae	1	0.22
Oxalidaceae	4	0.88	Gesneriaceae	1	0.22

Table 5.2 contd. ...

Table 5.2 contd. ...

Family	No of Species	%	Family	No of Species	%
Pasifloraceae	4	0.88	Iridaceae	1	0.22
Plantaginaceae	4	0.88	Juglandaceae	1	0.22
Ranunculaceae	4	0.88	Krameriacae	1	0.22
Rutaceae	4	0.88	Liliaceae	1	0.22
Scrophulariaceae	4	0.88	Linaceae	1	0.22
Urticáceae	4	0.88	Lobeliaceae	1	0.22
Anacardiaceae	3	0.66	Lythraceae	1	0.22
Apocynaceae	3	0.66	Monimiaceae	1	0.22
Bromeliaceae	3	0.66	Muntingiaceae	1	0.22
Cactaceae	3	0.66	Oleaceae	1	0.22
Capparaceae	3	0.66	Papaveraceae	1	0.22
Caryophyllaceae	3	0.66	Phyllanthaceae	1	0.22
Crassulaceae	3	0.66	Phytolaccaceae	1	0.22
Dryopteridaceae	3	0.66	Polygalaceae	1	0.22
Melastomataceae	3	0.66	Punicaceae	1	0.22
Pteridaceae	3	0.66	Salicaceae	1	0.22
Rubiaceae	3	0.66	Sapindaceae	1	0.22
Araliaceae	2	0.44	Smilacaceae	1	0.22
Basellaceae	2	0.44	Vitaceae	1	0.22
Berberidaceae	2	0.44	Zigophylaceae	1	0.22
Bignoniaceae	2	0.44			

of Cajamarca were native to the Americas—and mainly from the Andes—and 33% came from other continents. 45% of registered species were wild (most of them came from Jalca). The rest were domesticated in different degrees: cultivated, wild-cultivated, wild-arvense, arvense.

Diseases or conditions that involved a greater number of plant species were the woman's diseases, gynecological in general, diseases of the urogenital system and inflammatory, mainly of the respiratory and gastrointestinal system. The most used plant part was the leaf. It was followed by the leaf plus reproductive organs, the whole plant and the reproductive organs alone. The preponderant forms of preparation were infusion and cooking, and the main form of administration was as a beverage and to a lesser extent in plaster, poultice and rub.

The most important places for collecting medicinal species were Encañada, Otuzco, Cumbemayo, Combayo, Llacanora, San Juan and Jesus. The most exported plants to the coastal markets were manzanilla (*Matricaria chamomilla*), valeriana (*Valeriana pilosa*), eucalipto (*Eucalyptus globulus*), San Pedro (*Echinopsis pachanoi*), berro (*Roripa nasturtium-aquaticum*), romero de jalca (*Satureja weberbaueri*), papa madre (*Dioscorea mitoensis*) and andacushma (*Geranium ruizii*) (Aldave 2003).

It is necessary to study tendencies of the market of medicinal species in cities because it is very dynamic and subject to introduction of new species as a response to new conditions in users and globalization. It is also observed that certain species leave the market either because the supply fails or by replacement by other species. It is also convenient to use new tools in studies on medicinal plants in cities (Tinitana et al. 2016) to improve the reliability and validity of the information.

Commercialization of Plants by Herbalists in the Cajabamba's Market

Cajabamba is one of the thirteen provinces of the department of Cajamarca, located to the south, in the limit with the La Libertad department. Its capital is the district of Cajabamba. Castillo-Vera et al. (2017) made an inventory of food and medicinal species, which are sold every Sunday in the market of the capital city of the district. This market is supplied with plants coming from different ecological floors, from 1200 masl to 4496 masl, which include the Yunga, Quechua and Jalca or Suni regions. There were 85 medicinal species of 22 families, including Lamiaceae (15%) and Asteraceae (14%). According to the red book of Peruvian endemic plants (León et al. 2006), eight species are included in the list of endemic species of Peru and, according to the categorization of threatened species of wild flora of Peru (DS 043-2006-AG), one species is in danger (EN) and three are in critical danger (CR).

Market of Fresh Medicinal Plants and Cultivation of Medicinal Plants in Home Gardens of Cajamarca

The commercialization of fresh plants in one of the markets of Cajamarca (Revilla Perez) was studied by Bussmann et al. (2008). They found 42 species, the majority (81%) native, which was collected from the wild state. Only five of them were frequently grown in gardens and 11 were cultivated by vendors. The study shows that the increase in the demand for medicinal plants in the market does not affect the increase of their cultivation and, although the observation of two family gardens (one in Chigden and one in Higueron, San Juan district) showed the presence of 76 species, most of them behaved like spontaneous plants [probably as weeds] with few plants per species whose volumes were not significant for commercialization. These researchers base their explanation of little interest in cultivation, despite the demand, on the following points: (1) certain species grow abundantly in the wild state up in the highlands, so that harvesting is always efficient; (2) Other species, because of their narrow or rigid habitat requirement, do not thrive when they are displaced from their natural habitat. This would be the case of *V. pilosa* which, in its natural habitat, is dependent on ichu—*Calamagrostis tarmensis* and *Stipa ichu*—(Seminario et al. 2016); and (3) the crop demands time, land and other resources that families are not willing to invest.

Cajamarca as a Supplier of Medicinal Species for the Markets of the Peruvian Coast

Vásquez et al. (2010) carried out a study on medicinal plants of northern Peru and recognize that most of them come from the sierra and second from the jungle and that the contribution of the coast is small. In fact, the Cajamarca's region is an important center for collecting and exporting medicinal plants to the coastal markets, mainly for the large Moshoqueque market in the city of Chiclayo. The Aldave study (2003) showed that 64 of the 305 registered species had commercial demand in the city markets as well as in the coastal markets. Ten of them stood out for the volumes of commercialization: Manzanilla (*Matricaria chamonilla*), valeriana (*Valeriana pilosa*), San Pedro (*Echinopsis pachanoi*), eucalipto (*Eucalyptus globulus*), berro (*Roripa nasturtium-aquaticum*), Laurel (*Laurus nobilis*), romero de jalca (*Satureja sericea)*, papa madre (*Dioscorea* sp.), chinchimalí (*Gentianella graminea*) and andacushma (*Geranium ruizii*). Also, the study by Ramírez et al. (2006) indicated that more than 100 tons of fresh valeriana (*V. pilosa*) were collected annually; this volume would have increased in recent years due to the impulse of biotrade. On the other hand, Revene et al. (2008) followed a family that gathered—for 16 years —medicinal plants to take them to the Moshoqueque's market. These medicinal species came from six localities of Cajamarca (districts of San Juan and Cajamarca) and 40 of them were commercialized species, of which 23 were wild. Similarly, two to four families did the same activity, throughout the year.

The town of Combayo (district of Encañada, province of Cajamarca) until a few years ago was an important center for the gathering of medicinal plants from six other communities located in Jalca region. Plants collected by specialized merchants were destined to the local and coastal markets (mainly Chiclayo). Seminario and Sánchez (2010) showed that 15 species (out of 58) had a greater demand in local and regional markets (Table 5.3). Another important market of gathering of medicinal plants—not studied—is located in the capital of the district of San Marcos. In this regional market, every Sunday, the commercialization and gathering of various vegetable products and smaller and larger animals are carried out, which are then transported to coastal markets.

Studies on Particular Uses of Medicinal Species

Emollient and other out-of-home drinks in the city of Cajamarca

In 2004, results of an investigation that revealed a very common phenomenon in the cities of Peru were published: the sale of hot or cold drinks for nutraceutical purposes (Seminario 2004). One of them is emollient.

Table 5.3. Medicinal species of the Jalca of Cajamarca, with greater demand in the market (2010).

Species	Common name	Importance Order
Valeriana pilosa	Valeriana, coche coche	1
Gentianella graminea	Chinchimalí	2
Gentianella sp.	Amargón amarillo y morado	3
Perezia multiflora	Escorzonera	4
Puya fastuosa	Carnero, hierba del carnero	5
Senecio canescens	Vira vira	6
Huperzia crassum	Cóndor	7
Loricaria ferruginea	Maqui maqui dorado	8
Bejaria aestuans	Purunrosa	9
Satureja sericea	Romero de jalca o romero blanco	10
Satureja nubigena	Pachachamcua	11
Loricaria leptothamna	Maqui maqui chico	12
Clethra sp.	Murmum	13
Valeriana sp.	Órnamo morado	14
Valeriana sp.	Órnamo blanco	15

Source: Seminario and Sánchez (2010), with permission of the authors.

The emollient's origins are not clear. However, it is known that in 1927, there was already in Lima *the Society of Emollient Makers*, which would have been created with the support of the Japanese colony. Because of this, it was said that the Japanese people brought emollient to Peru.

Bussman et al. (2015) and Ríos et al. (2017) suggest that the emollient would be the same as *horchata*, a drink made from tubers of chufa (*Ciperus esculentus*) whose origins would be in ancient Egypt (2400 BC) from where it moved to Europe and then to America with the conquest. In Latin America, this drink has different compositions, depending on the country or region, and usually includes a grain or seed (barley, rice, almonds, more cinnamon and sugar). Ríos et al. (2017) indicate that in the south of Ecuador, the *horchata* is an infusion of herbal mixture prepared with 16 to 32 herbs plus sugar, honey or unrefined sugar cane and some drops of lemon. However, this drink is different from the Peruvian emollient, which is a drink formed by a syrup or concentrated juice of a plant—for example lemon, flaxseed, cat's claw, chicory, aloe, grade blood–plus a liquid, product of the cooking of several plants or parts thereof—for example grains of barley, chamomile, horsetail, dog foot, apple, pineapple, quince—which is drunk for food and therapeutic purposes. In total, it can involve up to 15 plants at the same time. Also, the purposes are different since you can approach the three-wheeled carriages and ask the seller for a preparation for the kidneys, liver or stomach, depending on the condition you have or to recover after a *bad night*. In each case, the base liquid is the same, but the syrup will be different.

Between the years 1995 and 2001, in the city of Cajamarca, the ambulatory sale of maca juice, aloe extract, quinoa *shampoo* and orange juice was established. Evaluations between the years 2001 and 2003 showed that these five drinks involved a variable number of plant species and three-wheeled carriages in the city (Table 5.4). This ambulatory activity employed 126 families and included 58 plant species (38 plant families and 57 genera)—trees, shrubs, grasses and lianas—cultivated and wild, 60% of them being of American origin.

Substantial changes in the system of these beverages, with respect to production and collection, preparation and the market, are likely to have occurred to date, and information needs to be updated. For example, in the last years the sale of sugar cane juice, pineapple, grapefruit and melon has been observed.

On the other hand, the Municipality of Cajamarca has paid attention to and trained business drivers on food handling. Moreover, on May 17, 2014, the Peruvian government promulgated the Emollient maker Law (N° 30198), which recognizes these ventures as self-productive employment microenterprises and establishes February 20 as the *day of emollient, quinoa, maca, kiwicha and other traditional natural drinks*. Another important issue is that more than 30% of Lima's emollient makers are people from Cajamarca (Cajamarquinos) (Bussmann et al. 2015). This confirms the role of Cajamarca's families in the diffusion of this drink. Likewise, preliminary observations indicate that people from Cajamarca, mainly from Bambamarca, have taken this business to Ecuador (e.g., Quito and Cuenca), where it is sold as an emollient—not as *horchata*—and has achieved remarkable development. Undoubtedly, the subject has ethnobotanical, economic, social, phytochemical, and public health perspectives that must be studied in detail.

Table 5.4. Number of three-wheeled carriages and species involved in the emollient, maca juice, aloe extract and orange juice, in the city of Cajamarca: 2001 and 2003.

Beberage	2001		2003	
	No three-wheeled carriages	No Species	No three-wheeled carriages	No Species used
Emollient	45	20	60	25
Maca juice	98	w.d.	45	14**
Aloe extract	2	2	8	29
Quinoa shampoo*	15	4	15	4
Orange juice	2	2	13	2

w.d.: Without data. *Was sold in some emollient three wheeled carriages. **Six species involved in the maca's adulteration (most common).
Source: Seminario (2004), with permission of the author.

Plant Species for Gynecological Use and for Birth Control

A pioneering study in this field was carried out by Castañeda and Vargas (1991). The research was carried out in the villages of Milpo and Shitabamba in the province of San Marcos. The use of plants and other resources in the birth control was investigated. They found that peasants of these localities used 14 plant species for this purpose, of which 10 were native—all wild—and four exotic (Table 5.5).

In four communities in the district of Huambos, province of Chota, an investigation was carried out to identify medicinal plants used for gynecological purposes in women—menstruation, fertility, delivery, postpartum and gynecological diseases (Ramos 2015). 39 plant species were used for this purpose, 28 of them were wild and native at the same time: 11 shrubs, 5 trees, 22 herbs and 1 succulent. Asteraceae was the most representative family with eight species, followed by Lamiaceae and Poaceae with three species each. Three species (*Gnaphalium dysodes* Spreng, *Juglans neotropica* Diels, *Mauria heterophylla* Kunth) are considered within one of the three IUCN and DS043-2006-AG categories of endangered species, one is endemic for Cajamarca and another is endemic for Cajamarca and La Libertad departments. The number of species according to 15 use categories is presented in Table 5.6.

Table 5.5. Plant species used in birth control in two communities of San Marcos, Cajamarca.

Common Name	Botanical Name	Common Name	Botanical Name
Granado	*Punica garnatum* L.	Quincetulpas	*Paranephilius uniflorus* (Poepp. & Endl.) H. Rob.
Huanga sola	*Hesperomeles* sp.	Ratanya	*Krameria triandra* (R. & P.)
Landa cushma	*Geranium* sp.	Ruda	*Ruta graveolens* L.
Oregano	*Origanum vulgare* L.	Tapa tapa	*Mimosa pectinata* Kunth
Pacharosa	*Lantana reptans* Hayer	Cascarilla	*Cinchona* sp.
Paja sola	*Aa paleacea* (HBK)	Corcho	*Quercus* suber L.
Papa madre	*Dioscorea* sp.	Sangre de Drago	*Croton* sp.

Source: Prepared with data from Castañeda and Vargas (1991).

The Arracacha (*Arracacia xanthorrhiza* Bancroft) and Its Wild Relatives in Traditional Medicine

The arracacha's tuberous root is rich in good quality starch (high in amylopectin) which is highly digestible, so it is recommended for children, as well as for the elderly and sedentary people. In addition, because of its richness in beta carotene, calcium and phosphorus, and because it is cultivated under organic conditions, it constitutes a food with perspective.

Arracacia xanthorrhiza **Bancroft** has a high morphological variation. The morphological characterization of 186 entries from northern Peru (La Libertad–Piura), through 17 standardized descriptors, indicate the presence of 76 groups or morphotypes (Seminario and Valderrama 2004). This species has, in Cajamarca, at least three wild relatives—*A. elata, A. equatorialis* and *A. incisa* (Blas et al. 2008) and both, the cultivated species and their relatives, are used in traditional medicine.

Table 5.6. Number of species, according to use categories.

Use Categories	Species	
	No	%
Vaginal infection *	21	42
Tear	5	10
Menstrual cramps	4	8
Analgesic	3	6
Abundant menstruation	3	6
Contraceptive	2	4
Dilator	2	4
Cleaning**	2	4
Regulation of the cycle	2	4
Cancer	1	2
Compose blood	1	2
Cut umbilical cord	1	2
Placental expulsion	1	2
Vaginal bleeding	1	2
Blow your hips	1	2

* Includes infections with white and yellow descents and vaginal inflammation.
** Includes vaginal and uterine cleansing.
Source: Ramos (2015, p.87), with permission of the author.

A study conducted in northern Peru registered 21 cases in which the arracacha and their wild relatives were used in therapeutic treatments, generally combined with other plants and ingredients. Nine refer to treatment and cure of the evil of fright or fright in its two forms: *evil of dry terror, evil of terror of water*. Six cases refer to the treatment of the woman's disorders: amenorrhea (suspension of the menstrual flow), delivery aid, and postpartum recovery, treatment of over part and retention of the placenta. Other uses are against the cleavage of the grandfather (the diseases that appear after having contact with places where the gentiles lived), shucaque or modesty (discomfort due to psychological causes), chirapa (a condition that is acquired when the person bathes in pools, ponds or dirty marshes) and purging in men (venereal disease). An interesting conclusion is that most uses of wild and cultivated arracachas (foliage) refer directly or indirectly to women and children (Valderrama and Seminario 2002).

Studies on yacon (Smallanthus sonchifolius (Poepp. & Endl.) H. Rob.)

The yacon until the 1990s was a plant at risk of extinction, grown in family gardens for consumption as fruit. However, as a result of Japanese studies, which showed that it contained high levels of phospho-oligosaccharides (FOS), it resurfaced in use and in the market. Today, it is cultivated throughout the country, it is in the market all year round and both consumers and researchers consider it as a medicinal plant or as a nutraceutical or functional food. The therapeutic uses collected from traditional knowledge and from initial scientific studies in humans were summarized by Lebeda et al. (2011), and there are studies—in laboratory animals—on their effects in the treatment of hyperglycemia, kidney problems, infertility, high cholesterol levels, immune system, cancer and as an antioxidant (Valentova and UlrichovÃ¡ 2003, Foy 2005, Alvarez et al. 2008, Lebeda et al. 2011, Choque et al. 2012, 2013, Satoh et al. 2013, Sook and Han 2013).

The yacon studies in Cajamarca began in 1993 through the Biodiversity Program of Roots and Andean Tubers, led by the International Potato Center (CIP), with funds from Swiss cooperation. They were basically focused on the diversity and variability of the plant in northern Peru, with an emphasis on Cajamarca. Products from these studies are a series of formal publications, frequently cited in the scientific literature (Seminario et al. 2003, Seminario and Valderrama 2004, Arnao et al. 2011, among others).

In 2002, the First National Course on Cultivation and Utilization of the Yacon was held in Cajamarca, whose reports were formally published (Seminario and Valderrama 2003). The theme was ethnobotany, biology, characterization and evaluation, management, harvesting, postharvest, uses and market.

The germplasm of yacon maintained by the National University of Cajamarca—Program of Roots and Andean Tubers—consists of 100 accessions. These were characterized by 20 standardized morphological descriptors and were grouped into eight morphotypes (Seminario et al. 2004). Agronomic and productivity studies are carried out with these materials (Seminario et al. 2017, Aguilar 2017).

Poisonous and Medicinal Plants and Their Use in Cajamarca

This study of Orozco (2003) was carried out in four places of Cajamarca (Contumazá, Chetilla, Sorochuco and Chilete), ranging from the upper limit of the Chala region (Chilete, 500 masl) to Jalca region (Chetilla, 4000 masl). Initially, it was focused on poisonous plants; however, the author says that when she raised the field inquiries, villagers were reluctant to address this issue because, culturally, this is a very sensitive aspect, poisonous plants are considered as *bad plants* or *devil's plants*. Because of this, the author had to change the focus, first covering the medicinal plants—*good plants*—or *plants of God* (of the amito) and around that subject, the poisonous plants were treated. For this reason, this work turned out to be a good inventory of medicinal plants in these areas. 95 species were registered. 57 species of this total were medicinal, 42 poisonous and within these, 22 useful. Twenty-six species were both medicinal and toxic (for humans, animals, and insects). Twenty-six were toxic to animals, of which seven were medicinal at the same time. Eight were toxic to humans, of which six were medicinal at the same time. The 16 toxic plants to the skin were separated from the rest, of which three were medicinal at the same time.

Studies on Valerian (*Valeriana pilosa* R. & P)

Valeriana pilosa is one of the main medicinal species collected in the Cajamarca's Jalca for the coastal markets. It is used as somnific, antispasmodic, soft anesthetic, anti-hysterical, anti-epileptic and sedative. It is also recommended to treat rheumatism, neurasthenia, insomnia and nervousness (Ramírez et al. 2006, Vásquez et al. 2010).

In its natural habitat, specifically in its first stage of life, it is dependent on the accompanying plants of the genera *Stipa* and *Calamagrostis*. Because of this, it does not thrive in the open field, without special care. Another considerable limitation is probably the excessive acidity of the soils where it grows (pH 3.4 to 3.9), which slows down its growth; hence, by the collectors' version, a new plant is able to be harvested for the market, after six to seven years after being seeded in its natural environment.

Studies on biodiversity and medicinal plants show Valerian as one of the most depredated species which would be at risk (Ramirez et al. 2006, Vásquez et al. 2010, Seminario et al. 2016), so it is proposed to deepen the studies leading to its incorporation into cropping systems. Valerian studies in the region date from 2006 and range from ethnobotany, plant biology in its natural environment, agronomic treatments for purposes of domestication, vegetative propagation and seed germination (Ramírez et al. 2006, Rojas and Seminario 2014, Rumay and Seminario 2015, Nazar and Alva 2015, Seminario et al. 2016, Valdez 2017). At the moment, basic knowledge is available to establish trials of response to growing condition.

Conclusions

The studies on medicinal plants in the Cajamarca's region had an ethnobotanical focus, mainly directed to know the diversity of the species, the market and the uses.

The studies cover the ecological regions of Yunga, Quechua and Jalca, with predominance towards the species of Jalca. They cover 16% of the districts of the Cajamarca's region and were developed mainly in the provinces of Cajamarca and Contumazá.

In 15 reviewed studies, 457 species in 105 families and 296 genera were recorded. Most species are wild and native to the Andes. The best represented families were Asteraceae, Lamiaceae, Fabaceae, Solanaceae, Rosaceae and Piperaceae.

The most frequently mentioned species in the studies were Ishpingo verde (*Achyrocline alata*), manzanilla (*Matricaria chamomilla*), cola de caballlo (*Equisetum bogotense*), pie de perro (*Desmodium mollicum*), chancua (*Minthostachys mollis*), hinojo (*Foeniculum vulgare*), ajenjo (*Artemisa absintium)*, carqueja (*Baccharis genistelloides*), escorzonera (*Perezia multiflora*), cerraja (*Sonchus oleraceus*), orégano (*Origanum vulgare*) and llantén (*Plantago major*).

Studies on the use of medicinal species as nutraceutical or functional foods and for the treatment of women's gynecological diseases and birth control are noteworthy.

There are gaps in information on medicinal plants in the rest of the territory and, in general, little is known about the state of conservation, species biology, local risk factors, demand-crop relationship and species response to Agronomic treatments for domestication purposes. There are also few studies on the content of the active ingredients and the pharmacological and medicinal effects.

Acknowledgments

To Rosel Orrillo for his support in the data collection and map elaboration.

To Juan Montoya (Herbarium Isidoro Sánchez Vega-UNC) for his collaboration with important documents.

References

Aguilar, I. 2017. Análisis estadístico de la productividad del germoplasma de yacón Smallanthus sonchifolius (Poepp. & Endl.) H. Robinson de la UNC. Tesis Ing. Agr. Cajamarca, PE, Facultad de Ciencias Agrarias, Universidad Nacional de Cajamarca. 127 p.

Aldave, M.A. 2003. Aspectos etnobotánicos de las plantas medicinales en la ciudad de Cajamarca. Tesis Ing. Agr. Cajamarca, PE, Facultad de Ciencias Agrarias, Universidad Nacional de Cajamarca. 72 p.

Alvarez, P., Jurado, B., Calixto, M., Incio, N. and Silva, J. 2008. Prebiótico inulina/oligofructosa en la raíz del yacón (*Smallanthus sonchifolius*) fitoquímica y estandarización como base de estudios preclínicos y clínicos. Rev. Gastroenterol. Perú 28: 22–27.

Alvitres, K., Huamán, I.M. and Vera, L. 2007. Plantas medicinales, biocidas y aromáticas del Distrito de La Encañada, Cajamarca. Tesis Licenciado en Educación. Cajamarca, PE. Facultad de Educación, UNC. 109 p.

Arnao, I., Seminario, J., Cisneros, R. and Trabuco, J. 2011. Potencial antioxidante de 10 accesiones de yacón (*Smallanthus sonchifolius* (Poepp. & Endl.) Robinson) procedentes de Cajamarca, Perú. An. Fac. Med. 72(4): 239–243.

Ayasta, D.M. 2012. Memoria e identidad: El caso de los cuspiniques. Librosperuanos.com. Consulta 22-11-2016. Disponible en http://www.librosperuanos.com/autores/articulo/00000002088/Memoria-e-Identidad-El-caso-de-los-Cupisnique.

Ayay, J.I. 2017. La agrobiodiversidad en la agricultura familiar del caserío Chilincaga, Centro Poblado Porcón Bajo, Cajamarca. Tesis Ing. Agr. Cajamarca, PE. Facultad de Ciencias Agrarias, Universidad Nacional de Cajamarca. 130 p.

Blas, R., Ghislain, M., Herrera, M.R. and Baudoin, J.P. 2008. Genetic diversity analysis of wild *Arracacia* species according to morphological and molecular markers. Genet. Resour Crop. Evol. 55: 625–642.

Bussmann, R.W. and Sharon, D. 2006. Traditional medicinal plant use in Northern Peru: Tracking two thousand years of healing culture. Journal of Ethnobiology and Ethnomedicine 2: 47. Consultado 29-08-2014. Disponible en: http://www.ncbi.nlm.nih.gov/pmc/articles/PMC1637095/.

Bussmann, R.W., Sharon, D. and Ly, J. 2008. From garden to market? The cultivation of native and introduced medicinal plant species in Cajamarca, Peru and implications for habitat conservation. Ethnobotany Research & Applications 6: 351–361. Consultado 27-07-2017. Disponible en: https://www.researchgate.net/publication/29744682. DOI: 10.17348/era.6.0.351-361.

Bussmann, R.W. and Sharon, D. 2009. Markets, healers, vendors, collectors: The sustainability of medicinal plant use in northern Peru. Mountain Research and Development (MRD) 29(2): 128–134.

Bussmann, R.W., Paniagua-Zambrana, N., Castañeda, R.Y., Prado, Y.A. and Mandujano, J. 2015. Healt in a pot—the ethnobotany of *emolientes* and *emolienteros* in Peru. Notes on economic plants. Economic Botany 69(1): 83–88.

Camino, L. 1992. Cerros, plantas y lagunas poderosas. La medicina al norte del Perú. Lluvia Editores, Lima, PE. 296 p.

Castañeda, D.T. and Vargas, M.L. 1991. Uso de plantas medicinales y otros métodos para el control de la natalidad: su conocimiento en los caseríos de Milco y Shitabamba de la provincia de San Marcos: 1987–1991. Tesis Lic. Sociología, Cajamarca, PE, Facultad de Ciencias Sociales, Universidad Nacional de Cajamarca. 110 p.

Castañeda, G.M. and Condori, E.M. 2010. Catálogo y estudio farmacognóstico de plantas medicinales del distrito de Llacanora, Provincia de Cajamarca, Departamento deCajamarca. Tesis Químico Farmacéutico. Lima, PE. Facultad de Farmacia y Bioquímica, UNMSM. 154 p.

Castillo-Vera, H., Cochachín, E. and Albán, J. 2017. Plantas comercializadas por herbolarios en el mercado del distrito de Cajabamba (Cajamarca, Perú). Blacpma 16(3): 303–318.

Castle, L.M., Leopold, S., Craft, R. and Kindscher, K. 2014. Ranking tool created for medicinal plants at risk of being overharvested in the wil. Ethonobioloy Letters 5: 77–88. Consultado 28-07-2017. Disponible en http://ojs.ethnobiology.org/index.php/ebl/article/view/169.DOI:10.14237/ebl.5.

Choque, G.T., Thome, R., Gabriel, D.L., Tamashiro, W.M.S.C. and Pastore, G.M. 2012. Yacon (*Smallanthus sonchifolius*)-derived fructooligosaccharides improves the immune parameters in the mouse. Nutrition Research 32: 884–892.

Choque, G.T., Silva, W.M.S.C., Maróstica, M.R. and Pastore, G.M. 2013. Yacon (*Smallanthus sonchifolius*): A funtional food. Plant Foods Hunt Nutr. 68: 222–228.

DS-043-2006-AG. 2006. Categorización de especies amenazadas de flora silvestre. DiarioOficial El Peruano, 13 julio 2006.

Foy, E. 2005. *Smallanthus sonchifolius* (llacón o yacón) en el tratamiento de hiperlipoproteinemias e hipercolesterolemia inducidas en ratas albinas. Rev. Fac. Mmed. Hum. 5(1): 27–31.

Iberico, L. 1881. El folklore agrario de Cajamarca. Universidad Nacional de Cajamarca, pp. 65–72.

Iberico, L. 1984. Folklore médico de Cajamarca. Universidad Nacional de Cajamarca. 238 p.

La Torre, M.A. 1998. Etnobotánica de los recursos vegetales silvestres del caserío de Yanacancha, distrito de Chumuch, provincia de Celendín, Cajamarca. Tesis Biólogo, Facultad de Ciencias, Universidad Nacional Agraria La Molina. 50 p.

Lebeda, A., Doležalová, I., Fernández, E. and Viehmannová, I. 2011. Yacon (Asteraceae; *Smallanthus sonchifolius*). Chapter 20. *In*: R.J. Singh (ed.). Genetic resources, chromosome engineering, and crop improvement: Medicnal plants. Vol. 6. Taylor & Francis Group, London.

León, B., Roque, J., Ulloa, C., Jorgenson, P.M., Pitman, N. and Cano, A. (eds.). 2006. El libro rojo de las plantas endémicas el Perú. Rev Peru Biol 13 (2). Número especial.

Ley 30198. Ley que reconoce la preparación y expendio o venta de bebidas elaboradas con plantas medicinales en la vía pública, como microempresas generadoras de autoempleo productivo. Diario Oficial El Peruano. Normas Legales. 17 de mayo, 2014.

Marcelo, J.L., Sánchez, I. and Millán, J. 2006. Estado de la diversidad florística del Páramo sectores: El Espino y Palanque, Sallique, Jaén, Cajamarca, Perú. Ecología Aplicada 5(1,2): 1–8.

Martínez Compañon, B.J. 1789. Codex Trujillo del Perú. Vol. V. Biblioteca Nacional de Colombia. Biblioteca digital. Disponible en: http://www.bibliotecanacional.gov.co/content/%E2%80%9Ccodex-trujillo-del-per%C3%BA%E2%80%9D.

Montoya, E. and Figueroa, G. 1991. Geografia de Cajamarca. Vol. II. Taller de Estudio Fanny Abanto Calle, Lima, PE. 359 p.

Nazar, J. and Alva, E. 2015. Efecto del encalado en el crecimiento de *Valeriana pilosa* R. & P. en Huanico, Cajamarca. Fiat Lux 11(2): 53–59.

Orozco, O. 2003. Poisons plants and type uses in Cajamarca, Peru. A dissertation submitted to the Graduate Faculty in Biology in partial fulfillment of the requirements for the degree of Doctor of Phylosophy, The City University of New York. 304 p.

Pulgar Vidal, J. 1996. Las ocho regiones naturales del Perú. PEISA, Lima, PE. 302 p.

Ramírez, J.P., Terán, R.M., Sánchez, I. and Seminario, J. 2006. Etnobotánica de la valeriana (*Valeriana* spp.) en la Jalca de Cajamarca, Perú. Arnaldoa 13(2): 368–379.

Ramos, G.E.V. 2015. Plantas medicinales de uso ginecológico de cuatro comunidades del distrito de Huambos, provincia de Chota, departamento de Cajamarca. Tesis biólogo. Lima, PE. Facultad de Ciencias, Universidad Nacional Agraria, La Molina. 169 p.

Revene, Z., Bussmann, R.W. and Sharon, D. 2008. From sierra to coast: Tracing the supply of medicinal plants in northern Peru—A plant collector's tale. Ethnobotany Research and Application 6: 15–22. Consultado 10-09-2014. Disponible en: www.ethnobotanyjournal.org/ vol6/i1547-3465-06-015. pdf.

Ríos, M., Tinitana, F., Jarrín, P., Donoso, N. and Romero-Benavides, J.C. 2017. "Horchata" drink in southern Ecuador: medicinal plants and peoples's wellbeing. Journal of Ethnobiology and Ethnomedicine 2017:13–18. Consultado 05-08-2017. Disponible https://link.springer.com/content/pdf/10.1186%2Fs13002-017-0145-z.pdf DOI:10.1186/S13002-017-0145-z.

Rojas, J.M. and Seminario, J. 2014. Método alométrico para estimar el área foliar de "valeriana" (*Valeriana pilosa* R. & P.) al estado silvestre. Arnaldoa 21(2): 305–316.

Roberson, E. 2008. Medicinal plants at risk. Tucson. Centre for Biological Diversity. Consultado 28-07-2017. Disponible en: http://www.biologicaldiversity.org/publications/papers/Medicinal_Plants_042008_lores.pdf.

Rumay, D. and Seminario, J. 2015. Respuesta de *Valeriana pilosa* R. & P a tres tratamientos agronómicos. Fiat. Lux. 11(2): 101–110.

Sagástegui, A. 1995. Diversidad florística de Contumazá. Universidad Privada del Norte, Fondo Editorial. Trujillo, Perú. 203 p.

Sagástegui, A., Dillon, M.O., Sánchez, I., Leiva, S. and Lezama, P. 1999. Diversidad florística del norte del Perú. Tomo I. WWF (World Wildlife Fund). Lima, PE. 228 p.

Sánchez, I. and Dillon, M. 2006. Jalcas. pp.77–90. *In*: Moraes, M., Øllarard, B., Kvist, P., Borchsenius, F. and Balslev, H. (eds.). Botánica Económica de los Andes. Universidad Nacional de San Andrés, La Paz, Bolivia.

Sánchez, I. 2011. Especies medicinales de Cajamarca I. Contribución etnobotánica, morfológica y taxonómica. Universidad Privada Antonio Guillermo Urrelo, Lumina Cooper Fondo Editorial. 228 p.

Sánchez, I. 2014. Plantas medicinales en los páramos de Cajamarca. pp. 175–194. *In*: Cuesta, F., Sevink, J., Llambi, L.D., de Bièvre, B. and Posner, J. (eds.). Avances en Investigación Para la conservación de los páramos andinos.

Satoh, H., Audrey, M.T., Kudoh, A. and Watanabe, T. 2013. Yacon diet (*Smallanthus sonchifolius*, Asteraceae) improves hepatic insulin resistance via reducing Trb3 expression in Zuker fa/fa rats. 2013. Nutrition & Diabetes 3. Consultado 27-07-2017. Disponible en http://www.nature.com/nutd/journal/v3/n5/full/nutd201311a.html?foxtrotcallback=true. DOI:10.1038/nutd.

Singh, R.J. 2012. Landmark research in medicinal plants (cap. 1). pp. 1–11. *In*: Singh, R.J. (ed.). Genetic Resources, Chromosome Engineering, and Crop Improvement. Medicinal Plants.

Singh, R.J., Lobeda, A. and Tucker, A.O. 2012. Medicinal plants—nature's pharmacy. pp. 14–51. *In*: Singh, R.J. (ed.). Genetic Resources, Chromosome Engineering, and Crop Improvement. Medicinal Plants.

Singh, R.D., Mody, S.K., Patel, H.B., Devi, S., Modi, C.M. and Kamani, D.R. 2014. Pharmaceutical biopiracy and protection of traditional knowledge. Int. J. Res. Dev. Pharm. L. Sci. 3(2): 866–871.

Seminario, J., Valderrama, M. and Manrique, I. 2003. El yacón. Fundamentos para el aprovechamiento de un recurso promisorio. CIP (Centro Internacional de la Papa), UNC (Universidad Nacional de Cajamarca), COSUDE (Cooperación Suiza al Desarrollo), Lima, PE. 57 p.

Seminario, J. and Valderrama, M. (eds.). 2003. I Curso Nacional Cultivo y Aprovechamiento del Yacón. Cajamarca, 26 al 29 agosto 2002. 113 p.

Seminario, J. and Valderrama, M. 2004a. Variabilidad morfológica y distribución geográfica de la colección de arracacha (*Arracacia xanthorrhiza* Bancroft) de la Universidad Nacional de Cajamarca, Perú. Arnaldoa 11(2): 79–104.

Seminario, J., Valderrama, M. and Romero, J. 2004b. Variabilidad morfológica y distribución geográfica del yacón, *Smallanthus sonchifolius* (Poepp. & Endl.) H. Robinson, en el norte peruano. Arnaldoa 11(1): 139–160.

Seminario, J. 2004. Etnobotánica del emoliente y otras bebidas de venta ambulatoria en la ciudad de Cajamarca. Caxamarca 12(1): 9–28.

Seminario, J., Rumay, L.D. and Seminario, A. 2016. Biología de *Valeriana pilosa* R. & P. (Valerianaceae): una especie en peligro de extinción de las altas montañas de Perú. Bol. Latinoam Caribe Plant Med. Aromat. 15(5): 337–351.

Seminario, J., Oblitas, I. and Escalante, B. 2017. Area foliar del yacón (*Smallanthus sonchifolius* (Poepp. & Endl.) H. Rob.) estimada mediante método indirecto. Agron. Mesoam. 28(1): 171–181.

Seminario, A. and Sánchez, I. 2010. Estado y factores de riesgo de la biodiversidad de especies vegetales medicinales en el Centro Poblado de Combayo, Cajamarca. Fiat. Lux. 6(1): 23–34.

Seminario, A. and Escalante, B. 2017. Potencial de la flora medicinal silvestre con fines de conservación en el distrito La Encañada, Cajamarca. Fiat. Lux. 12(1). In press.

Sook, J. and Han, K. 2013. The spermatogenic effect of yacon extract and its constituents and their inhibition effect of testosterone metabolism. Biomolecules & Therapeutics 21(2): 153–160.

Tinitana, F., Ríoss, M., Romero Benavides, J.C., De la Cruz, M. and Pardo, M. 2016. Medicinal plants sold at traditional markets in southern Ecuador. J. Ethnobiology and Ethnomedicine 2016: 12–29. Consultado 04-08-2017. Disponible en https://www.researchgate.net/publication/304915895_Medicinal_plants_sold_at_traditional_markets_in_southern_Ecuador. DOI:10.1186/s13002-016-0100-4.

Towle, M. 1961. The ethnobotany of pre-columbian Peru. Aldine publishing Company, Chicago, US. 180 p.

Ugent, D. and Ochoa, C.M. 2006. La etnobotánica del Perú. Desde la prehistoria al presente. Concytec. Lima, PE. 380 p.

Valderrama, M. and Seminario, J. 2002. Los parientes silvestres de la arracacha (*Arracacia xanthorrhiza* Bancroft) y su uso en medicina tradicional en el norte peruano. Arnaldoa 9(1): 67–91.

Valdez, H. 2017. Caracterización morfológica y germinación de la semilla de valeriana (*Valeriana pilosa* R. & P.). Tesis Ing. Agr, Cajamarca, PE. Facultad de Ciencias Agrarias, Universidad Nacional de Cajamarca. 62 p.

Valentova, K. and Ulrichová. 2003. *Smallanthus sonchifolius* and *Lepidium meyenii* prospective andean crops for the prevention of chronic diseases. Biomed. Papers 147(2): 119–130.

Vásquez, L., Escurra, J., Aguirre, R., Váquez, G. and Vásquez, L.P. 2010. Plantas medicinales del norte del Perú. FINCTYC, UNPRG, Lambayeque, PE. 382 p.

Watanabe, S. 2002. El reino de Cuismancu: orígenes y transformación en el Tahuantinsuyo. Boletín de Arqueología PUCP n° 6: 107–136.

Weberbauer, A. 1945. El mundo vegetal de los Andes peruanos. Estudio fitogeográfico. Estación Experimental Agrícola La Molina. Dirección de Agricultura. Ministerio de Agricultura. 2 ed. Talleres Gráficos de la Editorial Lumen S.A. Lima, PE.

Yacovleff, E. and Herrera, F.L. 1934. Botánica etnológica (3). El mundo vegetal de los antiguos peruanos. Revista del Museo Nacional 3(3): 243–322.

CHAPTER 6

Recent Reports on Ethnopharmacological and Ethnobotanical Studies of *Valeriana carnosa* Sm. (Valerianaceae)

Soledad Molares,[1,*] *Ana H. Ladio*[2] and *Nicolás Nagahama*[3]

Introduction

In southernmost South America, both in Argentinean and Chilean Patagonia, subterranean organs of numerous species of plants have long been recognized as being of great value to Mapuche and Tehuelche regional ethnic groups and rural Creole (Ladio and Lozada 2009, Molares and Ladio 2009a, Ochoa and Ladio 2011). These species also constitute an important part of many regional rites and legends (Ochoa and Ladio 2014).

From the perspectives of economic botany and ethnopharmacology, the main value of these species is based on the fact that their subterranean organs often contain starch and other carbohydrates of importance to the human diet, and also therapeutic compounds derived from plant secondary metabolism (Gurib-Fakim

[1] CIEMEP (Centro de Investigación Esquel de Montaña y Estepa Patagónica). Universidad Nacional de la Patagonia San Juan Bosco-CONICET. Roca 780 Esquel, Chubut. Argentina.
[2] Laboratorio Ecotono. INIBIOMA (Instituto de Biodiversidad y Medio Ambiente). Universidad Nacional del Comahue-CONICET. Quintral 1250-S.C. de Bariloche, Rio Negro. Argentina.
[3] Estación Experimental Agroforestal Esquel. Instituto Nacional de Tecnología Agropecuaria-CONICET. Chacabuco 513. Esquel, Chubut, Argentina.
* Corresponding author: smolares@gmail.com

2006). Amongst these species, some representatives of *Valeriana* L. genus have been used as medicinal plants, with high cultural and symbolic value. The underground organs of many *Valeriana* species contain numerous compounds and are used as a sedative and for treating insomnia, allowing the reduction of nervousness and agitation associated with stress (Thies and Funke 1966, Wagner et al. 1980, Nahrstedt 1984, Grusla et al. 1986, Upton 1999). This genus is widely studied, with special focus on anxiolytic properties (Hattesohl et al. 2008, Murphy et al. 2010).

For anxiolytic purposes, since ancestral times, extracts of rhizomes and roots of the Eurasian species *V. officinalis* L. have been used worldwide. Roots of *V. officinalis* are used for treatment of anxiety and mild sleep disorders. Studies indicate that flavonoids in this species have sedative activity in the central nervous system (Marder et al. 2003, Fernández et al. 2004, 2005, 2006, Lacher et al. 2007). Others authors suggests that biochemical composition and active constituents in *Valeriana* are valepotriates (Backlund and Moritz 1998), valeric acid and gamma-aminobutyric acid (Hallam et al. 2003, Nam et al. 2013). Besides, pre-clinical studies reported the antidepressant-like activity of *V. officinalis* (Hattesohl et al. 2008), *V. jatamansi* Jones (Subhan et al. 2010, Sah et al. 2011), *V. glechomifolia* Mey. (Müller et al. 2012) and *V. prionophylla* (Holzmann et al. 2011).

Currently, alternatives to *V. officinalis* are being sought in different countries for replacement by indigenous representatives, for example, in India with *V. jatamansi*, in Mexico with *V. edulis* Nutt. ex Torr. & A. Gray subsp. *procera* (Kunth) G.F. Mey and in Brazil with *V. glechomifolia* (Bos et al. 1999, Oliva et al. 2004).

In Patagonia, there are records of the use of at least five native species of *Valeriana* for medicinal purposes (Conticello et al. 1997, Molares and Ladio 2008). Among them, the most used *Valeriana* species by regional ethnic groups is *V. carnosa* Sm. ("Ñamkulawen", which means the white hawk medicine in Mapuzungun language, probably in reference to the high sites where the species grows and where the ñamku-*Buteo polyosoma*-can be seen in flight) and is considered a "sacred plant" (Estomba et al. 2005, Molares and Ladio 2008) (Fig. 6.1). This local name (Ñamkulawen) is shared with *V. clarionifolia* but this plant has different reputed attributes. Another local name is "Valeriana", which is used by some Creole settlers.

Valeriana carnosa stands out as one of the principal elements in the indigenous pharmacopoeias of Patagonia, and its roots and rhizomes have been used since ancient times (Molares and Ladio 2009b). The local perception of this plant is that it has wide-ranging curative powers: *"it's a cure-all"*. This attribute confers on the species high cultural and symbolic value for the Mapuche people, and its reputation and use has spread throughout the formal and informal medicinal herb market of Patagonian cities (Ladio 2006).

Fig. 6.1. General appearance of *V. carnosa* in a Patagonian forest-steppe ecotone habitat.

An "Appropriate" Taxonomy for a Cure-All Plant

Ñamkulawen (*Valeriana carnosa*, synonym: *Valeriana magellanica* Lam.), belonging to the Valerianaceae nom. conserv. family (currently considered within the Caprifoliaceae s.l.; APG III, 2009), has long been thought to represent a natural group of ca. 350 species distributed throughout much of the world, mainly found in the Northern Hemisphere and along the Andes mountain range, with the exception of Australia and New Zealand (Borsini 1966, Backlund 1996, Bell 2004, 2007).

South America is an important diversification center of Valerianaceae with approximately 250 species (Eriksen 1989, Bell 2004, Hidalgo et al. 2004, Bell and Donoghue 2005), of which the genus *Valeriana* is the most numerous with c.a. 200 species.

It has been suggested that Holartic *Valeriana* genera have been present on the South American continent for some time (> 13 MY), and have exploited new niche opportunities, migrating from a temperate to a more Mediterranean-style climate (Bell et al. 2012).

Despite Miller (1754) and de Candolle (1815) amendments, *Valeriana* remains a heterogeneous genus, especially due to the great diversity in species of South and Central America. Some authors suggest from phylogenetic studies that *Valeriana* is paraphyletic (Hidalgo et al. 2004). However, Bell (2004), based on phylogenetic analyzes combined with chloroplast and nuclear DNA sequences, observed a strongly supported clade that includes the South American *Valeriana* species.

In the southern Andes (Argentina and Chile), there are ~ 40 species of *Valeriana* that occur over a wide ecological as well as elevational gradient (Bell et al. 2012) and in the southernmost region (central Chile and the province of Neuquén in

Argentina), 25 endemic species have been identified and is considered an important center of secondary diversification for the genus (Kutschker and Morrone 2012).

The name of the genus stems from the latin *valere*, "to be healthy", a reference to the medicinal uses of its plants, particularly those associated with treating nervous conditions and hysteria (Borsini et al. in Correa 1999). Their epithet *carnosa* makes references to the consistency of the leaves (Ferreyra et al. 2006).

Ethnobotanical Reports About Ñamkulawen in Patagonia

The subterranean organs have been cited as a remedy used for hepatic, respiratory, circulatory, urinary and digestive disorders as well as having analgesic, anti-inflammatory, anti-tumoral, anti-depressive and wound-healing properties (e.g., Martinez Crovetto 1980, Estomba et al. 2006, Molares and Ladio 2009a,b, 2012, Richeri et al. 2013). It has also gained great prestige for its usefulness in treating cultural syndromes like the "*susto*", "*evil eye*" and "*frío*" (Molares 2010). *Valeriana carnosa* can also used in mixtures with other species, like "nalka" (*Gunnera tinctorea* (Molina) Mirb.) to strengthen its medicinal attributes (Molares 2010), or with "carqueja" (*Baccharis sagittalis* (Less.) DC.) and "palo piche" (*Fabiana imbricata* Ruiz et Pav.) to make "body cleansers" (Toledo and Kutschker 2012), which are used in a process which is both symbolic and practical, where the wellbeing of the person is sought by eliminating all the elements (physical, social and spiritual) which may be causing harm (Molares 2010). All these properties, grouped in seven ethnocategories according to the particular precepts of the Mapuche culture, have led to the plant also being recognized as "the remedy that cures the seven diseases" (Molares and Ladio 2012).

The local indigenous communities use the plant through decoction. They boil a piece of root, approximately 3 cm in length per liter of water, and then drink a cup each day until the liter is finished. According to our sources, ñamkulawen is "a powerful plant". Perception of the strong bitter taste of this decoction is an indicator of high therapeutic effectiveness, but also of potential danger, and because of this it is only consumed by adults and the dosages used are highly controlled and sporadic (Molares and Ladio 2009a). Traditionally, for those reasons its use is not recommended for children or pregnant women (Kutschker et al. 2002). In addition, the dosage must be small because it causes sleepiness (Weigandt et al. 2004) and an excessive dosage can even be fatal (Molares and Ladio 2009a).

Ecological and Sensorial Properties in the Local Criteria of Searching and Gathering Ñamkulawen

Valeriana carnosa is widely distributed and is common to the whole of Patagonia (Borsini et al. in Correa 1999). Usually, this plant grows in mid- to low elevation habitats, with a few occurring at higher elevations. In Chile, it inhabits the southern mountain range, in the VI, VII, VIII, IX, X, XI and XII regions; in Argentina, it inhabits the Mendoza, Neuquén, Río Negro, Chubut, Santa Cruz and Tierra del

Fuego provinces. Its altitudinal range is from 0 m.a.s.l. to 2,700 m.a.s.l. (Zuloaga et al. 2008). In phytogeographic terms, it is found in the Sub-Antarctic, Patagonian and High-Andean provinces (Borsini et al. in Correa 1999).

The species flourishes in xeric, open, sunny environments in the rocky soils of the forest, steppe and the Patagonian-Andean forest-steppe ecotone. It is also found in sandy sites, on low, sunny slopes or even in rocky sites of the Patagonian Andes. It flowers from October to December and fruits during the months of January and February (Borsini et al. in Correa 1999, Kutschker 2011).

In spite of the wide geographic distribution of this species, gathering carried out by the settlers is characterized by the search for specimens in stony areas with a high level of light exposure, preferably at the highest altitude possible. The underground part is collected with the help of simple tools like knives and spades. In the process of identification and selection of specimens, cultural practices of sensory perception come into play. These include the recognition of organoleptic qualities directly associated with this species, such as its bitter and unpleasant smell (*"like dirty feet"*) and its strong, bitter, repulsive flavor (*"füre"*), which is rather spicy (*"trapi"*) and astringent (*"seco"*) (Molares and Ladio 2009a).

Various studies indicate that the collection of this species is associated with the care of livestock. People take advantage of the time during which their animals are grazing to look for the plant in places far from their dwellings (Estomba et al. 2006, Richeri et al. 2013). With regard to the identification and collection of *V. carnosa* and *V. clarionifolia* by Patagonian inhabitants, studies reveal levels of organoleptic differentiation between the two species, which are of great cultural and ethnopharmacological value. For example, it was discovered that locals are capable of differentiating between *Valeriana* species, and that even though they recognize them as related (which can be deduced by the fact that both have the same common name), they can tell them apart by their smell and taste, which consequently determine their different uses and value (Molares and Ladio 2012). Unlike *V. carnosa*, *V. clarionifolia* is used for a limited number of ailments, mainly to relieve lower back pain and treat kidney and bladder disorders and cultural syndromes. In a curiously similar way, by means of laboratory tests with electronic noses, differences between the aromatic profiles of *V. carnosa* and *V. clarionifolia* have been found, which are determined by the chemical differences between the species (Baby et al. 2005).

The collected pieces of *V. carnosa* are usually taken to the dwellings where they are dried in the open air and in the shade, under cover, to be preserved later in mesh or paper bags. This practice ensures availability of the dried resource all year round, and is particularly useful in winter when the search for medicinal herbs on the mountains can become difficult due to the accumulation of snow (Molares and Ladio 2012).

Although *V. carnosa* gathering is very important and its commercialization has increased rapidly over the last decades (Cuassolo 2009), this species can be regarded as not threatened. However, settlers say that in some regions it is increasingly difficult to find plants, and that longer distances must be travelled in the search for them (Estomba et al. 2005, 2006). For this reason, the study of this plant's cultivation requirements must be encouraged (Cuassolo 2009). Currently, keeping in view the importance of this species, a simple low cost technique using rooting hormones for vegetative production of *V. carnosa* through macro-propagation has been developed (Fig. 6.2) (Nagahama et al. 2016). Additionally, these authors identified potential habitats of *V. carnosa* along the Argentine Patagonian with predicting models for its cultivation in order to obtain potential marketable font of phytomedicines and strengthen the non-conventional productive development in Patagonia (Nagahama et al. 2016).

Fig. 6.2. Vegetative production of *V. carnosa* through macro-propagation techniques.

What do we Know About *V. carnosa* Ethnopharmacology?

Research carried out on *Valeriana carnosa* reveals the presence of active ingredients similar to those of *V. officinalis*, which is present in many pharmacopoeias for oral consumption as a sedative and sleep inducer for humans (Gratti et al. 2010).

Several studies on the *Valeriana* genus indicate that the main active ingredients are the valepotriates, lignans, flavonoids, tannins, phenolic acids and essential oils (Kutschker et al. 2010). In particular, the essential oils have been researched; they primarily consist of elemol, bornyl-acetate, bornyl-isovalerate, isovalerate, and valerenone (Baby et al. 2005). Of all the Patagonian species belonging to this genus, the dry extract of the whole *V. carnosa* plant has been most studied (Cuadra and Fajardo 2002). It has been found that its valepotriate composition pattern, and especially its valtrates, is similar to *V. officinalis*, which is known for its tranquilizing and sleep inducing effect (Kutschker et al. 2010). However, according to Castillo and Martínez (2007), the chemical composition of *V. carnosa* varies according to time of collection, preparation and packaging. In addition, Cuadra and Fajardo (2002) have isolated caffeoyl methyl ester and two pinoresinol-type lignans. Fajardo et al. (2010) have also suggested that in terms of its biological activity, it would present cytotoxic activity and negative toxicological activity. Guajardo et al. (2018) suggests that phenolic compounds vary quantitatively and qualitatively between populations as well as among plants' phenological stages. Regarding the total phenol content, the values obtained in *V. carnosa* (from ethanolic root extracts) varied between 3.3–14.3 mg eq GAE/g dry material, being similar to those reported for root extracts in other species of medicinal use within the family Valerianaceae such as *V. officinalis* (14.2 mg eq GAE/g dry material; Surveswaran et al. 2007) and *V. jatamansi* (8.7 to 14.6 mg eq GAE/g dry material; Jugran et al. 2013).

Taking account of the above information, the traditional and recommended method of use, which consists of the decoction of a handful of the material, followed by ingestion of one cupful, orally, over a variable timeframe (Cuassolo 2009, Cuassolo et al. 2011), Kutschker et al. (2002) propose uses of the plant in modern medicine by means of the preparation of tinctures. The crude drug consists of dried pieces of the roots and rhizomes. Kutschker et al. (2002) describes a dosage of a daily cupful drunk on an empty stomach for a week. The roots are placed in a jar with 300 ml of alcohol, left for 15 days and then filtered. The recommended dosage is 1 to 2 ml as a sedative.

Morphological and Anatomical Description: Very Important Information in Quality Control of Commercial Samples

Valeriana carnosa is an evergreen herb of up to 80 cm in height, simple or branching from the base. Fleshy rhizome is up to 50 cm long, with weak branches. Basal leaves are 6–21 × 3–7 cm, obovate or elliptic, smooth edged or coursely toothed, glabrous and fleshy; petioles are 3–12 cm long. Upper leaves are sessile or petiolate, 0.6–4.5 cm, obovate, triangular or lanceolate, smooth edged or toothed. Inflorescenses are axillary or terminal, paniculiform and lax. Bracts are 3–9 mm in length, whole, oblong-lanceolate, ovate. Bracteoles are 2.5–4 mm in length, entire or auriculate, oblong-lanceolate, acute, glabrous or have long hairs on the edges, at the base. Flowers are hermaphrodite, the corollas are 4 mm long, bell-shaped or funnel-like, gibbous at the base with oblong lobes, and included stamens. The female flowers are 2–3 mm long, bell-shaped with ovate lobes. The styles are exerted and thickened at the tip. The fruit measures 5–7 × 2–3.5 mm, and is pyriform, with thick veins, and is glabrous; pappus is formed by 14–15 feathery setae (Borsini et al. in Correa 1999) (Fig. 6.3).

Even if *Valeriana carnosa* is the most used *Valeriana* in the Patagonian region, other species, mainly *V. clarionifolia,* are used in similar ways in traditional and nontraditional medicine, which has led to the need for comparative anatomical studies between the two species in order to avoid confusion in relation to the raw material. Likewise, the quality control works are very important considering that "Ñamkulawen" is sold in bulk or hand packed in paper or cellophane bags for sale

Fig. 6.3. Diagram of the aerial parts of the plant (a), floral structures (b and c) and fruit (d) of *V. carnosa* Sm. (Taken from Borsini et al. in Correa 1999).

in drugstores and herbalist's shops, at different degrees of fragmentation and only under their common name, from which the botanical identification is difficult.

According to diagnostic anatomical data provided by Bach et al. (2014), *V. carnosa* showed a primary pentarch aktinostele root, pith in the secondary structure and a rhizome with anomalous structure. *V. clarionifolia,* in contrast, has no rhizome and showed a protostele as a primary root structure and a secondary structure without pith. During the maceration process, the *V. carnosa* rhizome presented cork with irregular polygonal cells with acute and obtuse angles, while in *V. clarionifolia* rectangular cork cells with right angles were observed. Starch grains are simple, spherical in *V. carnosa* and polyhedral in *V. clarionifolia*. In addition, Molares and Ladio (2012) studied cross sections of *V. carnosa* primary root and observed a well-developed periderm consisting of cells with thickened, birefringent walls, from irregular to polygonal; cells of this tissue and phloem parenchyma have essential oils in the form of droplets (Sudan IV+); cortex has large air spaces between oval cells with brown contents.

Conclusions

Valeriana carnosa is one of the most prominent medicinal plants in the Mapuche tradition, and from an ethnopharmacological viewpoint, one of the most versatile promising medicinal plants in Patagonia, when taking into account the wide range of therapeutic alternatives it can offer for the treatment of the different ailments of the region (Molares and Ladio 2009b, Richeri et al. 2013). However, *V. carnosa* is not included in the Argentine Pharmacopoeia (http://www.anmat.gov.ar), nor does it appear on the list of toxic species not recommended for consumption.

The similarity between the active compounds found in *V. carnosa* and *V. clarionifolia* and those of *V. officinalis* is promising, since this species is included worldwide in many pharmacopoeias and consumed orally as a sedative and sleep inducer in humans. However, little conclusive evidence for the efficacy of the other local uses can be provided. The key problem of various investigations has been the emphasis on very few compounds rather than traditional preparations. Much more research is required to evaluate the actual efficacy of the preparations. The scientific research and cultural revalorization of the role played by *V. carnosa* in local herbal medicines is of considerable ethnopharmacological interest and highly relevant to the medicinal security of Patagonian communities. However, there is evidence to indicate that the abundance of this species in natural environments is decreasing, mainly due to disturbance of the environments (Estomba et al. 2006, Ladio et al. 2007) and lack of regulation of its commercialization in Patagonian cities (Cuassolo 2009). Given that the roots are the organs of medical interest in this valuable species, the establishment of conservation strategies *in situ* and studies that provide guidelines for its cultivation and preservation *ex situ* are of the utmost importance (Nagahama et al. 2016).

Acknowledgements

We are profoundly grateful to the inhabitants of the rural and urban communities where we learned about the ethnobotany of *V. carnosa*. This study was funded by Consejo Nacional de Investigaciones Científicas y Técnicas (CONICET).

References

Baby, R.E., Cabezas, M., Kutschker, A., Messina, V. and Walsöe de Reca, N.E. 2005. Discrimination of different valerian types with an electronic nose. J. Arg. Chem. Soc. The Journal of the Argentine Chemical Society 93: 43–50.

Bach, H.G., Varela, B.G., Fortunato, R.H. and Wagner, M.L. 2014. Pharmacobotany of two *Valeriana* species (Valerianaceae) of argentinian Patagonia known as "Ñancolahuen." Lat. Am. J. Pharm. 33: 891–896.

Backlund, A.A. 1996. Phylogeny of the Dipsacales. Doctoral Dissertation. Dept. Systematic Botany, Uppsala University, Sweden.

Backlund, A. and Moritz, T. 1998. Phylogenetic implications of an expanded valepotriate distribution in the Valerianaceae. Biochem. Syst. Ecol. 26: 309–335.

Bell, C.D. 2004. Preliminary phylogeny of Valerianaceae (Dipsacales) inferred from nuclear and chloroplast DNA sequence data. Mol. Phylogenet. Evol. 31: 340–350.

Bell, C.D. and Donoghue, M.J. 2005. Phylogeny and biogeography of Valerianaceae (Dipsacales) with special reference to the South American valerians. Org. Divers. Evol. 5: 147–159.

Bell, C.D. 2007. Phylogenetic placement and biogeography of the North American species of *Valerianella* (Valerianaceae: Dipsacales) based on chloroplast and nuclear DNA. Mol. Phylogenet. Evol. 44: 929–941.

Bell, C.D., Kutschker, A. and Arroyo, M.T.K. 2012. Phylogeny and diversification of Valerianaceae (Dipsacales) in the southern Andes. Mol. Phylogenet. Evol. 63: 724–737. doi:10.1016/j. ympev.2012.02.015.

Borsini, O.E. 1966. Valerianáceas de Chile. Lilloa 32: 375–476.

Borsini, O.E., Rossow, R.A. and Correa, M.N. 1999. Valerianaceae. pp. 449–468. *In*: Correa, M.N. (ed.). Parte VI. Dicotyledones Gamopetalas. Flora Patagónica. Buenos Airess: INTA.

Bos, R., Woerdenbag, H.J., Hendriks, H., Smit, H.F., Wikström, H.V. and Scheffer, J.J. 1997. Composition of the essential oil from roots and rhizomes of *Valeriana wallichii* DC. Flavour Frag. J. 12: 123–131.

Castillo García, E. and Martínez Solís, I. 2007. Manual de Fitoterapia. Elsevier, España, 536 pp.

Conticello, L., Gandullo, R., Bustamante, A. and Tartaglia, C. 1997. El uso de plantas medicinales por la comunidad Mapuche de San Martín de los Andes, Provincia de Neuquén. Parodiana 10: 165–180.

Cuadra, P. and Fajardo, V. 2002. A new lignan from the Patagonian *Valeriana carnosa* Sm. Bol. Soc. Chil. Quim. 47: 361–366.

Cuassolo, F. 2009. Estudio Etnobotánico de las plantas medicinales nativas y exóticas comercializadas en la Ciudad de Bariloche, Patagonia, Argentina. Universidad Nacional del Comahue.

Cuassolo, F., Ladio, A.H. and Ezcurra, C. 2011. Aspectos de la comercialización y control de calidad de las plantas medicinales más vendidas en una comunidad urbana del NO de la Patagonia Argentina. Bol. Lat. y del Caribe, Pl. Medic. y Arom. 9: 166–176.

De Candolle, A.P. 1815. Flore française. V6. Paris.

Eriksen, B. 1989. Notes on generic and infrageneric delimitation in the Valerianaceae. Nordic J. Bot. 9: 179–187.

Estomba, D., Ladio, A.H. and Lozada, M. 2005. Plantas medicinales utilizadas por una comunidad Mapuche en las cercanías de Junín de los Andes, Neuquén. Bol. Lat. y del Caribe, Pl. Medic. y Arom. 4: 107–112.

Estomba, D., Ladio, A. and Lozada, M. 2006. Medicinal wild plant knowledge and gathering patterns in a Mapuche community from North-western Patagonia. J. Ethnopharmacol. 103: 109–119. doi:10.1016/j.jep.2005.07.015.

Fajardo, V., Gallardo, A., Araya, M., Joseph-Nathan, P., Oyarzún, A., Cuadra, P., Sanhueza, V., Manosalva, L., Villarroel, L. and Darias, J. 2010. Química y algunos antecedentes y ensayos simples de la actividad biológica de plantas de la zona austral de chile. Dominguezia 26: 4–5.

Fernández, S., Wasowski, C., Paladini, A.C. and Marder, M. 2004. Sedative and sleep-enhancing properties of linarin, a flavonoid-isolated from *Valeriana officinalis*. Pharmacol. Biochem. Behav. 77: 399–404.

Fernández, S., Wasowski, C., Paladini, A.C. and Marder, M. 2005. Synergistic interaction between hesperidin, a natural flavonoid, and diazepam. Eur. J. Pharmacol. 512: 189–198.

Fernández, S., Wasowski, C., Loscalzo, L., Granger, R.E., Johnston, G.A.R., Paladini, A.C. and Marder, M. 2006. Central nervous system depressant action of flavonoid glycosides. Eur. J. Pharmacol. 539:168–176.

Ferreyra, M., Ezcurra, C. and Clayton, S. 2006. Flores de alta montaña de los Andes patagónicos. "High Mountain Flora of the Patagonian Andes". Editorial L.O.L.A., Buenos Aires.

Gratti A., Peneff, R. and Freile, M.L. 2010. *Valeriana* en la estepa patagonica, Argentina: Aportes al conocimiento fitoquimico. Dominguezia 26: 19–20.

Grusla, D., Holzl, J. and Krieglstein, J. 1986. Effect of valerian on rat brain. Dtsch. Apoth. Ztg. 126: 2249–2253.

Guajardo, J., Gastaldi, B., González, S. and Nagahama, N. 2018. Variability of phenolic compounds at different phenological stages in two populations of *Valeriana carnosa* Sm. (Valerianoideae, Caprifoliaceae) in Patagonia. Bol. Lat. y del Caribe, Pl. Medic. y Arom. 17 (4): 381–393.

Gurib-Fakim, A. 2006. Medicinal plants: traditions of yesterday and drugs of tomorrow. Mol. Aspects Med. 27: 1–93. doi:10.1016/j.mam.2005.07.008.

Hallam, K.T., Olver, J.S., McGrath, C. and Norman, T.R. 2003. Comparative cognitive and psychomotor effects of single doses of *Valeriana officinalis* and triazolam in healthy volunteers. Hum. Psychopharmacol. 18: 619–625.

Hattesohl, M., Feistel, B., Sievers, H., Lehnfeld, R., Hegger, M. and Winterhoff, H. 2008. Extracts of *Valeriana officinalis* L. sl show anxiolytic and antidepressant effects but neither sedative nor myorelaxant properties. Phytomedicine 15: 2–15.

Hidalgo, O., Garnatje, T., Susanna, A. and Mathez, J. 2004. Phylogeny of Valerianaceae based on matK and ITS markers, with reference to matK individual polymorphism. Ann. Bot. 93: 283–293.

Holzmann, I., Cechinel Filho, V., Mora, T.C., Cáceres, A., Martínez, J.V., Cruz, S.M. and de Souza, M.M. 2011. Evaluation of behavioral and pharmacological effects of hydroalcoholic extract of *Valeriana prionophylla* Standl. from Guatemala. Evid. Based Complement. Alternat. Med. http://dx.doi.org/10.1155/2011/312320.

http://www.anmat.gov.ar/webanmat/fna/fna.asp. 2013. Farmacopea Argentina. 7ed. Consultado: Febrero de 2015.

Jugran, A., Rawat, S., Dauthal, P., Mondal, S., Bhatt, I.D. and Rawal, R.S. 2013. Association of ISSR markers with some biochemical traits of *Valeriana jatamansi* Jones. Ind. Crops Prod. 44: 671–676.

Kutschker A., Menoyo, H. and Hechem, V. 2002. Plantas medicinales de uso popular en comunidades del oeste del Chubut. Ed. Bavaria. INTA-UN de la Patagonia S.J.B.-GTZ, Bariloche.

Kutschker, A. 2011. Revisión del género *Valeriana* (Valerianaceae) en Sudamérica austral. Gayana Bot. 68: 244–296.

Kutschker, A., Ezcurra, C. and Balzaretti, V. 2010. *Valeriana* (Valerianaceae) de los Andes australes: biodiversidad y compuestos químicos. pp. 219–224. *In*: Pochettino, M.L., Ladio, A.H. and Arenas, P.M. (eds.). Tradiciones y Transformaciones en Etnobotánica, La Plata: CYTED.

Kutschker, A. and Morrone, J.J. 2012. Distributional patterns of the species of *Valeriana* (Valerianaceae) in southern South America. Pl. Syst. Evol. 298: 535–547.

Lacher, S.K., Mayer, R., Sichardt, K., Nieber, K. and Müller, C.E. 2007. Interaction of valerian extracts of different polarity with adenosine receptors: Identification of isovaltrate as an inverse agonist at A1 receptors. Biochem. Pharmacol. 73: 248–258.

Ladio, A. 2006. Gathering of wild plant foods with medicinal use in a Mapuche community of Northwest Patagonia. pp. 297–321. *In*: Pieroni, A. and Price, L.L. (eds.). Eating and Healing: Traditional Food… . USA: Harworth Press.

Ladio, A., Lozada, M. and Weigandt, M. 2007. Comparison of traditional wild plant knowledge between aboriginal communities inhabiting arid and forest environments in Patagonia, Argentina. J. Arid Environ. 69: 695–715. doi:10.1016/j.jaridenv.2006.11.008.

Ladio, A.H. and Lozada, M. 2009. Human ecology, ethnobotany and traditional practices in rural populations inhabiting the Monte region: Resilience and ecological knowledge. J. Arid Environ. 73: 222–227. doi:10.1016/j.jaridenv.2008.02.006.

Marder, M., Violab, H., Wasowski, C., Fernández, S., Medinab, J.H. and Paladini, A.C. 2003. 6-Methylapigenin and hesperidin: new valeriana flavonoids with activity on the CNS. Pharmacol. Biochem. Behav. 75: 537–545.

Martínez Crovetto, R. 1980. Apuntes sobre la vegetación de los alrededores del Lago Cholila. Publicación Técnica de la Facultad de Ciencias Agrarias 1: 1–22.

Miller, P. 1754. The gardeners dictionary, abr. Self published, London. Vol. 3, 584 pp.

Molares, S. and Ladio, A. 2008. Plantas Medicinales en una comunidad mapuche del NO de la Patagonia Argentina: clasificación y percepciones organolépticas relacionadas con su valoración. Bol. Lat. y del Caribe, Pl. Medic. y Arom. 7(3): 149–155.

Molares, S. and Ladio, A. 2009a. Chemosensory perception and medicinal plants for digestive ailments in a Mapuche community in NW Patagonia, Argentina. J. Ethnopharmacol. 123: 397–406. doi:10.1016/j.jep.2009.03.033.

Molares, S. and Ladio, A. 2009b. Ethnobotanical review of the Mapuche medicinal flora: use patterns on a regional scale. J. Ethnopharmacol. 122: 251–60. doi:10.1016/j.jep.2009.01.003.

Molares, S. 2010. Flora medicinal aromática de la Patagonia: características anatómicas y propiedades organolépticas utilizadas en el reconocimiento por parte de la terapéutica popular. Tesis Doctoral. Universidad Nacional del Comahue. Bariloche.

Molares, S. and Ladio, A.H. 2012. Plantas aromáticas con órganos subterráneos de importancia cultural en la patagonia argentina: una aproximación a sus usos desde la etnobotánica, la percepción sensorial y la anatomía. Darwiniana 50: 7–24.

Müller, L.G., Salles, L.A., Stein, A.C., Betti, A.H., Sakamoto, S., Cassel, E. and Rates, S.M. 2012. Antidepressant-like effect of *Valeriana glechomifolia* Meyer (Valerianaceae) in mice. Prog. Neuropsychopharmacol. Biol. Psychiatry 36: 101–109.

Murphy, K., Kubin, Z.J., Shepherd, J.N. and Ettinger, R. 2010. *Valeriana officinalis* root extracts have potent anxiolytic effects in laboratory rats. Phytomedicine 17: 674–678.

Nagahama, N., Bach, H.G., Opazo, W., Miserendino, E., Arizio, C., Manifesto, M. and Fortunato, R. 2016. Aprovechamiento sustentable de recursos genéticos nativos: *Valeriana carnosa* Sm., estudio de caso de una planta medicinal patagónica. Dominguezia 32: 68–69.

Nam, S.M., Choi, J.H., Yoo, D.Y., Kim, W., Jung, H.Y., Kim, J.W. and Yoon, Y.S. 2013. *Valeriana officinalis* extract and its main component, valerenic acid, ameliorate D-galactose-induced reductions in memory, cell proliferation, and neuroblast differentiation by reducing corticosterone levels and lipid peroxidation. Exp. Gerontol. 48: 1369–1377.

Nahrstedt, A. 1984. Drugs and phytopharmaca having sedative activity. Dtsch. Apoth. Ztg. 124: 1213–1216.

Ochoa, J.J. and Ladio, A.H. 2011. Pasado y presente del uso de plantas silvestres con órganos de almacenamiento subterráneos comestibles en la Patagonia. Bonplandia 20: 265–284.

Ochoa, J.J. and Ladio, A.H. 2014. Ethnoecology of *Oxalis adenophylla* Gillies ex Hook. & Arn. J. Ethnopharmacol. 155: 533–542.

Oliva, I., González-Trujano, M., Arrieta, J., Enciso-Rodríguez, R. and Navarrete, A. 2004. Neuropharmacological profile of hydroalcohol extract of *Valeriana edulis* ssp. *procera* roots in mice. Phytother. Res. 18: 290–296.

Richeri, M., Cardoso, M.B. and Ladio, A.H. 2013. Soluciones locales y flexibilidad en el conocimiento ecológico tradicional frente a procesos de cambio ambiental: estudios de caso en Patagonia. Ecología Austral. (23): 184–193.

Sah, S.P., Mathela, C.S. and Chopra, K. 2011. Antidepressant effect of *Valeriana wallichii* patchouli alcohol chemotype in mice: Behavioural and biochemical evidence. J. Ethnopharmacol. 135: 197–200.

Subhan, F., Karim, N., Gilani, A.H. and Sewell, R.D.E. 2010. Terpenoid content of *Valeriana wallichii* extracts and antidepressant-like response profiles. Phytother. Res. 24: 686–91.

Surveswaran, S., Cai, Y.Z., Corke, H. and Sun, M. 2007. Systematic evaluation of natural phenolic antioxidants from 133 Indian medicinal plants. Food Chem. 102: 938–953.

Thies, P.W. and Funke, S. 1966. Über die wirkstoffe des baldrians: 1. Mitteilung Nachweis und isolierung von sedativ wirksamen isovaleriansäureestern aus wurzeln und rhizomen von verschiedenen valeriana-und kentranthus-arten. Tetrahedron Letters 7: 1155–1162.

Toledo, C. and Kutschker, A. 2012. Plantas Medicinales en el Parque nacional los alerces, chubut, Patagonia Argentina. Bol. Soc. Argent. Bot. 47: 461–470.

Upton, R., Petrone, C., Swisher, D., Goldberg, A., Mc Guffin, M. and Pizzorno, N.D.J. 1999. Valerian Root *Valeriana officinalis* Analytical, Quality Control, and Therapeutic Monograph. Santa Cruz, CA: American Herbal Pharmacopoeia (AHP).

Wagner, H., Jurcic, K. and Schaette, R. 1980. Comparative studies on the sedative action of *Valeriana* extracts, valepotriates and their degradation products. Planta Med. 39: 358–365.

Weigandt, M., Ladio, A. and Lozada, M. 2004. Plantas medicinales utilizadas en la comunidad Mapuche Curruhuinca. Ediciones Imaginaria. Bariloche. Argentina. 75 pp.

Zuloaga, F.O., Morrone, O. and Belgrano, J.M. 2008. Catálogo de las plantas vasculares del Cono Sur. Monographs in systematic botany from the Missouri Botanical Garden. Ed. Missouri Botanical Garden Press. http://www2.darwin.edu. ar.

CHAPTER 7

Traditional Knowledge of Antivenom Plants

Bioactive Compounds and Their Antiophidic Properties

Carolina Alves dos Santos,[1,4] *Marco V. Chaud,*[1,4] *Valquíria Miwa Hanai Yoshida,*[3] *Raksha Pandit,*[2] *Mahendra Rai*[2,]* and *Yoko Oshima-Franco*[4]

Introduction

The medicinal plants and their secondary metabolites are capable of promoting different pharmacological and biological responses, and hence they are important source of pharmaceuticals. In Brazil, natural and animal resources favour the study and discovery of new plant activities, and hence the use of plants with antivenom property is an interesting area of research. In developing countries, communities in rural and urban areas are still heavily dependent on herbal medicines as primary health care (Maroyi 2017). In tropical and subtropical countries, such as Africa, Latin America, Asia and Oceania, snakebite poisoning is a real public health issue (Gomes et al. 2016).

[1] LaBNUS – Biomaterials and Nanotechnology Laboratory, University of Sorocaba, Sorocaba/SP, Brazil.

[2] Department of Biotechnology, SGB Amravati University, Amravati-444 602, Maharashtra, India.

[3] Department of Technology & Environmental Process, University of Sorocaba UNISO, Sorocaba, SP – Brazil.

[4] Post Graduate Program in Pharmaceutical Sciences, Department of Development and Evaluation of Bioactive Substances, University of Sorocaba, UNISO, Sorocaba, SP, Brazil.

* Corresponding author: pmkrai@rediffmail.com

Serum therapy is one of the measure for the treatment of poisoning caused by snakebites. The main aim of serum therapy is that it can reverse the effects of the complex and varied composition of substances and proteins present in the venom. The pathophysiological process of poisoning involves toxins such as metalloproteins, proteinases, phospholipases, hyaluronidases, pharmacological enzymes and mediators (Gomes et al. 2016). Poisoning may result in different degrees of toxicity such as neuroparalysis, multiple organ failure, and death (Naik and Sadananda 2017). The pathophysiological effects induced by snake venom are tissue necrosis, inhibition of platelet aggregation, hemorrhage, edema, etc. (Fig. 7.1).

Serum therapy consists of a mixture of antivenom antibodies isolated from animals that aim to reverse the systemic effects of poisoning such as cardiac changes, clotting disorders, renal and tissue damage, etc. (Félix-Silva et al. 2014). However, serum therapy is often not able to reverse tissue damage caused by the venom such as edema, injury, hemorrhage and necrosis. Such damages are very debilitating and often generate permanent sequelae. Therefore, the strategies to reduce these toxic effects are necessary, which can prevent the toxic effects associated with snake venom. The neutralization of the local effects of venom is a frequent problem and the use of therapeutic compounds is an alternative, which is capable of reversing or minimizing tissue damage. In this context, the use of plants with antiophidic properties is interesting which can be used in the treatment and it may enhance the benefits obtained from serum therapy. The major disadvantages of serum therapy is that it is toxic, costly and the accessibility limitations in rural regions and distant urban centers (Tianyi et al. 2017).

The present chapter discusses about the significance of plants with antivenom properties. A few therapies are known, which can be used in the treatment of snake

Fig. 7.1. The pathophysiological effects of snake venom.

bites and the available therapies have many drawbacks. The secondary metabolites present in traditionally used plants can be used against snake venom.

The Global Scenerio

Globally, there are alarming reports of snake-bite. Naik and Sadananda (2017) reported that more than 5 million people are bitten by venomous snakes in a year with fatal case in 100,000 of them and that in India nearly 45,000 people died in 2005 due to snake bites.

According to Chippaux (2017), two families of snakes are responsible for snake envenomations in the America: The Viperidae and the Elapidae. The other types represent less than 1% of envenomations. Chippaux (2017) also represent the statistics for snakebite, about 27,200 per year in Brazil and more than 115 deaths during the period of 2001–2012. The main genus responsible for the bites in Brazil is *Bothrops* with a high incidence of bites in northern Amazon, in male, children or people older than 40 years. Feitosa et al. (2013) carried out epidemiological surveillance of annual snakebite count in the State of Amazonas and reported 200 cases/100,000 inhabitants in some areas, which is among the highest annual snakebite rate in the world.

In US (United States), 4735 native venomous snakebites are reported every year and approximately half of these are from crotalids. The majority of the victims are males less than 19 years. Recently, it was reported that the death rate of snake bitten patients has decreased since the last 5–6 years (Corbett and Clark 2017).

Schiermeier (2015) reported that according to the Médecins Sans Frontiéres (MSF), health care workers responsible for treating snakebites in Africa and Sudan report an estimated 30,000 deaths every year with amputations for around 8,000 people due to snakebite.

In Africa, the problem is more critical because the main company responsible for serum production stopped its production and the other types of antivenom product around the world are not totally effective for some species of snakes present in Africa (Schiermeier 2015).

Plants with Antivenom Properties

The main therapeutic strategy to solve snakebite envenoming is the use of antiserum therapy that is able to avoid the lethal effects of venom. However, this main therapy is not able to minimize the local effects of venom enzymes (phospholipases, proteases and hyaluronidases) that are mainly responsible for necrosis and hemorraghic conditions observed after a snakebite. To minimize the effects of these enzymes present in the local injury, plants with antivenom properties have been used with good results since antiquity (Table 7.1).

The list of compounds related with antivenom properties is wide. In fact, the bioactive compounds responsible for antivenom activity are a combination between alkaloids, flavonoids, saponins, tanins and others.

Table 7.1. Plants species, main parts of this plants studied, experimental design and inhibition of snakes' poison.

Family	Plant	Part of Plant used/Bioactive Compounds	Snake Venoms/Isolated Compounds	Experimental Design	References
Acanthaceae/ Aristolochiaceae	*Andrographis paniculata* (Burm. f.) Wall. ex Nees	Polarity-based stem and leaf extracts	*Naja naja*	*In vivo*	Gopi et al. 2011
		Methanol whole plant extract	*Daboia russelli*	*In vitro* and *in vivo*	Meenatchisundaram et al. 2009a
	Aristolochia indica	Roots/Aristolochic acid	*Vipera russelli*	*In vitro* and *in vivo*	Bhattacharjee and Bhattacharyya 2013
		Methanol whole plant extract	*Daboia russelli*	*In vitro* and *in vivo*	Meenatchisundaram et al. 2009
	Aristolochia cymbifera L.	Aqueous-isopropanol green-leaf extract	*Bothrops alternatus*	*In vitro* and *in vivo*	Melo et al. 2007
Amaryllidaceae	*Crinum jagus* L.	Methanol bulb extract	*Echis ocellatus* *Bitis arietans* *Naja nigricollis*	*In vitro* and *in vivo*	Ode and Asuzu 2006
Anacardiaceae	*Annacardium occidentale* L.	Methanol bark extract	*Vipera russelli*	*In vitro* and *in vivo*	Ushanandini et al. 2009b
	Mangifera indica L. cv. 'Fahlun'	Aqueous stem bark extract	Group IA sPLA2 (i.e., purified NN-XIaPLA2 phospholipase A2 enzyme from *Naja naja* venom)	*In vitro* and *in vivo*	Dhananjaya and Shivalingaiah 2016
			NN-XIb-PLA2 from *Naja naja*	*In vitro* and *in vivo*	Dhananjaya et al. 2016
			VRV-PL-VIIIa from *Daboia russelli*	*In vitro* and *in vivo*	Dhananjaya and Sudarshan 2015

		Ethanol seed kernel extract/ pentagalloyl glucopyranose	*Calloselasma rhodostoma; Naja naja kaouthia*	*In vitro*	Pithayanukul et al. 2009
			Calloselasma rhodostoma; Naja naja kaouthia	*In vitro*	Leanpolchareanchai et al. 2009
Annonaceae	*Annona senegalensis* Pers	Methanol root extract	*Naja nigricollis nigricollis* Wetch	*In vivo*	Adzu et al. 2005
Apiaceae	*Eryngium yuccifolium* Michx.	aqueous extract of the flowers, rootlets, leaves and seeds		*In vitro*	Price 2016
Apocynaceae	*Hemidesmus indicus* (L.) R.Br.	HI-RVIF compound from root extract	*Daboia russelli*	*In vitro* and *in vivo*	Alam et al. 1994
		2-hydroxy-4-methoxy benzoic acid from root extract		*In vitro* and *in vivo*	Alam and Gomes 1998a and 1998b
		lupeol acetate from root extract	*Daboia russellii Naja kaouthia*	*In vitro* and *in vivo*	Chatterjee et al. 2006
	Mandevilla velutina K. Schum	Aqueous subterranean system extract	*Crotalus durissus terrificus*; CB; *crotoxin*; *Bothrops jararacussu*; BthTX-I; BthTX-II; BthA-I-PLA2; *B. alternatus*; *B. moojeni*; *B. pirajai*; PrTX-I.	*In vitro* and *in vivo*	Biondo et al. 2003
	Mandevilla illustris (Vell) Woodson	Aqueous subterranean system extract	*Crotalus durissus terrificus* snake venom, isolated basic phospholipase A_2 (CB) and crotoxin	*In vitro* and *in vivo*	Biondo et al. 2004

Table 7.1 contd. ...

...Table 7.1 contd.

Family	Plant	Part of plant used/bioactive compounds	Snake venoms/isolated compounds	Experimental design	References
	Rauwolfia serpentina L.	Ethanol whole plant extract	*Naja naja*	*In vitro* and *in vivo*	Rajasree et al. 2013
	Tabernae montana catharinensis A. DC.	Aqueous root barks extract/ fractions	*Crotalus durissus terrificus*	*In vitro* and *in vivo*	Batina et al. 2000
		Eight fractions, PI to PVIII	*Crotalus durissus terrificus*	*In vitro* and *in vivo*	De Almeida et al. 2004
		Aqueous root barks extract	*Bothrops jararacussu*; Bothropstoxin-I (BthTX-I); Bothropstoxin-II (BthTX-II)	*In vitro* and *in vivo*	Veronese et al. 2005
Araceae	*Dracontium croatii* Zhu.	Ethanol leaf and branches extract	*Bothrops asper*	*In vitro* and *in vivo*	Nuñez et al. 2004
Asteraceae	*Bidens pilosa* L.	Plant extracts	*Dendroaspis jamesoni* *Echis oceliattus*	*In vivo*	Chippaux et al. 1997
	Calendula officinalis L.	Ointment containing 10% extract	*Bothrops alternatus*	*In vitro* and *in vivo*	Melo et al. 2005
	Eclipta prostrata (L.) L.	Ethanol aerial parts extract; Wedelolactone; sitosterol; stigmasterol	*Crotalus durissus terrificus*	*In vitro* and *in vivo*	Mors et al. 1989
		Wedelolactone	*Crotalus viridis viridis*; *Agkistrodon contortrix laticinctus*; CVV and ACL PLA2-myotoxins	*In vitro* and *in vivo*	Melo and Ownby 1999
		Aqueous extract/ wedelolactone, stigmasterol, sitosterol	*Bothrops jararaca, B. jararacussu, Lachesis muta* and myotoxins	*In vitro* and *in vivo*	Melo et al. 1994

Family	Species	Extract	Venom	Assay	Reference
		Butanol extract of plant/ demethylwedelolactone	*Calloselasma rhodostoma*	*In vitro* and *in vivo*	Pithayanukul et al. 2004
		Plant extract	*Gloydius brevicaudus, G. shedaoensis, G. ussuriensis* and *Deinagkistrodon acutus*	*In vivo*	Chen et al. 2005
	Mikania laevigata Sch. Bip. ex Baker	Ethanol leaf extract	*Philodryas olfersii*	*In vitro* and *ex vivo*	Collaço et al. 2012
	Pluchea indica (L.) Less.	Root extract/beta-sitosterol, stigmasterol	*Daboia russelli; Naja naja*	*In vitro* and *in vivo*	Gomes et al. 2007
Balanitaceae/ Zygophyllaceae	*Balanites aegyptiaca* (L.) Del.	Polarity-based stem barks extract	*Echis carinatus*	*In vivo*	Wufem et al. 2007
		Seed oil	*Vipera russelli*	*In vivo*	Mishal et al. 2014
Bignoniaceae	*Tabebuia rosea* (Bertold.) DC.	Ethanol leaf and branches extract	*Bothrops asper*	*In vitro* and *in vivo*	Nuñez et al. 2004
Bixaceae	*Bixa orellana* L.	Ethanol leaf and branches extract	*Bothrops asper*	*In vitro* and *in vivo*	Nuñez et al. 2004
Boraginaceae	*Argusia argentea* (L.f.) H. Heine	Methanol extract/Rosmarinic acid	*Trimeresurus flavoviridis; Crotalus atrox, Gloydius blomhoffii, Bitis arietans and mettalloproteinases HT-b (C. atrox),* bilitoxin 2 (*Agkistrodon bilineatus*), HF (*B. arietans*), and Ac1-proteinase (*Deinagkistrodon acutus*)	*In vivo*	Aung et al. 2010

Table 7.1 contd. ....

...*Table 7.1 contd.*

Family	Plant	Part of plant used/bioactive compounds	Snake venoms/isolated compounds	Experimental design	References
	Cordia verbanaceae DC.	Methanol extract/Rosmarinic acid	*B. jararacussu* venom and its basic PLA2s	*In vitro* and *in vivo*	Ticli et al. 2005, Soares et al. 2005
		Rosmarinic acid	PrTX-I (*Bothrops pirajai*)	*In vitro* and *ex vivo*	Santos et al. 2011
Burseraceae	*Boswellia dalzielli* Hutch.	Aqueous stem barks extract	*Naja nigricollis*	*In vitro* and *in vivo*	Goje et al. 2013
	Commiphora africana A. Rich.	Methanol stem bark extract/ fractions	*Naja nigricollis*	*In vitro*	Isa et al. 2015
Caesalpiniaceae/ Fabaceae	*Bauhinia forficata* Link. (Fabaceae)	Aqueous aerial parts extract	*Bothrops jararacussu; B. moojeni; Crotalus durissus terrificus* and PLA2s	*In vitro*	Oliveira et al. 2005
	Brownea rosademonte Berg.	Ethanol leaf and branches extract	*Bothrops asper*	*In vitro* and *in vivo*	Nuñez et al. 2004
Combretaceae	*Combretum leprosum* Mart.	Root extract; Arjunolic acid	*Bothrops jararaca; B. jararacussu*	*In vitro* and *in vivo*	Fernandes et al. 2014
	Combretum molle R. Br. ex G. Don	Aqueous extract	*Bitis arietans; Naja nigricollis*	*In vitro*	Molander et al. 2014
	Guiera senegalensis J.F.Gmel.	Leaf extract	*Echis carinatus; Naja nigricollis*	*In vitro* and *in vivo*	Abubakar et al. 2000
	Terminalia fagifolia Mart.	Ethanol stem barks extract	*Bothrops jararacussu*	*In vitro* and *ex vivo*	Tribuiani et al. 2017
Clusiaseae	*Hypericum brasiliense* Choisy	Ethanol whole plant extract	*Bothrops jararaca*	*In vitro* and *in vivo*	Assafim et al. 2011
		Ethanol leaf extract	*Crotalus durissus terrificus;* crotamine and crotoxin	*In vitro* and *ex vivo*	Dal Belo et al. 2013

Family	Plant	Extract/Compound	Snake	Study type	Reference
Crassulaceae	*Kalanchoe brasiliensis* Camb.; *Kalanchoe pinnata* (Lam.) Pers.	Hydroalcoholic leaf extract of *K. brasileinsis* and flavonoid glycosides derived from patuletin and eupafolin; Hydroalcoholic leaf extract of *K. pinnata* and flavonoids glycosides derived from quercetin and kaempferol.	*Bothrops jararaca*	*In vitro* and *in vivo*	Fernandes et al. 2016
Crysobalanaceae	*Parinari curatellifolia* Planch. ex Benth.	Ethanol extract	*Naja nigricollis*	*In vitro* and *in vivo*	Omale et al. 2012
Dillenaceae	*Davilla elliptica* St. Hill.	Methanol leaf extract	*Bothrops jararaca*	*In vivo*	Nishijima et al. 2009
Ebenaceae	*Diospyros kaki* Thunb.	Condensed tannins from fruits	*Laticauda semifasciata* Reinwardt; *Laticauda semifasciata*; *Trimeresurus flavoviridis* Hallowell	*In vitro* and *in vivo*	Okonogi et al. 1979
	Diospyros kaki L.	Fractionation of lyophilized persimmon tannin/isolated compounds	*Naja naja atra* PLA2	*In vitro* and *in vivo*	Xu et al. 2012
		Polymeric persimmon proanthocyanidin fraction		*In vitro*	Li et al. 2013
		epigallocatechin-3-gallate; epigallocatechin-3-gallate (A-type EGCG dimer) and epicatechin-3-gallate-epicatechin-3-gallate (A-type ECG dimer)		*In vitro* and *in vivo*	Zhang and Li 2017
Ehretiaceae	*Cordia macleodii* (Griff.) Hook. f. & Thoms.	Ethanol bark extract	*Naja naja*	*In vitro* and *in vivo*	Soni and Bodakhe 2014
Euphorbiaceae	*Acalypha indica* L.	Ethanol leaf extract	*Vipera russelli russelli*	*In vivo*	Shirwaikar et al. 2004

Table 7.1 contd. ...

...Table 7.1 contd.

Family	Plant	Part of plant used/bioactive compounds	Snake venoms/isolated compounds	Experimental design	References
	Bridelia ferruginea Benth.	Ethanol leaf extract	*Naja nigricollis*	*In vitro* and *in vivo*	Momoh et al. 2012.
	Croton urucurana Baillon	Aqueous extract/ proanthocyanidins	*Bothrops jararaca*	*In vitro* and *in vivo*	Esmeraldino et al. 2005
	Emblica officinalis Linn. Gaertn.	Methanol root extract	*Vipera russelli Naja kaouthia*	*In vitro* and *in vivo*	Alam and Gomes 2003
		Root extract/Phtalate compound	*Naja Kaouthia; Vipera russellii*	*In vitro, ex vivo, in vivo*	Sarkhel et al. 2011
	Jatropha elliptica (Pohl) Oken.	Ethanol ryzomes extract	*Bothrops jararacussu*	*In vitro* and *in vivo*	Ferreira-Rodrigues et al. 2016
	Jatropha gossypiifolia L.	Aqueous leaf extract	*Bothrops jararaca*	*In vitro* and *in vivo*	Felix-Silva et al. 2014
	Jatropha mollissima (Pohl) Bail	Aqueous leaf extract	*Bothrops erythromelas; B. jararaca*	*In vitro* and *in vivo*	Gomes et al. 2016
	Dipteryx alata Vogel	Polarity-based barks extracts	*Bothrops jararacussu; Crotalus durissus terrificus*	*In vitro* and *ex vivo*	Nazato et al. 2010
		Polarity-based barks fractions	*Bothrops jararacussu*	*In vitro* and *ex vivo*	Puebla et al. 2010
		Lupeol; lupenone; 28-OH-lupenone; betulin	*Bothrops jararacussu; Crotalus durissus terrificus*	*In vitro* and *ex vivo*	Ferraz et al. 2012
		7,8,3'-trihydroxy-4'-methoxyisoflavone	*Bothrops jararacussu; Bothropstoxin-I (BthTX-I)*	*In vitro* and *ex vivo*	Ferraz et al. 2014
		Betulin	*Bothrops jararacussu*	*In vivo*	Ferraz et al. 2015
		Protocatechuic acid; Vanillic acid; Vanillin		*In vitro* and *ex vivo*	Yoshida et al. 2015

Erythrina senegalensis DC.				
Mucuna pruriens (L.) DC	Aqueous seed extract	Family Elapidae *Ophiophagus hannah*; *Naja sputatrix*; *Bungarus candidus*; *Notechis scutatus*; *Pseudechis australis*; Family Viperidae *Trimeresurus purpureomaculatus*; *Naja nigricollis*; *Bothrops asper*; *Agkistrodon piscivorus*; *Vipera russelli russelli*.	*In vitro* and *in vivo*	Tan et al. 2009
		Echis carinatus		Guerranti et al. 2002 Guerranti et al. 2004
Pentaclethra macroloba (Willd.) Kuntze	Aqueous extract	*Bothrops jaracussu*; *B. atrox*; metalloproteases	*In vitro* and *in vivo*	Da Silva et al. 2005
	Triterpenoid saponin inhibitors, named macrolobin-A and B	*B. neuwiedi*; *B. jararacussu*; metlloproteases	*In vitro* and *in vivo*	Da Silva et al. 2007
Tamarindus indica L.	Seed extract	*Vipera russelli*	*In vitro* and *in vivo*	Ushanandini et al. 2006
Flacourtiaceae	*Casearia gossypiosperma* Briquet	Hexane leaf extract	*Bothrops jararacussu*	*In vitro* and *ex vivo*	Soares-Silva et al. 2014
	Hydroalcoholic leaf extract			Camargo et al. 2010
Casearia sylvestris Sw.	Aqueous leaf extract	*Bothrops moojeni*; *B. jararacussu*; *B. neuwiedi*; *Crotalus durissus terrificus*; *Micrurus frontalis*	*In vitro* and *in vivo*	Borges et al. 2000

Table 7.1 contd. ...

...Table 7.1 contd.

Family	Plant	Part of plant used/bioactive compounds	Snake venoms/isolated compounds	Experimental design	References
			Bothrops asper; *B. jararacussu;* *B. moojeni;* *B. neuwiedi;* *B. pirajai;* and metalloproteinases.	*In vitro* and *in vivo*	Borges et al. 2001
		Aqueous/Hydroalcoholic leaf extract	Bothropstoxin-I (BthTX-I) from *Bothrops jararacussu*	*In vitro* and *ex vivo*	Oshima-Franco et al. 2005
		Polarity-based leaf extracts	*Bothrops jararacussu;* Bothropstoxin-I (BthTX-I)	*In vitro* and *ex vivo*	Cintra-Francischinelli et al. 2008
Heliconiaceae	*Heliconia curtispatha* Petersen	Ethanol leaf and branches extract	*Bothrops asper*	*In vitro* and *in vivo*	Nuñez et al. 2004
Hymenophyllaceae	*Trichomanes elegans* L.C. Rich	Ethanol leaf and branches extract	*Bothrops asper*	*In vitro* and *in vivo*	Nuñez et al. 2004
Icacinaceae	*Humirianthera ampla* (Miers) Baehni	Extracts and constituents	*Bothrops jararacussu,* *B. atrox; B. jararaca*	*In vitro* and *in vivo*	Strauch et al. 2013
Lamiaceae	*Peltodon radicans* Pohl.	Polarity-based leaf, flowers stem and barks extracts; phytochemical isolated	*Bothrops atrox*	*In vitro* and *in vivo*	Costa et al. 2008
	Vitex negundo L.	Methanol root extract	*Vipera russelli* *Naja kaouthia*	*In vitro* and *in vivo*	Alam and Gomes 2003
Leguminosae	*Brongniartia intermedia* Moric.; *B. podalyrioides* Kunth.	Isolated (-)-edunol from roots	*Bothrops atrox*	*In vivo*	Reyes-Chilpa et al. 1994

Family	Species	Extract	Snake venom	Assay	Reference
	Galactia glaucecens Kunth.	Ethanol leaf extract	*Crotalus durissus terrificus*	*In vitro* and *in vivo*	Dal Belo et al. 2008
Loganiaceae	*Strychnos nux vomica* L.	Ethanol seed extract	*Daboia russelli; Naja kaouthia*	*In vitro* and *in vivo*	Chatterjee et al. 2004
	Strychnos pseudoquina St. Hill.	Methanol leaf extract	*Bothrops jararaca*	*In vivo*	Nishijima et al. 2009
	Strychnos spinosa Lam.	Aqueous stem bark extract	*Echis carinatus*	*In vivo*	
Loranthaceae	*Struthanthus orbicularis*	Ethanol leaf and branches extract	*Bothrops asper*	*In vitro* and *in vivo*	Nuñez et al. 2004
Malpighiaceae	*Byrsonima crassa* Niendezu	Methanol leaf extract; Amenthoflavone; Quercetin	*Bothrops jararaca*	*In vivo*	Nishijima et al. 2009
Malvaceae	*Hibiscus aethiopicus* L.	Aqueous whole plant extract	*Echis ocellatus; Naja n. nigricollis*	*In vitro* and *in vivo*	Hasson et al. 2010
	Hibiscus subdariffa L.	Ethanol leaf and barks extract	*Echis carinatus* Sochureki	*In vivo*	Hasson et al. 2012
Melastomataceae	*Bellucia dichotoma* Cogn.	Aqueous barks extract	*Bothrops atrox*	*In vitro* and *in vivo*	Moura et al. 2014
	Mouriri pusa Gardn.	Methanol leaf extract; Quercetin; Myricetin	*Bothrops jararaca*	*In vivo*	Nishijima et al. 2009
Meliaceae	*Azadirachta indica* A. Juss.	Purified compound (AIPLAI) from methanol leaf extract	*Naja naja; Naja kaouthia; Daboia russelli*	*In vitro*	Mukherjee et al. 2008
Mimosaceae	*Abarema cochliacarpos* (Gomes) Barneby & J.W.Grimes	Hydroethanol stem barks extract	*Bothrops leucurus*	*In vitro* and *in vivo*	Saturnino-Oliveira et al. 2014

Table 7.1 contd. ...

...Table 7.1 contd.

Family	Plant	Part of plant used/bioactive compounds	Snake venoms/isolated compounds	Experimental design	References
	Dichorastachys cinerea W. & A.	Methanol and etherial root extracts	*Vipera russelli*	*In vivo*	Mishal 2002
	Mimosa pudica L.	Aqueous and alcoholic roots extracts	*Naja kaouthia*	*In vitro* and *in vivo*	Mahanta and Mukherjee 2001
		Tannin isolate	*Naja kaouthia*	*In vitro* and *in vivo*	Sia et al. 2011
					Ambikabothy et al. 2011
		Aqueous extract	*Naja naja kaouthia; Ophiophagus hannah; Bungarus candidus; B. fasciatus; Calloselasma rhodostoma*	*In vitro* and *in vivo*	Vejayan et al. 2007
	Parkia biglobosa	Water-methanol stem bark extract	*Echis ocellatus; Naja nigricollis*	*In vitro, ex vivo* and *in vivo*	Asuzu and Harvey 2003
Mimosoideae	*Plathymenia reticulata* Benth	Polarity-based barks extracts	*Bothrops jararacussu*	*In vitro* and *ex vivo*	Farrapo et al. 2011
Moraceae	*Ficus nymphaeifolia* Miller	Ethanol leaf and branches extract	*Bothrops asper*	*In vitro* and *in vivo*	Nuñez et al. 2004
	Morus alba L.	Leaf extract	*Daboia russelli*	*In vitro* and *in vivo*	Chandrashekara et al. 2009
Musaceae	*Musa paradisiaca* L.	Exudates or juice from liberian vessels	*Bothrops jararacussu; B. neuwiedi*	*In vitro* and *in vivo*	Borges et al. 2005
Piperaceae	*Piper longum* L.	Ethanolic fruits extract/ piperine	*Daboia russelli*	*In vitro* and *in vivo*	Shenoy et al. 2013, 2014
Polygalaceae	*Securidada longepedonculata* Fresen.	Aqueous leaf and root bark	*Naja nigricollis*	*In vitro* and *in vivo*	Sanusi et al. 2014

Polypodiaceae	*Pleopeltis percussa* (Cav.) Hook & Grev.	Ethanol leaf and branches extract	*Bothrops asper*	*In vitro* and *in vivo*	Nuñez et al. 2004
Rubiaceae	*Gonzalagunia panamensis* (Cav.) Schumm.	Ethanol leaf and branches extract	*Bothrops asper*	*In vitro* and *in vivo*	Nuñez et al. 2004
Rutaceae	*Citrus limon* L.	Ethanol leaf and branches extract	*Bothrops asper*	*In vitro* and *in vivo*	Nuñez et al. 2004
Salicaceae	*Salix nigra* Marshall	Methanol leaf extract	*Bungarus sindarus*	*In vitro*	Ahmed et al. 2016
Sapindaceae	*Sapindus saponaria* L.	Callus culture extracts (polarity-based solvents)	*Bothrops jararacussu; B. moojeni; B.alternatus; Crotalus durissus terrificus;* Myotoxins and phospholipases A2	*In vitro* and *in vivo*	da Silva et al. 2012
Sapotaceae	*Manilkara subsericea* (Mart.) Dubard	Polarity-based leaf and stems extracts	*Lachesis muta*	*In vitro* and *in vivo*	De Oliveira et al. 2014
Solanaceae	*Whitania somnifera* (L.) Dunal	NN-XIa-PLA2	*Naja naja*	*In vitro, ex vivo, in vivo*	Machiah and Gowda 2006
Theaceae	*Camellia sinensis* (L.) Kuntze	Methanol leaf extract	*Naja naja kaouthia* Lesson (Elapidae); *Calloselasma rhodostoma* Kuhl (Viperidae)	*In vitro* and *in vivo*	Pithayanukul et al. 2010
		Hydroalcohol leaf extract; theaflavin and epigallocatechin	*Crotalus durissus terrificus*	*In vitro* and *ex vivo*	de Jesus Reis Rosa et al. 2010
		Hydroalcohol leaf extract	*Bothrops jararacussu;* Bothropstoxin-I (BthTX-I)	*In vitro* and *ex vivo*	Oshima-Franco et al. 2012

Table 7.1 contd. ...

...*Table 7.1 contd.*

Family	Plant	Part of plant used/bioactive compounds	Snake venoms/isolated compounds	Experimental design	References
	Thea sinensis Linn	Melanin extrated from black tea	*Agkistrodon contortrix laticinctus; Agkistrodon halys blomhoffii; Crotalus atrox*	*In vivo*	Hung et al. 2004
Urticaceae	*Pouzolzia indica* (L.) G. Benn.	Polatity-based aerial parts extracts	*Viper russelli*	*In vitro* and *in vivo*	Ahmed et al. 2010
Velloziaceae	*Vellozia flavicans* Mart. ex Schult.	Hydroalcohol leaf extract	*Bothrops jararacussu*	*In vitro* and *ex vivo*	Tribuiani et al. 2014
Verbenacea	*Clerodendrum viscosum* Vent.	Alcohol root extract	*Naja naja*	*In vitro* and *in vivo*	Lobo et al. 2006
Vitaceae	*Vitis vinifera* L.	Methanol extract of seed grapes	*Duboia/Vipera russelli*	*In vitro* and *in vivo*	Mahadeswaraswamy et al. 2009
Zingiberaceae	*Curcuma longa* L.	Ar-turmerone fraction	*Bothrops jararaca; Crotalus durissus terrificus*	*In vitro* and *in vivo*	Ferreira et al. 1992
		Rhyzomes extracts (polarity-based solventes)/ ar-turmerone	*Bothrops alternatus*	*In vitro* and *in vivo*	Melo et al. 2005
		Aqueous extract	*Naja naja sinensis*	*In vitro* and *ex vivo*	Daduang et al. 2005
		Aqueous-isopropanol green leaves extract	*Bothrops alternatus*	*In vitro* and *in vivo*	Melo et al. 2007
	Curcuma sp.	Aqueous extract	*Naja naja siamensis*	*In vitro*	Cherdchu and Karlsson 1983
	Curcuma parviflora Wall.	Aqueous extract	*Naja naja sianensis*	*In vitro* and *ex vivo*	Daduang et al. 2005
	Renealmia alpinia (Rottb.)	Ethanol leaf and branches extract	*Bothrops asper*	*In vitro* and *in vivo*	Nuñez et al. 2004

Another advantage of using these plants in association with serum therapy is low cost and accessibility of this treatment in distant and poor areas, facilitating the access of treatment in remote areas, a problem common and frequent in developing countries.

Plant secondary metabolites or their analogues have proven effective and safe in the antivenom activity. Plant metabolities can be used indirectly as supplement to conventional antivenom therapy. It was revealed that bioprospecting of plants with antiophidic activity can be used in the inhibition of snake venom. Félix-Silva et al. (2014) reported antiophidic activity in plants of Fabaceae, Euphorbiaceae, Apocynaceae and Sapindaceae (Félix-Silva et al. 2014). Topical application of plant based material on the bitten area, by chewing leaves or barks, and decoctions of plant extract can counteract snake envenomation (Sebastin Santhosh et al. 2013).

Triterpenoids saponin isolated from *Pentaclothra macroloba* such as macrolobin A and B can inhibit the hemorrhagic activity of *Bothrops* venom, *B. neuwiedi* and *B. jararacussu* (Da Silva et al. 2007). Clerodane is a diterpenoid obtained from Brazilian plant *Baccharis trimera*. It was found that the obtained bioactives demonstrated anti-hemorrhagic and anti-proteolytic activity against bothrops snake venom (Januario et al. 2007).

Gomes et al. (2007) reported isolation of plant metabolites such as β sitosterol, stigmasterol from the root extract of *Plucha indica*. The plant extract neutralizes venom induced lethal defibrinogenation and edema. Stigmasterol neutralizes snake venom induced pharmacological effect. Wedelolactone, stigmasterol and sitosterol were isolated from *Eclipta prostrata*, and their antihemorrhagic and antimytotoxic effect was studied against *Bothrops jararacussu*, *B. jararica* and *Lachesis muta* (Melo et al. 1994). Pithayanukul et al. (2009) investigated that phenolic compound from the seed kernels of Thai mango plants such as pentagallegalloylglucopyranose, methyl gallate and gallic acid showed proteolytic activity against *Naja naja* kaouthia venom and *Calloselasma rhodostoma*. Bioactive compounds such as quercetin, myricetin, catechin and gallocatechin were isolated from the extract of *Schizolobium parahyla* leaves, and fibrinogenolytic and hemorrhagic activity of bothrops crude venoms was evaluated (Vale et al. 2011).

Main Bioactive Compounds Present in Plants with Antivenom Properties

The complexity of the damage resulting from ophidian accidents and the necessity of alternative and complementary therapeutic sources able to improve serum therapy for this neglected disease is urgent and necessary to minimize the sequelaes and damages.

Among the variety of compounds that constitute these plants with antivenom properties, some seem to play a key role in the antiofidic properties described (Table 7.2). Torres et al. (2011) reported that for bioactive compounds from the genus *Solanum*, three new solanidae steroidal alkaloids were studied with regard to myotoxicity, hemorrhage and necrosis induced by *Bothrops pauloensis* venom. In

Table 7.2. Some examples of plants with antiophidic properties with mechanism of inhibition.

Pharmaceutical form Plant Species Phytochemical	Property Studied	Species of Snake/ Poison	References
Aqueous extract *Casearia sylvestris* (Flacourtiaceae)	Inhibit myotoxic, anticoagulant, and edema	*Bothrops moojeni, B. pirajai, B. neuwiedi,* and *B. jararacussu* venom and its Asp49 and Lys49-PLA2 isolated toxins	Borges et al. 2000
Aqueous extract *Casearia sylvestris* (Flacourtiaceae)	Neutralize hemorrhagic activity	*B. pirajai, B. jararacussu, B. asper, B. moojeni,* and *B. neuwiedi* venom	Borges et al. 2001
Aqueous extract *Casearia sylvestris* (Flacourtiaceae)	Protective effects against muscle damage induced by two Lys49-PLA2 toxins and prevented the neuromuscular blockage induced by all PLA2 toxins	PrTX-I from *B. pirajai* and BthTX-I from *B. jararacussu* snake venom	Cavalcante et al. 2007
Eclipta alba (Asteraceae) was genetically engineered using *Agrobacterium rhizogenes* LB9402 to enhance the production of secondary wedelolactone metabolites	Reduce the PLA2 activities and myotoxic and neurotoxic effects of the *C.d. terrificus* and *B. jararacussu* snake venom		Mors et al. 1989, Diogo et al. 2009
Aqueous extract *Mandevilla velutina* (Apocynaceae)	Inhibitor of PLA2 activity and some toxic effects	*Bothrops* and *Crotalus* genus	Biondo et al. 2003
Aqueous extract *Mandevilla velutina* (Apocynaceae)	Inhibit the activity of the Crotoxin B, the basic Asp49-PLA2	*Crotalus durissus terrificus* venom	Biondo et al. 2004

this study, three alkaloids were isolated and characterized. The results showed that alkaloids studied could inhibit myotoxicity of *B. pauloensis* by inhibition of Asp49 activity. For antihemorrhagic and antinecrotizing activity, the hemorrhage provoque by venom are clearly decreased when alkaloid 1 and 2 are previously administered. The necrotic effect was also affected by alkaloid 1 and 2 administration. Authors also emphasized that snake venom metalloproteases (SVMP) is the main toxin responsible for hemorrhagic and necrosis and the antihemorrhagic and antinecrosis properties found in alkaloids 1 and 2 are due to the capacity of these compounds to interact with metals and metalloproteases.

Silva et al. (2007) studied the capacity of plants' mixture used in indigenous medicinal practice in Sri Lanka against venom toxins of *Naja naja* and *Daboia russelii* (Russell's viper). These plants mixture was composed by *Sansevieria cylindrica*, *Jatropha podagrica* and *Citrus aurantiifolia*. The study was performed by evaluation of the potential of herbal mixture which inhibits venom toxicity in chicken embryo model. The results showed the potential of herbal mixture that neutralizes action of *D. russeli* venom. In the presence of herbal mixture, the activity of phospholipase A2 (PLA2), of *Daboia russelii* venom was reduced from 9.2 10^3 mM min^{-1} to 8.0 10^3 mM min^{-1} and from *N. naja* from 2.92 10^2 mM min^{-1} to 0.188 × 10^{-2} mM min^{-1}. With *Naja* venom, herbal mixture was also able to reduce metalloprotease activity. Shabbir et al. (2014) reported that the main components involved in neutralization properties against venom of snakes (*Naja* species) include lupeol acetate, β-sitosterol, stigmasterol, rediocides A and G, quercertin, aristolochic acid, and curcumin, as well as the broad chemical groups of tannins, glycoproteins, and flavones. The authors emphasize that the chemical compounds of plants enhance the survival of snakebite victims by decreasing the severity of toxins, protect cardiovascular system, and enhance diaphragm contraction and others.

Khrishnan et al. (2013) studied the capacity of *Ophiorrhiza* mungos root extact that neutralizes the venom of *Daboia russelii* snake. The neutralization/inhibition of venom-induced lethality or hemorrhage was achieved by incubating venom in chick embryo. The results showed 100% recovery of embryos with 10 mg/mL of extract which was effective in neutralizing the hemorrhage induced by viper. Extracts of *Ophiorrhiza* mungos revealed the presence of terpenes, phenols, flavonoids, alkaloids, quinones, tannins, glycosides and saponins.

Nazato et al. (2010) evaluated the antiophidian properties of *Dipteryx alata* Vogel bark extract against *Bothrops jararacussu* and *Crotalus durissus terrificus* snake venom. Authors reported the neurotoxic and myotoxic inhibition of methanolic extracts against *Bothrops jararacussu* relating this inhibitory capacity with the presence of phenolic acids and flavonoids contained in the methanolic extract plus tannins, terpenoids from dicloromethane extract seems to play an important role in *Bothrops jararacussu* and *Crotalus durissus terrificus*. Important observations discussed in this study are related with the potential of some extract that do not inhibit venom from different species or even for the same species, a fact that makes use of "antiophidic plant" many times premature. Consequently, the antivenom activity of plants also depend on bioactive compounds and venom pharmacology.

Main classes of PLA2 inhibitors are the phenolic compounds, which include flavonoids, coumestans and alkaloids, steroids and terpenoids (mono-, di-, and triterpenes), and polyphenols (vegetable tannins) (Carvalho et al. 2013).

Identification of bioactive compounds present in plants extracts can be performed using analytical techniques that allow identifying the profiles and elucidating molecular characteristics of the main compounds enrolled in antivenom activity.

Liu et al. (2015) reported the use of high-performance liquid chromatography, high-resolution mass spectrometry, solid-phase extraction and nuclear magnetic resonance spectroscopy (HPLC–HRMS–SPE–NMR). These techniques were used to measure 88 plants known by antinecrosis properties that are traditionally used in Chinese medicine against snakebite. These results showed that 61 plant extracts could inhibit the hyaluronidase against at least one venom while 35 plants exhibited more than 50% in the PLA2 against at least one venom. After that, the most potent extract was submitted to high-resolution hyaluronidase inhibition profiling with three steps: (i) chromatographic separation, (ii) hyaluronidase inhibition assaying of the material, and (iii) plotting hyaluronidase inhibition for each well against its respective retention time to construct a biochromatogram that can be compared with the HPLC chromatogram.

Moura et al. (2016) studied the inhibitory potential of condensed tannin of *Plathymenia reticulata* against *Bothrops atrox*. In this study, the chemical composition of the aqueous extract was first isolated by TLC chromatography and afterwards, fractionated using gel filtration sephadex LH-20 chromatography. After a colorimetric solution containing the differents extracts was obtained, it was analysed using UV/VIS spectrophotometer at 510 nm.

Conclusion and Future Perspectives

Indigenous populations since antiquity have been using plants with antivenom properties in therapy against snake bites. The compounds with antisnake venom properties are related with a mix of molecules as alkaloids, tannins, flavonoids and others, which are able to promote local effects in affected area. The local effects promoted by serum therapy are limited, reforcing the necessity to combine different therapies in a snakebite treatment. However, the use and denomination of a plant as "antiophidic plant" seems to be premature due to the capacity of some plants unable to inhibit the venom of all different snakes or even from the same snake species. This fact brings out the link between the mixed substances present in the plant extract and its interaction with the pharmacological properties related with the snake venom composition. In necrosis and hemorrhagic conditions promoted by snakebite venom inoculation, the presence of enzymes such as phospholipases, proteases and hyaluronidases play an important role in tissue damage, injury, necrosis, bleeding and permanent sequels. The study and relation between phytochemistry and snake venom compound is determinant for the success and achievement of combination therapy in snake bite treatment. The variety in fauna and flora present in most of the countries with a high incidence of antiophidic accidents can be a key therapy, more effective, with less adverse effects and with accessible price.

References

Abubakar, M.S., Sule, M.I., Pateh, U.U., Abdurahman, E.M., Haruna, A.K. and Jahun, B.M. 2000. *In vitro* snake venom detoxifying action of the leaf extract of *Guiera senegalensis*. J. Ethnopharmacol. 69(3): 253–257.

Adzu, B., Abubakar, M.S., Izebe, K.S., Akumka, D.D. and Gamaniel, K.S. 2005. Effect of *Annona senegalensis* rootbark extracts on *Naja nigricotlis nigricotlis* venom in rats. J. Ethnopharmacol. 96(3): 507–513.

Ahmed, A., Rajendaran, D., Jaiswal, D., Singh, H.P., Mishra, A., Chandra, D., Yadav, I.K. and Jain, D.A. 2010. Anti-snake venom activity of different extracts of *Pouzolzia indica* against russel viper venom. Int. J. ChemTech Res. 2(1): 744–751.

Ahmed, W., Ahmad, M., Khan, R.A. and Mustaq, N. 2016. Promising inhibition of krait snake's venom acetylcholinesterase by *Salix nigra* and its role as anticancer, antioxidante agent. Indian J. Anim. Res. 50(3): 317–323.

Alam, M.I., Auddy, B. and Gomes, A. 1994. Isolation, purification and partial characterization of viper venom inhibiting factor from the root extract of the Indian medicinal plant sarsaparilla (*Hemidesmus indicus* R. Br.). Toxicon 32(12): 1551–1557.

Alam, M.I. and Gomes, A. 1998. Adjuvant effects and antiserum action potentiation by a (herbal) compound 2-hydroxy-4-methoxy benzoic acid isolated from the root extract of the Indian medicinal plant "sarsaparilla" (*Hemidesmus indicus* R. Br.). Toxicon 36(10): 1423–1431.

Alam, M.I. and Gomes, A. 1998. Viper venom-induced inflammation and inhibition of free radical formation by pure compound (2-hydroxy-4-methoxy benzoic acid) isolated and purified from anantamul (*Hemidesmus indicus* R.Br) root extract. Toxicon 36(1): 207–215.

Alam, M.I. and Gomes, A. 2003. Snake venom neutralization by Indian medicinal plants (*Vitex negundo* and *Emblica officinalis*) root extracts. J. Ethnopharmacol. 86: 75–80.

Ambikabothy, J., Ibrahim, H., Ambu, S., Chakravarthi, S., Awang, K. and Vejayan, J. 2011. Efficacy evaluations of *Mimosa pudica* tannin isolate (MPT) for its anti-ophidian properties. J. Ethnopharmacol. 137(1): 257–262.

Assafim, M., de Coriolano, E.C., Benedito, S.E., Fernandes, C.P., Lobo, J.F.R., Sanchez, E.F., Rocha, L.M. and Lopes, A. 2011. *Hypericum brasiliense* plant extract neutralizes some biological effects of *Bothrops jararaca* snake venom. J. Venom Res. 2: 11–16.

Aung, H.T., Nikai, T., Niwa, M. and Takaya, Y. 2010. Rosmarinic acid in *Argusia argentea* inhibits snake venom-induced hemorrhage. J. Nat. Med. 64(4): 482–486.

Batina, M., De, F., Cintra, A.C., Veronese, E.L., Lavrador, M.A., Giglio J.R., Pereira, O.S., Dias, D.A., França, S.C. and Sampaio, S.V. 2000. Inhibition of the lethal and myotoxic activities of *Crotalus Durissus Terrificus* Venom By *Tabernaemontana Catharinensis*: Identification of one of the active components. Planta Med. 66(5): 424–428.

Barma, A.D., Mohanty, J.P. and Bhuyan, N.R. 2014. A review on anti-venom activity of some medicinal plants. IJPSR 5(5): 1612–1615.

Bhattacharjee, P. and Bhattacharyya, D. 2013. Characterization of the aqueous extract of the root of *Aristolochia indica*: evaluation of its traditional use as an antidote for snake bites. J. Ethnopharmacol. 145(1): 220–226.

Biondo, R., Pereira, A.M., Marcussi, S., Pereira, O.S., França, S.C. and Soares, A.M. 2003. Inhibition of enzymatic and pharmacological activities of some snake venoms and toxins by *Mandevilla velutina* (Apocynaceae) aqueous extract. Biochimie. 85(10): 1017–1025.

Biondo, R., Soares, A.M., Bertoni, B.W., França, S.C. and Pereira, M.A.S. 2004. Direct organogenesis of *Mandevilla illustris* (Vell) Woodson and effects of its aqueous extract on the enzymatic and toxic activities of *Crotalus durissus terrificus* snake venom. Plant Cell. Rep. 22(8): 549–552.

Borges, M.H., Soares, A.M., Rodrigues, V.M., Andrão-Escarso, S.H., Diniz, H., Hamaguchi, A., Quintero, A., Lizano, S., Gutiérrez, J.M., Giglio, J.R. and Homsi-Brandeburgo, M.I. 2000. Effects of aqueous extract of *Casearia sylvestris* (Flacourtiaceae) on actions of snake and bee venoms ando n activity of phspholipases A_2. Comp. Biochem. Physiol. Part B 127: 21–30.

Borges, M.H., Soares, A.M., Rodrigues, V.M., Oliveira, F., Frannsheschi, A.M., Rucavado, A., Giglio, J.R. and Homsi-Brandeburgo, M.I. 2001. Neutralization of proteases from *Bothrops* snake venoms by the aqueous extract from *Casearia sylvestris* (Flacourtiaceae). Toxicon 39: 1863–1869.

Borges, M.H., Alves, D.L.F., Raslan D.S., Piló-Veloso, D., Rodrigues, V.M., Homsi-Brandeburgo, M.I. and de Lima, M.E. 2005. Neutralizing properties of *Musa paradisiaca* L. (Musaceae) juice on phospholipase A_2, myotoxic, hemorrhagic and lethal activities of crotalidade venoms. J. Ethnopharmacol. 98: 21–29.

Camargo, T.M., Nazato, V.S., Silva, M.G., Cogo, J.C., Groppo, F.C. and Oshima-Franco, Y. 2010. *Bothrops jararacussu* venom-induced neuromuscular blockade inhibited by *Casearia gossypiosperma* Briquet hydroalcoholic extract. J. Venom Anim. Toxins Incl. Trop. Dis. 16(3): 432–441.

Carvalho, B.M.A., Santos, J.D.L., Xavier, B.M., Almeida, J.R., Resende, L.M., Martins, W., Marcussi, S., Marangoni, S., Stábeli, R.G., Calderon, L.A., Soares, A.M., Silva, S.L. and Marchi-Salvador, D.P. 2013. Snake venom PLA2s inhibitors isolated from Brazilian plants: synthetic and natural molecules. BioMed. Res. Int. (ID 153045): 8.

Cavalcante, W.L.G., Campos, T.O., Dal, Pai-Silva, M., Pereira, P.S., Oliveira, C.Z., Soares, A.M. and Gallacci, M. 2007. Neutralization of snake venom phospholipase A2 toxins by aqueous extract of *Casearia sylvestris* (Flacourtiaceae) in mouse neuromuscular preparation. J. Ethnopharmacol. 112(3): 490–497.

Chacko, N., Ibrahim, M., Shetty, P. and Shastry, C.S. 2012. Evaluation of antivenom activity of *Calotropis gigantea* plant extract against *Vipera russelli* snake venom. IJPSR 3(7): 2272–2279.

Chandrashekara, K.T., Nagaraju, S., Nandini, S.U., Basavaiah and Kemparaju, K. 2009. Neutralization of local and systemic toxicity of *Daboia russelli* venom by *Morus alba* plant leaf extract. Phytother. Res. 23(8): 1082–1087.

Chatterjee, I., Chakravarty, A.K. and Gomes, A. 2004. Antisnake venom activity of ethanolic seed extract of *Strychnos nux vomica* Linn. Indian J. Exp. Biol. 42(5): 468–475.

Chatterjee, I., Chakravarty, A.K. and Gomes, A. 2006. *Daboia Russellii* and *Naja Kaouthia* venom neutralization by lupeol acetate isolated from the root extract of indian sarsaparilla *Hemidesmus Indicus* R.Br. J. Ethnopharmacol. 106(1): 38–43.

Chen, J.J., Shi, D.J., Li, K.H., Liu, G.F. and Wang, Q.C. 2005. Effect of Ecliptaprostrate on inflammation and hemorrhage induced by the snake venom. J. Snake 17(2): 65–68.

Cherdchu, C. and Karlsson, E. 1983. Proteolytic-independent cobra neurotoxin inhibiting activity of *Curcuma* sp. (Zingiberaceae). Southeast Asian J. Trop. Med. Public Health 14(2): 176–180.

Chippaux, J.P., Rakotonirina, V.S., Rakotonirina, A. and Dzikouk, G. 1997. Drug or plant substances which antagonize venoms or potentiate antivenins. Bull. Soc. Pathol. Exot. 90(4): 282–285.

Chippaux, J.P. 2017. Incidence and mortality due to snakebite in the Americas. Plos Negl. Trop. Dis. 11(6): e0005662. DOI: 10.1371/journal.pntd.0005662.

Cintra-Francischinelli, M., Silva, M.G., Andréo-Filho, N., Gerenutti, M., Cintra, A.C., Giglio, J.R., Leite, G.B., Cruz-Höfling, M.A., Rodrigues-Simioni, L. and Oshima-Franco, Y. 2008. Antibothropic action of Casearia sylvestris Sw. (Flacourtiaceae) extracts. Phytother. Res. 22(6): 784–790.

Collaço R de, C., Cogo, J.C., Rodrigues-Simioni, L., Rocha, T., Oshima-Franco, Y. and Randazzo-Moura, P. 2012. Protection by Mikania laevigata (guaco) extract against the toxicity of Philodryas olfersii snake venom. Toxicon. 60(4): 614–622.

Corbett, B. and Clark, R.F. 2017. North American snake envenomation. Emerg. Med. Clin. North Am. 35(2): 339–354.

Costa, H.N.R., Santos, M.C., Alcântara, A.F.C., Silva, M.C., França, R.C. and Piló-Veloso, D. 2008. Chemical constituents and antiedematogenic activity of *Peltodon radicans* (Lamiaceae). Quím Nova 31(4): 744–750.

da Silva, J.O., Coppede, J.S., Fernandes, V.C., Sant'ana, C.D., Ticli, F.K., Mazzi, M.V., Giglio, J.R., Pereira, P.S., Soares, A.M. and Sampaio, S.V. 2005. Antihemorrhagic, antinucleolytic and other antiophidian properties of the aqueous extract from *Pentaclethra macroloba*. J. Ethnopharmacol. 100(1-2): 145–152.

Da Silva, J.O., Fernandes, R.S., Ticli, F.K., Oliveira, C.Z., Mazzi, M.V., Franco, J.J., Giuliatti, S., Pereira, O.S., Soares, A.M. and Sampaio, S.V. 2007. Triterpenoids saponins, new metalloprotease snake venom inhibitors isolated from *Pentaclethra macroloba*. Toxicon. 50(2): 283–291.

Da Silva, M.L., Marcussi, S., Fernandes, R.S., Pereira, P.S., Januário, A.H., França, S.C., Da Silva, S.L., Soares, A.M. and Lourenço, M.V. 2012. Anti-snake venom activities of extracts and fractions from callus cultures of *Sapindus saponaria*. Pharm. Biol. 50(3): 366–375.

da Silva, J.O., Fernandes, R.S., Ticli, F.K., Oliveira, C.Z., Mazzi, M.V., Franco, J.J., Giuliatti, S., Pereira, P.S., Soares, A.M. and Sampaio, S.V. 2007. Triterpenoid saponins, new metalloprotease snake venom inhibitors isolated from *Pentaclethra macroloba*. Toxicon 50(2): 283–291.

Daduang, S., Sattayasai, N., Sattayasai, J., Tophrom, P., Thammathaworn, A., Chav eerach, A. and Konkchaiyaphum, M. 2005. Screening of plants containing Naja naja siamensis cobra venom inhibitory activity using modified Elisa technique. Anal. Biochem. 341(2): 316–325.

Dal Belo, C.A., Colares, A.V., Leite, G.B., Ticli, F.K., Sampaio, S.V., Cintra, A.C., Rodrigues-Simioni, L. and dos Santos, M.G. 2008. Antineurotoxic activity of Galactia glaucescens against *Crotalus durissus* terrificus venom. Fitoterapia 79(5): 378–80.

Dal Belo, C.A., Lucho, A.P.B., Vinadé, L., Rocha, L., França, H.S., Marangoni, S., Rodrigues-Simioni, L. 2013. *In vitro* antiophidian mechanisms of *Hypericum brasiliense* Choisy standardized extract: quercetin-dependent neuroprotection. BioMed. Res. Int. 943520: DOI: 10.1155/2013/943520.

De Almeida, L., Cintra, A.C., Veronese, E.L., Nomizo, A., Franco, J.J., Arantes, E.C., Giglio, J.R. and Sampaio, S.V. 2004. Anticrotalic and antitumoral activities of gel filtration fractions of aqueous extract from *Tabernaemontana catharinensis* (Apocynaceae). Comp. Biochem. Physiol. C Toxicol. Pharmacol. 137(1): 19–27.

De Jesus Reis Rosa, L., Silva, G.A., Filho, J.A., Silva, M.G., Cogo, J.C., Groppo, F.C. and Oshima-Franco, Y. 2010. The inhibitory effect of *Camellia sinensis* extracts against the neuromuscular blockade of Crotalus durissus terrificus venom. J. Venom Res. 1: 1–7.

De Oliveira, E.C., Fernandes, C.P., Sanchez, E.F., Rocha, L. and Lopes, A. 2014. Inhibitory effect of plant *Manilkara subsericea* against biological activities of Lachesis muta snake venom. BioMed. Res. Int. 2014: Article ID 408068, 7 pages.

Dey, A. and De, J.N. 2012. Phytopharmacology of antiophidian botanicals: a review. Int. J. Pharmacol. 8(2): 62–79.

Dhananjaya, B.L. and Shivalingaiah, S. 2016. The anti-inflammatory activity of standard aqueous stem bark extract of *Mangifera indica* L. as evident in inhibition of Group IA sPLA2. Ann. Braz. Acad. Sci. 88(1): 197–209.

Dhananjaya, B.L., Sudarshan, S., Dongol, Y. and More, S.S. 2016. The standard aqueous stem bark extract of Mangifera indica L. inhibits toxic PLA2 – NN-XIb-PLA2 of Indian cobra venom. Saudi Pharm. J. 24(3): 371–378.

Diogo, L.C., Fernandes, R.S., Marcussi, S., Menaldo, D.L., Roberto, P.G., Matrangulo, P.V., Pereira, P.S., França, S.C., Giuliatti, S., Soares, A.M. and Lourenço, M.V. 2009. Inhibition of snake venoms and phospholipases A (2) by extracts from native and genetically modified *Eclipta Alba*: Isolation of active coumestans. Basic Clin. Pharmacol. Toxicol. 104(4): 293–299.

Domingos, T.F.S., Vallim, M.A., Carvalho, C., Sanchez, E.F., Teixeira, V.L. and Fuly, A.L. 2011. Anti-snake venom effect of secodolastane diterpenes isolated from Brazilian marine brown alga *Canistrocarpus cervicornis* against *Lachesis muta* venom. Revista Brasileira de Farmacognosia 21(2): 234–238.

Emmanuel, N. and Mamoudou. 2015. Some investigations on the traditional Pharmacopoeia about venomous bite and stings from Scorpions, snakes and spiders, in the Hina subdivision, Far-North Cameroon. British J. Pharma. Res. 5(5): 344–358.

Esmeraldino, L.E., Souza, A.M. and Sampaio, S.V. 2005. Evaluation of the effect of aqueous extract of *Croton urucurana* Bailon (Euphorbiaceae) on the hemorrhagic activity induced by the venom of *Bothrops jararaca* using new techniques to quantify hemorrhagic activity in rat skin. Phytomed. 12(8): 570–576.

Farrapo, N.M., Silva, G.A.A., Costa, K.N., Silva, M.G., Cogo, J.C., Dal Belo, C.A., dos Santos, M.G., Groppo, F.C. and Oshima-Franco, Y. 2011. Inhibition of Bothrops jararacussu venom activities by *Plathymenia reticulata* Benth extracts. J. Venom Res. 2: 52–58.

Feitosa, E.S., Sampaio, V., Sachett, J.,Castro, D.B., Noronha, M.D.N., Lozano, J.L.L, Muniz, E., Ferreira, L.C.L Lacerda, M.V.G. and Monteiro, W.M. 2013. Snakebites as a largely neglected problem in the Brazilian Amazon: highlights of the epidemiological trends in the State of Amazonas. http://dx.doi.org/10.1590/0037-8682-0105-2015.

Félix-Silva, J., Souza, T., Menezes, Y.A., Cabral, B., Câmara, R.B., Silva-Junior, A.A., Rocha, H.A., Rebecchi, I.M., Zucolotto, S.M. and Fernandes-Pedrosa, M.F. 2014. Aqueous leaf extract of *Jatropha gossypiifolia* L. (Euphorbiaceae) inhibits enzymatic and biological actions of Bothrops jararaca snake venom. PLoS One 9(8): e104952. DOI: 10.1371/journal.pone.0104952.

Fernandes, J., Félix-Silva, J., Cunha, L., Gomes, A., Siqueira, M., Gimenes, L., Lopes, N., Soares, L., Fernandes-Pedrosa, M. and Zucolotto, S. 2016. Inhibitory effects of hydroethanolic leaf extracts of *Kalanchoe brasiliensis* and *Kalanchoe pinnata* (Crassulaceae) against local effects induced by Bothrops jararaca snake venom. PLoS One 11(12): e0172598.

Fernandes, F.F., Tomaz, M.A., El-Kik, C.Z., Monteiro-Machado, M., Strauch, M.A., Cons, B.L., Tavares-Henriques, M.S., Cintra, A.C., Facundo, V.A. and Melo, P.A. 2014. Counteraction of Bothrops snake venoms by *Combretum leprosum* root extract and arjunolic acid. J. Ethnopharmacol. 155(1): 552–562.

Ferraz, M.C., Parrilha, L.A.C., Moraes, M.S.D., Amaral Filho, J., Cogo, J.C., dos Santos, M.G., Franco, L.M., Groppo, J.C., Puebla, P., San Feliciano, P. and Oshima-Franco, Y. 2012. The effect of lupane triterpenoids (Dipteryx alata Vogel) in the in vitro neuromusuclar blockade and myotoxicity of two snake venoms. Curr. Org. Chem. 16(22): 2717–2723.

Ferraz, M.C., Yoshida, E.H., Tavares, R.V., Cogo, J.C., Cintra, A.C., Dal Belo, C.A., Franco, L.M., dos Santos, M.G., Resende, F.A., Varanda, E.A., Hyslop, S., Puebla, P., San Feliciano, A. and Oshima-Franco, Y. 2014. An isoflavone from Dipteryx alata Vogel is active against the in vitro neuromuscular paralysis of *Bothrops jararacussu* snake venom and bothropstoxin-I, and prevents venom-induces myonecrosis. Molecules 19(5): 5790–5805.

Ferraz, M.C., de Oliveira, J.L., de Oliveira, J.R., Cogo, J.C., dos Santos, M.G., Franco, L.M., Puebla, P., Ferraz, H.O., Ferraz, H.G., da Rocha, M.M., Hyslop, S., San Feliciano, A. and Oshima-Franco, Y. 2015. The tripernoid betulin protects against the neuromuscular effects of *Bothrops jararacussu* snake venom *in vivo*. Evid. Based Complement Alternat. Med. 939523.

Ferreira, L.A., Henriques, O.B., Andreoni, A.A., Vital, G.R., Campos, M.M., Habermehl, G.G. and de Moraes, V.L. 1992. Antivenom and biological effects of ar-turmerone isolated from *Curcuma longa* (Zingiberaceae). Toxicon 30(10): 1211–1218.

Ferreira-Rodrigues, S.C., Rodrigues, C.M., Dos Santos, M.G., Gautuz, J.A., Silva, M.G., Cogo, J.C., Batista-Silva, C., Dos Santos, C.P., Groppo, F.C., Cogo-Müller, K. and Oshima-Franco, Y. 2016. Anti-inflammatory and antibothropic properties of Jatropha elliptica, a plant from Brazilian Cerrado Biome. Adv. Pharm. Bull. (4): 573–579.

Goje, L.J., Aisami, A., Maigari, F.U., Ghamba, P.E., Shuaibu, I. and Goji, A.D.T. 2013. The Anti snake venom effects of the aqueous extracts of Boswellia delzielli stem bark on the parameters of hepatic functions and energy metabolism of Naja nigricollis (Spitting Cobra) envenomed albino rats. Res. J. Chem. Env. Sci. 1(4): 61–68.

Gomes, A., Saha, A., Chatterjee, I. and Chakravarty, A.K. 2007. Viper and cobra venom neutralization by beta-sitosterol and stigmasterol isolated from the root extract of *Pluchea indica* Less. (Asteraceae). Phytomedicine 14(9): 637–643.

Gomes, J.A., Félix-Silva, J., Morais Fernandes, J., Geraldo Amaral, J., Lopes, N.P., Tabosa do Egito, E.S., da Silva-Júnior, A.A., Maria Zucolotto, S. and Fernandes-Pedrosa, M.F. 2016. Aqueous leaf extract of *Jatropha mollissima* (Pohl) Bail decreases local effects Induced by bothropic venom. Biomed. Res. Int. 6101742.

Gopi, K., Renu, K., Raj, M., Kumar, D. and Muthuvelan, B. 2011. The neutralization effect of methanol extract of Andrographis paniculata on Indian cobra Naja naja snake venom. J. Pharm. Res. 4(4): 1010–1012.

Guerranti, R., Aguiyi, J.C., Neri, S., Leoncini, R., Pagani, R. and Marinello, E. 2002. Proteins from Mucuna pruriens and enzymes from Echis carinatus venom: characterization and cross-reactions. J. Biol. Chem. 277(19): 17072–17078.

Guerranti, R., Aguiyid, J.C., Oguelia, I.G., Onoratia, G., Neria, S., Rosatib, F., Del Buonob, F., Lampariellloc, R., Pagania, R. and Marinello, E. 2004. Protection of Mucuna pruriens seeds against Echis carinatus venom is exerted through a multiform glycoprotein whose oligosaccharide chains are functional in this role. Biochem. Biophys. Res. Comm. 323: 484–490.

Hasson, S.S., Al-Jabri, A.A., Sallam, T.A., Al-Balushi, M.S. and Mothana, R.A.A. 2010. Antisnake venom activity of *Hibiscus aethiopicus* L. against *Echis ocellatus* and Naja n. nigricollis. J. Toxicol. 2010: Article ID 837864, doi: 10.1155/2010/837864.

Hasson, S.S., Al-Balushi, M.S., Said, E.A., Habbal, O., Idris, M.A., Mothana, R.A.A., Sallam, T.A. and Al-Jabri, A.A. 2012. Neutralisation of local haemorrhage induced by the saw-scaled viper

Echis carinatus sochureki venom using ethanolic extract of Hibiscus aethiopicus L. Evid. Based Complement Alternat. Med. 2012: Article ID 540671, 8 pages doi: 10.1155/2012/540671.

Hung, Y.C., Sava, V., Hong, M.Y. and Huang, G.S. 2004. Inhibitory effects on phospholipase A2 and antivenin activity of melanin extracted from Thea sinensis Linn. Life Sci. 74(2004): 2037–2047.

Isa, H.I., Ambali, S.F., Suleiman, M.M., Abubakar, M.S., Kawu, U.M., Shittu, M., Yusuf, P.O. and Habibu, B. 2015. *In vitro* neutralization of Naja nigricollis venom by stem-bark extracts of *Commiphora africana* A. Rich. (Burseraceae). IOSR J. Environ Sci. Toxicol. Food Technol. 9(12): 100–105.

Kadali, V.N., Kameswara, R.K. and Sandeep, B.V. 2015. Medicinal plants with anti-snake venom property—a review. Pharma Innov. J. 4(7): 11–15.

Krishnan, S.A., Dileepkumar, R., Nair, A.S. and Oommen, O.V. 2013. Studies on neutralizing effect of *Ophiorrhiza mungos* root extract against Daboia russelii venom. http://dx.doi.org/10.1016/j.jep.2013.11.010.

Leanpolchareanchai, J., Pithayanukul, P., Bavovada, R. and Saparpakorn, P. 2009. Molecular docking studies and anti-enzymatic activity of Thai mango seed kernel extract against snake venoms. Molecules 14(4): 1404–1422.

Li, C.M., Zhang, Y., Yang, J., Zou, B., Dong, X.Q. and Hagerman, A.E. 2013. The interaction of persimmon proanthocyanidin fraction with Chinese cobra PLA2 and BSA. Toxicon. 67: 71–79.

Liu, Y., Staerk, D., Nielsen, M.N., Nyberg, N. and Jäger, A.K. 2015. High-resolution hyaluronidase inhibition profiling combined with HPLC–HRMS–SPE–NMR for identification of anti-necrosis constituents in Chinese plants used to treat snakebite. Phytochem. 112: 62–69.

Lobo, R., Punitha, I.S.R., Rajendran, K., Shirwaikar, A. and Shirwaikar, A. 2006. Preliminary study on the anti-snake venom activity of alcoholic root extract of *Clerodendrum viscosum* (Vent.) in Naja naja venom. Nat. Prod. Sci. 12(3): 153–156.

Machiah, D.K. and Gowda, T.V. 2006. Purification of a post-synaptic neurotoxic phospholipase A2 from Naja naja venom and its inhibition by a glycoprotein from *Withania somnifera*. Biochimie 88(6): 701–710.

Mahadeswaraswamy, Y.H., Devaraja, S., Kumar, M.S., Goutham, Y.N. and Kemparaju, K. 2009. Inhibition of local effects of Indian Daboia/Vipera russelli venom by the methanolic extract of grape (*Vitis vinifera* L.) seeds. Indian J. Biochem. Biophys. 46(2): 154–160.

Mahanta, M. and Mukherjee, A.K. 2001. Neutralization of lethality, myotoxicity and toxic enzymes of Naja kaouthia venom by *Mimosa pudica* root extracts. J. Ethnopharmacol. 75: 55–60.

Maroyi, A. 2017. Kirkia Acuminata oliv.: a review of its ethnobotany and pharmacology. Afr. J. tradit. complement Altern. Med. 14(2): 217–226.

Meenatchisundaram, S., Parameswari, G. and Michael, A. 2009. Studies on antivenom activity of *Andrographis paniculata* and *Aristolochia indica* plant extracts against Daboia russelli venom by *in vivo* and *in vitro* methods. Indian J. Sci. Technol. 2(4): 76–79.

Meenatchisundaram, S., Priyagrace, S., Vijayaraghavan, R., Velmurugan, A., Parameswari, G. and Michael, A. 2009. Antitoxin activity of Mimosa pudica root extracts against Naja naja and Bangarus caerulus venoms. Bangladesh J. Pharmacol. 4(2): 105–109.

Melo, M.M., Habermehl, G.G., Oliveira, N.J.F., Nascimento, E.F., Santos, M.M.B. and Lúcia, M. 2005. Treatment of Bothrops alternatus envenomation by *Curcuma longa* and *Calendula officinalis* extracts and ar-turmerone. Arq. Bras. Med. Vet. Zootec. 57(1): 7–17.

Melo, M.M., Lúcia, M. and Habermehl, G.G. 2007. Plant extracts for topic therapy of *Bothrops alternatus* envenomation. Braz. J. Pharmacogn. 17(1): 29–34.

Melo, P.A. and Ownby, C.L. 1999. Ability of wedelolactone, heparin, and para-bromophenacyl bromide to antagonize the myotoxic effects of two crotaline venoms and their PLA2 myotoxins. Toxicon 37(1): 199–215.

Melo, P.A., do Nascimento, M.C., Mors, W.B. and Suarez-Kurts, G. 1994. Inhibition of the myotoxic and hemorrhagic activities of crotalid venoms by Eclipta prostrata (Asteraceae) extracts and constituents. Toxicon. 32(5): 595–603.

Mishal, H.B. 2002. Screening of anti-snake venom activity of Dichrostachys cinerea W.A.J. Nat. Remedies 2(1): 92–95.

Mishal, R.H., Vadnere, G.P. and Mishal, H.B. 2014. Screening of anti snake venom potential of the seed oil of Balanites aegyptiaca L. Del. J. Nat. Remedies 14(2): 126–131.

Molander, M., Nielsen, L., Sogaard, S., Staerk, D., Rosnted, N., Diallo, D., Chifundera, K.Z., van Staden, J. and Jäger, A.K. 2014. Hyaluronidase, phospholipase A2 and protease inhibitory activity of plants used in traditional treatment of snakebite-induced tissue necrosis in Mali, DR Congo and South Africa. J. Ethnopharmacol. 18(157): 171–180.

Momoh, S., Friday, E.T., Raphael, E., Stephen, A. and Umar, S. 2012. Anti-venom activity of ethanolic extract of *Bridelia ferruginea* leaves against Naja nigricollis venom. E3 J. Med. Res. 1(5): 69–73.

Mors, W.B., do Nascimento, M.C., Parente, J.P., da Silva, M.H., Melo, P.A. and Suarez-Kurtz, G. 1989. Neutralization of lethal and myotoxic activities of South American rattlesnake venom by extracts and constituents of the plant *Eclipta prostrata* (Asteraceae). Toxicon. 27(9): 1003–1009.

Moura, V.M., Serra Bezerra, A.N., Mourão, R.H.V., Varjão Lameiras, J.L., Almeida Raposo, J.D., Sousa, R.L., Boechat, A.L., Oliveira, R.B., Chalkidis, H.M. and Dos-Santos, M.C. 2014. A comparison of the ability of *Bellucia dichotoma* Cogn. (Melastomataceae) extract to inhibit the local effects of Bothrops atrox venom when pre-incubated and when used according to traditional methods. Toxicon. 85: 59–68.

Moura, V.M., Silva, W.C.R., Raposo, J.D.A., Freitas-de-Sousa, L.A., Dos-Santos, M.C., Oliveira, R.B. and Mourão, R.H.V. 2016. The inhibitory potential of the condensed-tannin-rich fraction of *Plathymenia reticulata* Benth. (Fabaceae) against Bothropsatrox envenomation. http://dx.doi.org/10.1016/j.jep.2016.02.047.

Mukherjee, A.K., Doley, R. and Saikia, D. 2008. Isolation of a snake venom phospholipase A2 (PLA2) inhibitor (AIPLAI) from leaves of *Azadirachta indica* (Neem): mechanism of PLA2 inhibition by AIPLAI *in vitro* condition. Toxicon. 51(8): 1548–1553.

Naik, B. and Sadananda. 2017. "Dry bite" in venomous snakes: a review. DOI: 10.1016/j.toxicon.2017.04.015.

Nazato, V.S., Rubem-Mauro, L., Vieira, N.A., Rocha-Junior, D., dos, S., Silva, M.G., Lopes, P.S., Dal-Belo, C.A., Cogo, J.C., dos Santos, M.G., da Cruz-Höfling, M.A. and Oshima-Franco, Y. 2010. *In vitro* antiophidian properties of *Dipteryx alata* Vogel bark extracts. Molecules 15(9): 5956–5970.

Nishijima, C.M., Rodrigues, C.M., Silva, M.A., Lopes-Ferreira, M., Vilegas, W. and Hiruma-Lima, C.A. 2009. Anti-hemorrhagic activity of four Brazilian vegetable species against *Bothrops jararaca* venom. Molecules 14(3): 1072–1080.

Núñez, V., Otero, R., Barona, J., Saldarriaga, M., Osorio, R.G., Fonnegra, R., Jiménez, S.L., Díaz, A. and Quintana, J.C. 2004. Neutralization of the edema-forming, defibrinating and coagulant effects of Bothrops asper venom by extracts of plants used by healers in Colombia. Braz. J. Med. Biol. Res. 37(7): 969–977.

Ode, O.J. and Asuzu, I.U. 2006. The anti-snake venom activities of the methanolic extract of the bulb of *Crinum jagus* (Amaryllidaceae). Toxicon. 48(3): 331–342.

Okonogi, T., Hattori, Z., Ogiso, A. and Mitsui, S. 1979. Detoxification by persimmon tannin of snake venoms and bacterial toxins. Toxicon. 17: 324–327.

Oliveira, C.Z., Maiorano, V.A., Marcussi, S., Sant'ana, C.D., Januário, A.H., Lourenço, M.V., Sampaio, S.V., França, S.C., Pereira, O.S. and Soares, A.M. 2005. Anticoagulant and antifibrinogenolytic properties of the aqueous extract from *Bauhinia forficata* against snake venoms. J. Ethnopharmacol. 98(1-2): 213–216.

Omale, S., Aguiyi, J.C., Wannang, N.N., Ogbole, E., Amagon, K.I., Banwat, S.B. and Auta, A. 2012. Effects of the ethanolic extract of *Parinari curatellifolia* on blood clotting factors in rats pretreated with venom of Naja nigricolis. Drug Invention Today 4(4): 363–364.

Oshima-Franco, Y., Alves, C.M.V., Andréo Filho, N., Gerenutti, M., Cintra, A.C.O., Leite, G.B., Rodrigues-Simioni, L. and Silva, M.G. 2005. Neutralization of the neuromuscular activity of bothropstoxin-I, a myotoxin from Bothrops jararacussu snake venom, by a hydroalcoholic extract of *Casearia sylvestris* sw. (guaçatonga). J. Venom Anim. Toxins incl. Trop. Dis. 11(4): 465–478.

Oshima-Franco, Y., Rosa, L.J.R., Silva, G.A.A., Amaral Filho, J., Silva, M.G., Lopes, P.S., Cogo, J.C., Cintra, A.C.O. and Cruz-Höfling, M.A. 2012. Antibothropic action of *Camellia sinensis* extract against the neuromuscular blockade by Bothrops jararacussu snake venom and its main toxin, Bothropstoxin-I. Pharmacology, Dr. Luca Gallelli (ed.). ISBN: 978-953-51-0222-9, InTech,

Available from: www.intechopen.com/books/pharmacology/antibothropic-action-of-camellia-sinensis-extract-against-theneuromuscular-blockade-of-bothrops-jar.

Pithayanukul, P., Laovachirasuwan, S., Bavovada, R., Pakmanee, N. and Suttisri, R. 2004. Antivenom potential of butanolic extract of *Eclipta prostrata* against Malayan pit viper venom. J. Ethnopharmacol. 90(2-3): 347–352.

Pithayanukul, P., Leanpol chareanchai, J., Saparpakorn, P., Devaraja, S., Kumar, M.S. and Goutham, Y.N. 2009. Molecular docking studies and anti-snake venom metallo proteinase activity of Thai mango seed kernel extract. Molecules 14(9): 3198–3213.

Pithayanukul, P., Leanpol chareanchai, J. and Bavovada, R. 2010. Inhibitory Effect of tea polyphenols on local tissue damage induced by snake venoms. Phytother. Res. 24(Suppl 1): S56–S62.

Price, J.A. 2016. An *in vitro* evaluation of the nativeamerican ethnomedicinal plant *Eryngium yuccifolium* as a treatment for snakebite envenomation. J. Intercult. Ethnopharmacol. 5(3): 219–225.

Puebla, P., Oshima-Franco, Y., Franco, L.M., Santos, M.G., Silva, R.V., Rubem-Mauro, L. and Feliciano, A.S. 2010. Chemical constituents of the bark of *Dipteryx alata* vogel, an active species against Bothrops jararacussu venom. Molecules 15(11): 8193–8204.

Rajasree, P.H., Singh, R. and Sankar, C. 2013. Anti-venom activity of ethanolic extract of *Rauwolfia serpentina* against Naja Naja (Cobra) Venom. IJDDHR 3(1): 521–524.

Reyes-Chilpa, R., Gómez-Garibay, F., Quijano, L., Magos-Guerrero, G.A. and Ríos, T. 1994. Preliminary results on the protective effect of (−)-edunol, a pterocarpan from *Brongniartia podalyrioides* (Leguminosae), against Bothrops atrox venom in mice. J. Ethnopharmacol. 42(3): 199–203.

Samy, R.P., Thwin, M.M., Gopalakrishnakone, P. and Ignacimuthu, S. 2008. Ethnobotanical survey of folk plants for the treatment of snakebites in Southern part of Tamilnadu India. J. Ethnopharmacol. 115: 302–312.

Santos, J.I., Cardoso, F.F., Soares, A.M., dal Pai Silva, M., Gallacci, M. and Fontes, M.R.M. 2011. Structural and functional studies of a bothropic myotoxin complexed to rosmarinic acid: new insights into Lys49-PLA2 inhibition. PLoS One 6(12): 1–11.

Sanusi, J., Shehu, K., Jibia, A.B., Mohammed, I. and Liadi, S. 2014. Anti snake venom potential of *Securidaca longepedunculata* leaf and root bark on spitting cobra (*Naja nigricollis* Hallowel) in envenomed Wistar rats. IOSR J. Pharm. Biol. Sci. 9(6): 92–96.

Sarkhel, S., Chakravarty, A.K., Das, R., Gomes, A. and Gomes, A. 2011. Snake venom neutralising factor from the root extract of *Embilica officinalis* Linn. Orient. Pharm. Exp. Med. 11(1): 25–33.

Saturnino-Oliveira, J., Santos, D., do, C., Guimarães, A.G., Santos Dias, A., Tomaz, M.A., Monteiro-Machado, M., Estevam, C.S., De Lucca Júnior, W., Maria, D.A., Melo, P.A., Araújo, A.A., Santos, M.R., Almeida, J.R., Pereira de Oliveira, A. and Quintans Júnior, L.J. 2014. *Abarema cochliacarpos* extract decreases the inflammatory process and skeletal muscle injury induced by *Bothrops leucurus* venom. Biomed. Res. Int. 2014: 820761. DOI: 10.1155/2014/820761.

Schiermeier. Africa braced for snakebite crisis. Nature, 525, 2015. http://www.nature.com/news/africa-braced-for-snakebite-crisis-1.18357.

Sebastin Santhosh, M., Hemshekhar, M., Sunitha, K., Thushara, R.M., Jnaneshwari, S., Kemparaju, K. and Girish, K.S. 2013. Snake venom induced local toxicities: plant secondary metabolites as an auxiliary therapy. Mini-Rev. Med. Chem. 13: 106–123.

Shabbir, A., Shahzad, M., Masci, P. and Gobe, G.C. 2014. Protective activity of medicinal plants and their isolated compounds against the toxic effects from the venom of Naja (Cobra) species. DOI: 10.1016/j.jep.2014.09.039.

Shenoy, P.A., Nipate, S.S., Sonpetkar, J.M., Salvi, N.C., Waghmare, A.B. and Chaudhari, P.D. 2013. Anti-snake venom activities of ethanolic extract of fruits of *Piper longum* L. (Piperaceae) against Russell's viper venom: characterization of piperine as active principle. J. Ethnopharmacol. 147(2): 373–382.

Shenoy, P.A., Nipate, S.S., Sonpetkar, J.M., Salvi, N.C., Waghmare, A.B. and Chaudhari, P.D. 2014. Production of high titre antibody response against Russell's viper venom in mice immunized with ethanolic extract of fruits of *Piper longum* L. (Piperaceae) and piperine. Phytomedicine 21(2): 159–163.

Shirwaikar, A., Rajendran, K., Bodla, R. and Kumar, C.D. 2004. Neutralization potential of *Viper russelli russelli* (Russell's viper) venom by ethanol leaf extract of *Acalypha indica*. J. Ethnopharmacol. 94(2-3): 267–273.

Sia, F.Y., Vejayan, J., Jamuna, A. and Ambu, S. 2011. Efficacy of tannins from *Mimosa pudica* and tannic acid in neutralizing cobra (*Naja kaouthia*) venom. J. Venom. Anim. Toxins incl. Trop. Dis. 27(1): 42–48.

Silva, J.F., Silva-Junior, A.A., Zucolotto, S.M. and Fernandes-Pedrosa, M. 2017. Medicinal plants for the treatment of local tissue damage induced by snake venoms: An overview from traditional use to pharmacological evidence. Evid Based Complementary Alt. Med. 2017, Article ID 5748256, 52 pages DOI: 10.1155/2017/5748256.

Silva, M.M., Seneviratne, S.S., Weerakoon, D.K. and Goonasekara, C.L. 2017. Characterization of Daboia russelii and Naja naja venom neutralizing ability of an undocumented indigenous medication in Sri Lanka. J. Ayurveda Integr. Med. 8(1): 20–26.

Soares, A.M., Ticli, F.K. and Marcussi, S. 2005. Medicinal plants with inhibitory properties against snake venoms. Curr. Med. Chem. 12(22): 2625–2641.

Soares-Silva, J.O., Oliveira, J.L., Cogo, J.C., Tavares, R.V.S. and Oshima-Franco, Y. 2014. Pharmacological evaluation of hexane fraction of *Casearia gossypiosperma* Briquet: Antivenom potentiality. J. Life Sci. 8(4): 306–315.

Soni, P. and Bodakhe, S.H. 2014. Antivenom potential of ethanolic extract of *Cordia macleodii* bark against *Naja* venom. Asian Pac. J. Trop. Biomed. 4 (Suppl 1): S449–S454.

Strauch, M.A., Tomaz, M.A., Monteiro-Machado, M., Ricardo, H.D., Cons, B.L., Fernandes, F.F., El-Kik, C.Z., Azevedo, M.S. and Melo, P.A. 2013. Antiophidic activity of the extract of the Amazon plant *Humirianthera ampla* and constituents. J. Ethnopharmacol. 145(1): 50–58.

Tan, N.H., Fung, S.Y., Sim, S.M. and Marinello, E. 2009. The protective effect of *Mucuna pruriens* seeds against snake venom poisoning. J. Ethnopharmacol. 123: 356–358.

Tareen, R.B., Bibi, T., Khan, M.A., Ahmad, M. and Zafar, M. 2010. Indigenous knowledge of folk medicine by the women of Kalat and Khuzdar regions of Balochistan, Pakistan. Pak J. Bot. 42(3): 1465–1485.

Tianyi, F.L., Dimala, C.A. and Feteh, V.F. 2017. Shortcomings in snake bite management in rural Cameroon: a case report. BMC Res. Notes (2017) 10: 196. DOI: 10.1186/s13104-017-2518-8.

Ticli, F.K., Hage, L.I.S., Cambraia, R.S., Pereira, P.S., Magro, A.J., Fontes, M.R., Stábeli, R.G., Giglio, J.R., França, S.C., Soares, A.M. and Sampaio, S.V. 2005. Rosmarinic acid, a new snake venom phospholipase A2 inhibitor from *Cordia verbenacea* (Boraginaceae): antiserum action potentiation and molecular interaction. Toxicon 46(3): 318–327.

Torres, M.C., Jorge, R.J., Ximenes, R.M., Alves, N.T.Q., Santos, J.V., Marinho, A.D., Monteiro, H.S.A., Toyama, M.H., Braz-Filho, R., Silveira, E.R. and Pessoa, O.D. 2011. Solanidane and iminosolanidane alkaloids from *Solanum campaniforme*. Phytochem. 97: 457–464.

Tribuiani, N., da Silva, A.M., Ferraz, M.C., Silva, M.G., Bentes, A.P., Graziano, T.S., dos Santos, M.G., Cogo, J.C., Varanda, E.A., Groppo, F.C., Cogo, K. and Oshima-Franco, Y. 2014. *Vellozia flavicans* Mart. ex Schult. Hydroalcoholic extract inhibits the neuromuscular blockade induced by *Bothrops jararacussu* venom. BMC Complement Altern. Med. 14: 48. DOI: 10.1186/1472-6882-14-48.

Tribuiani, N., Tavares, M.O., Santana, M.N., Fontana Oliveira, I.C., Amaral Filho, J.D., Silva, M.G., Dos Santos, M.G., Cogo, J.C., Floriano, R.S., Cogo-Müller, K. and Oshima-Franco, Y. 2017. Neutralising ability of *Terminalia fagifolia* extract (Combretaceae) against the *in vitro* neuromuscular effects of *Bothrops jararacussu* venom. Nat Prod. Res. 2: 1–5.

Ushanandini, S., Nagaraju, S., Harish Kumar, K., Vedavathi, M., Machiah, D.K., Kemparaju, K., Vishwanath, B.S., Gowda, T.V. and Girish, K.S. 2006. The anti-snake venom properties of *Tamarindus indica* (Leguminosae) seed extract. Phytother. Res. 20(10): 851–858.

Ushanandini, S., Nagaraju, S., Nayaka, S.C., Kumar, K.H., Kemparaju, K. and Girish, K.S. 2009. The anti-ophidian properties of *Anacardium occidentale* bark extract. Immunopharmacol. Immunotoxicol. 31(4): 607–615.

Vale, F., Mendes, L.H., Fernandes, M.M., Costa, T.R., S. Hage-Melim, L.I., Sousa, A., Hamaguchi, M., Homsi- Brandeburgo, A., Franca, M.I., Silva, S.C., Pereira, P.S., Soares, A.M. and Rodrigues, V.M.

2011. Protective effect of *Schizolobium parahyba* flavonoids against snake venoms and isolated toxins. Curr. Top Med. Chem. 11(20): 2566–2577.

Veronese, E.L.G., Esmeraldino, L.E., Trombone, A.P.F., Santana, A.E., Bechara, G.H., Kettelhut, I., Cintra, A.C.O., Giglio, J.R. and Sampaio, S.V. 2005. Inhibition of the myotoxic activity of *Bothrops jararacussu* venom and its two major myotoxins, BthTX-I and BthTX-II, by the aqueous extract of *Tabernaemontana catharinensis* A. DC. (Apocynaceae). Phytomed. 12(1-2): 123–130.

Wufem, B.M., Adamu, H.M., Cham, Y.A. and Kela, S.L. 2007. Preliminary studies on the anti-venim potential and phytochemical analysis of the crude extracts of *Balanites aegyptiaca* (Linn.) Delile on albino rats. Nat. Prod. Radiance 6(1): 18–21.

Xu, S., Zou, B., Yang, J., Yao, P. and Li, C. 2012. Characterization of a highly polymeric proanthocyanidin fraction from persimmon pulp with strong Chinese cobra PLA_2 inhibition effects. Fitoterapia 83(1): 153–160.

Yoshida, E.H., Ferraz, M.C., Tribuiani, N., Tavares, R.V.S., Cogo, J.C., dos Santos, M.G., Franco, L.M., Dal Belo, C.A., de Grandis, R.A., Resende, F.A., Varanda, E.A., Puebla, P., San Feliciano, A., Groppo, F.C. and Oshima-Franco, Y. 2015. Evaluation of the safety of three phenolic compounds from *Dipteryx alata* Vogel with antiophidian potential. Chinese Medicine 6: 1–12.

CHAPTER 8

Ethnobotanical Study of Dakshin Dinajpur District of West Bengal

An Overview

Tanmay Chowdhury,[1] *Subhas Chandra Roy*[2] *and Dilip De Sarker*[3,*]

Introduction

The records of conscious use of plants other than food by human races have been obliterated in remote historical past. When the "Vedas" appeared in written form, the use of plants appeared to be the first documentary evidence as curative agent. The subject ethnobotany gained importance at the beginning of 20th century. In those days, it was thought that the knowledge of plants of different ethnic people is the major source of ethnobotany. Those people had no written alphabet. It was further conceived that discourses about the uses of plants by these aborigines and ethnic people is ethnobotany. But the area of ethnobotany is deeper and wider.

Ethnobotany, as it stands now, is the domain of knowledge which researches all sorts of plant uses including medicinal, cultural, primitive, agricultural or other forms of anthropogenic uses.

Plants have been used as medicine from the beginning of civilization to present day. Perhaps since Stone Age, plants are believed to have healing powers on man. Ancient Vedas dating back between 3500 BC and 800 BC reveal many references on medicinal plants. One of the remotest works in traditional herbal medicine is

[1] Department of Sericulture, Raiganj University, Raiganj, Uttar Dinajpur, West Bengal-733134, India.
[2] Plant Genetics and Molecular Breeding Laboratory, Department of Botany, University of North Bengal, Raja Rammohunpur, Darjeeling, West Bengal-734013, India.
[3] College para, Raiganj, Uttar Dinajpur, West Bengal-733134, India.
* Corresponding author: dilipdesarker.rnj@gmail.com

"Virikshayurveda", compiled even before the beginning of the Christian era. World's most ancient literary work, the "Rig Veda" was composed around 2000 BC (Bently and Trimen 1980). About 200 years ago, Indian pharmacopoeia was dominated by herbal medicines (Ernst 2005) and almost 25% of the drugs prescribed worldwide were obtained from plants. Of the 252 drugs considered as basic and essential by the WHO, 11% are exclusively of plant origin and a significant number are synthetic drugs obtained from natural precursors (Rates 2001).

Dakshin Dinajpur, consisting of eight developmental blocks, is primarily an agriculturally sustained district. The principal tribal communities are—Santal, Munda, Oraon, along with Scheduled Caste communities like Rajbanshis. This district has an old folk culture of using herbal medicines. However, the importance and such use of medicinal plants/plant parts are being lost due to rapid urbanization and deforestation. As a result, many useful medicinal plants are becoming threatened and precious knowledge is lost. An attempt has been made to accumulate the previous ethnobotanical data in a nutshell so that the knowledge of various ethnic people never get lost and the knowledge be used in different pharmaceutical industry to exploit the active principal constituted of different plants. It may give a scientific base on the ethnobotanical study.

Ethnobotany in West Bengal

In West Bengal, some of the remarkable studies in the field of ethnobotany were on lesser known plant food among the tribals of Purulia (Jain and De 1964) which was the first publication on ethnobotany from West Bengal. Later on, De (1965, 1969) worked on ethnobotanical study of Purulia district. Subsequently, Jain and Tarafder (1970) studied the medicinal plant-lore of Santals. Chaudhuri and Pal (1976) made a preliminary study on ethnobotany of Medinipur district. Ghosh (1986) studied the ethnobotanical survey of Cooch Behar district, use of plants by Lodha tribe in Midnapur District (Pal and Jain 1989, Pant et al. 1993), ethnobotanical study of Purulia district (Sur et al. 1992), etc. Molla et al. (1984, 1985, 1996) studied the ethnobotany of Jalpaiguri district. Mitra and Mukherjee (2005a,b, 2010, 2013) studied the ethnobotanical works of undivided Dinajpur district. De Sarker and his associates have made a significant contribution in the ethnomedicine and ethnobotany of Malda, Uttar and Dakshin Dinajpur district. De Sarker et al. (2011) have recorded good number of medicinal plants in the districts of Malda and Uttar and Dakshin dinajpur and published them as a book. Review work on the plants of Malda district has been done by Saha et al. (2013). Preparation and use of native drink by Oraon tribe has been worked out by Saha et al. (2015) and healthcare management of tribal people of Malda district has been done by Saha et al. (2014a). Interestingly, ethnoveterinary medicine works have been done by Saha et al. (2014b). Specific use of *Acacia nilotica* for prevention of diabetes has been elucidated by Saha et al. (2017).

Trends in Herbal Medicine

Herbal medicine is still the mainstay of about 75–80% of the whole population, and the major part of traditional therapy involves the use of plant extract and their active constituents (Kamboj 2000). All the developing countries put greater value towards traditional healing treatments. Consequently, there always remains a need to find an effective and safety drug molecule because of the inefficiency to cure the disease completely and number of serious adverse effects associated with the existing synthetic drug (Yamashita et al. 2002, Pagan et al. 2005).

During the last few decades, there has been an increasing interest in the study of medicinal plants and their traditional use in different parts of India. Survival of tribals and backward class communities depend upon the use of these useful plants. Plants are one of the most important sources of medicines and play a significant role in the survival of the tribal and ethnic communities as well as common Indian people. Tribal people as well as traditional practitioners widely use the medicinal plants in their every day practice for safer, low cost, efficient and locally availability. India has more than 427 tribal communities with rich diversity of indigenous tradition (Xavier et al. 2014). However, traditional knowledge based practices have been marginalized due to political and socio-economic reasons. The advantage of medicinal plants in various treatments is relatively safer than synthetic drugs, besides being less expensive, having greater efficacy and availability throughout the world (Sirkar 1989, Siddiqui 1993). Therefore, collection of ethnobotanical information and documentation of traditional knowledge has gained prominence from the prospective of drug development (Ragupathy et al. 2008). Since interest in traditional medicine has been increasing world over, ethnobotanical studies have gained prominence to explore the traditional knowledge, particularly in developing countries (Joshi and Joshi 2000). Recently, Schmidt et al. (2009) experimentally proved that plants offer immense scope for researchers engaged in validation of traditional claims for the development of novel drugs.

The Study Area

Dakshin Dinajpur of West Bengal is a small agriculturally active district having eight blocks and two sub-divisions, namely, Balurghat (Hili, Balurghat, Kumarganj and Tapan) and Gangarampur (Gangarampur, Bangshhari, Harirampur and Kushmandi). The district Dakshin Dinajpur lies between 26° 35' 15" N to 25° 10' 55" N latitude and 89° 00' 30" E - 87° 48' 37" E longitude, covering an area of 2162 Sq. Km. The district is situated between Bangladesh on the East and South, Uttar Dinajpur district on the North and West and some southern part lies adjacent to Malda district (Fig. 8.1). The study site has even topography, old alluvial soil and a range in altitude from 25 to 40 m. Annual temperature generally ranges from 7.2 to 33.1°C. Annual maximum rainfall received was 445.9 mm during August, 2015 (Data collected from the Office of the Additional Director of Agriculture, North Bengal Regional Office, Jalpaiguri, Govt. of West Bengal). The main lakes

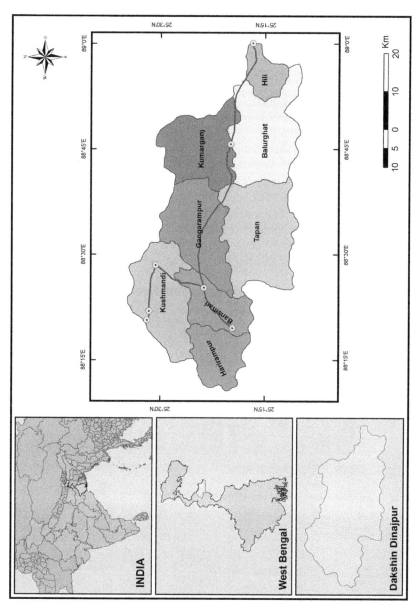

Fig. 8.1. Map of the study area.

(dighi) of the district are Kaldighi, Dhaldighi, Pransagar in Gangarampur block, Altadighi, Maliandighi, Gourdighi, Hatidobadighi in Banshihari block, Tapandighi in Tapan block, Mahipaldighi in Kushmandi block, etc. and rivers of the study area are Jamuna, Atryee, Punarbhava and Tangon, which flows across the blocks of Hili, Kumarganj and Balurghat, Gangarampur and Tapan, Kushmandi and Banshihari, respectively.

Culture and People of the Study Area

The major ethnic groups of the district are Santal, Munda, Oraon, Rajbanshi, Lodha, Sabar, etc. which are about (18.61%) of the total population. Besides their mother tongues, the tribal communities are very fluent in Bengali language. According to the census 2011, along with these tribes, the other non-tribal population (30.57%) belongs to other category such as Rajbanshis (18.4%), Hindus, Muslims and other minorities. The major ethnic groups of the district have an old tradition to use natural resources for their day to day life to cure different type of ailments. This district also has a very old tradition of practicing Kabiraji, Ayurveda and Unani (Fig. 8.2).

The district has various interesting and pleasurable folk culture. '*Khan*' is a unique age old folk culture performed mainly by the scheduled caste and scheduled tribe communities of the district. Some of the notable '*Khan*' palas are—'Cyclesari',

Fig. 8.2. Traditional healers and their medicinal preparations of Dashin Dinajpur district.

'Budhasari', etc. Besides '*Khan*', some of the other important folk culture of the district are Natua, Jang Gan, Mokha dance (Mask Dance), Halua-Haluani, Bislihava, Saitpir (Satyapir Gan), Jalmanga Gan, Khaja Gan, Chorchunni, etc. In addition to folk culture, the district has some historical and tourist sports like Dhaldighi, Dumduma, Bangarh, Mahipaldighi, Bairhatta and Kichaka Kunda, Tapan Dighi, District Library, College Museum, Pancha Ratna Temple, etc.

Ethnomedicine of Dakshin Dinajpur

Dakshin Dinajpur is a small and diversified ethnic culture district of West Bengal. The tribal people of this district are still using plants for their medicinal purposes (Table 8.1). However, few studies have been reported which showed the ethnobotanical prospects and how the medicinal plants have been incorporated into the cultural tradition of local ethnic people of Dakshin Dinajpur district. Sur along with his co-workers (1987) had studied ethno-economic importance of 73 plant species under 67 genera in the district Malda and West Dinajpur. Immediately after completion of the work again in the year 1990, they documented 52 plants under 48 genera for ethno-economic importance in the same study area. Both the ethno-economic works have documented the methods of application of plants in treating certain diseases and their distribution. Similarly, Banerjee and Ghora (1996) documented some domestic uses of plants which were not reported from West Dinajpur district. In this study, a total of 29 common wild plants were documented highlighting their use as food and as vegetable-ingredients, mostly during drought/flood, by the common people. The study of Mitra and Mukherjee (2005a) showed that 16 grass taxa were found to have 27 ethnobotanical uses and *Vetiveria zizanioides* was the mostly used taxon. The study demands that the Santals are more ethnomedicinally sound than the other tribal communities such as the Mundas and Oraons. Mitra and Mukherjee (2005b) reported 107 lesser known uses of the root and rhizome drugs of 71 species of angiosperms belongings to 68 genera of 45 families grown in West Dinajpur district for the treatment of different common human diseases. Mitra and Mukherjee (2007) had studied the ethnoveterinary medicine in Uttar and Dakshin Dinajpur districts of West Bengal to treat the ailments of cattle, to promote better lactation and also to improve the quality of meat, egg, etc., which are being traditionally used till date. In this study, total 23 different prescriptions of ethno-veterinary usages of 21 plant species covering 21 genera belonging to 17 angiospermic families were recorded. Out of these, nine are for healing of wounds, followed by eight preparations for the improvement of lactation and three for the healing of sores. De Sarker et al. (2011) carried out a documentation work on medicinal plants, their uses and availability in Uttar Dinajpur, Dakshin Dinajpur and Malda district. In this study, a total of 610 plant species including their medicinal importance were well documented. It is found from the study of Kundu and Bag (2012) that the Rajbanshis of Dakshin Dinajpur district use plants as preventive and curative health measures. Talukdar and Talukdar (2012) had studied the floral diversity and its folk uses in the banks of Atreyee River at Balurghat,

Table 8.1. Ethnomedicinal plants used by the people of Dakshin Dinajpur district.

Scientific Name and Voucher Specimen Number	Vernacular Name	Family	Parts used	Medicinal uses and preparation
Abroma augusta (L.) L.f. DD-12	Ulatkambol	Sterculiaceae	Petiole	Young petiole cut into small pieces and kept in a glass of water for overnight and infusion is used early morning on an empty stomach to cure "meho" (a sexual disease) and physical weakness.
Abrus precatorius L. DD-10	Kanch	Papilionaceae	Root	Roots are used for "Meho" (a sexual disease), jaundice and also used as abortifacient.
Acacia nilotica (L.) Delile * DD-130	Babla	Mimosaceae	Stem bark	Stem bark decoction used to treat cough, diarrhoea, dysentery, indigestion, acidity, "Meho" (a sexual disease) and diabetes.
Achyranthes aspera L. DD-78	Apang/Chatchota	Amaranthaceae	Root	Stem and root (2–3 piece) decoction mixed with "Ada" (*Zingiber officinale*) and used in jaundice.
Acmella oleracea (L.) R.K.Jansen* DD-25	Rasun sag	Asteraceae	Whole plant	Whole plant eaten as vegetables to treat body pain, especially after child birth.
Acorus calamus L. DD-23	Bach	Acoraceae	Rhizome	Fresh rhizome juice clears the vocal cord.
Aegle marmelos (L.) Corr. DD-58	Bel	Rutaceae	Fruit	One teaspoonful of young dried fruits powder mixed with water is given early morning on an empty stomach to cure dysentery and gastric problems.
Alocasia macrorrhizos (L.) G.Don* DD-39	Man-kachu	Araceae	Peliole	Petiole ash mixed with coconut oil and applied as emollient on carbuncle.
Alstonia scholaris (L.) R. Br.* DD-27	Chatim/Chatan	Apocyanaceae	Stem bark	Stem bark paste applied on breast for better lactation.
Amaranthus spinosus L. DD-16	KantaKhuria	Amaranthaceae	Root	Mature root decoction gives physical strength and used in indigestion problem, dysentery and diphtheria.
Ampelocissus latifolia (Roxb.) Planch.* DD-11	Goalilata	Vitaceae	Root	Root paste mixed with 12 "Golmarich" (*Piper nigrum*) and applied to cure gout and rheumatism.

Plant	Local name	Family	Part used	Uses
Andrographis paniculata (Burm. f.) Wall. Ex Nees DD-24	Kalmegh	Acanthaceae	Leaf	Leaf juice used for cold and cough, diabetes and leaf paste applied on hair before 30 minutes of bath to control dandruff.
Argemone mexicana L.* DD-01	Siyalkanta	Papaveraceae	Root	Root juice used for piles.
Aristolochia indica L.* DD-26	Iswarmul	Aristolochiaceae	Root	Root and leaf juice is given early morning to cure stomachache, fever, indigestion problem and diarrhoea.
Artemisia vulgeris L. DD-35	Nagdona	Asteraceae	Leaf	Leaf paste applied externally on the forehead to reduce headache and sinus problem.
Averrhoa carambola L.* DD-33	Kamranga	Oxalidaceae	Fruit	roasted fruit juice mixed with a pinch of sugar is given early morning to cure cough and bronchitis.
Azadirachta indica A. Juss. DD-103	Neem	Meliaceae	Stem bark	Stem bark and leaves boiled with water and applied on carbuncle, boil and skin diseases for quick healing.
Baccharoides anthelmintica (L.) Moench.* DD-61	Somraji	Asteraceae	Seed	Seeds are crushed with cream of milk and applied to cure carbuncle.
Bambusa tulda L.* DD-21	Bansh	Poaceae	Stem	Juice of roasted young shoot is applied to cure earache.
Basella alba L.* DD-77	Pui	Basellaceae	Stem	Dried stem ash used as tooth paste to cure pyorrhoea.
Bauhinia acuminata L.* DD-59	SwetKancan	Caesalpiniaceae	Flower	Dried flowers crushed and then lightly wormed with mustered oil are applied externally to cure skin diseases.
Blumea lacera (Burm. f.) DC. DD-22	Kukurmuta	Asteraceae	Root	Root (2–3 pieces) decoction mixed with "Ada" (*Zingiber officinale*) is given to cure flatulence and indigestion problem.
Boerhavia diffusa L. DD-74	Punarnaba	Nyctaginaceae	Whole plant	Whole plant juice is given to cure burning sensation during urination.
Bombax ceiba L. DD-66	Shimul	Bombacaceae	Root	Fresh root (1–2 years old plant) used in "Meho" (a sexual disease) and physical weakness.

Table 8.1 contd. ...

...Table 8.1 contd.

Scientific Name and Voucher Specimen Number	Vernacular Name	Family	Parts used	Medicinal uses and preparation
Borassus flabellifer L.* DD-72	Tal	Palmae	Root	Root mixed with Churchuri roots (*Achyranthes aspera*) are boiled and used for gargling to cure toothache.
Bryophyllum pinnatum (Lam.) Oken DD-41	Patharkuchi	Crassulaceae	Leaf	Fresh leaf juice is used for dissolving kidney stones and used to treat cold and cough, flatulence and acidity. Leaf paste applied also on burns for quick healing from burning sensation.
Butea monosperma (Lam.) Taub. DD-20	Palash	Papilionaceae	Leaf	Leaf juice used as aphrodisiac and enhances sperm count.
Cajanus cajan (L.) Millsp. DD-68	Arhar	Papilionaceae	Leaf	Fresh leaf juice is given early morning to treat jaundice.
Calotropis gigantea (L.) Dryand. DD-105	Akanda	Asclepiadaceae	Leaf	A leaf warm with "Ghee" is applied to get relief from paralysis, rheumatism and body pain.
Canna indica L.* DD-54	Kalabati	Cannaceae	Root	Root paste applied on the upper surface of the cheek to reduce toothache.
Cassia fistula L. DD-87	Bandarlathi	Caesalpiniaceae	Leaf	Tender leaf juice is given to cure constipation.
Centella asiatica (L.) Urban DD-32	Thankuni/ Dholamonia	Umbelliferae	Leaf	Fresh leaves (5–6) chewed early morning preferably on empty stomach to control diabetes, chronic dysentery, blood stool and diarrhoea.
Cheilocostus speciosus (J. Koenig) C. D. Specht DD-49	Kuttus/Jangliada	Costaceae	Rhizome	Rhizome juice mixed with a pinch of salt and "Ada" (*Zingiber officinale*) is given to cure indigestion and flatulence.
Chromolaena odorata (L.) R. M. King & H. Rob.* DD-50	Assam Lata	Asteraceae	Leaf	Leaf paste applied on cuts and wound to stop bleeding.
Cissus quadrangularis L. DD-108	Harjora	Vitaceae	Whole plant	Stem paste warmed with "Ghee" is applied on the fractured bones. Whole plant eaten as vegetable to reduce constipation problem.

Clerodendrum indicum (L.) Kuntze* DD-83	Bhamot	Verbenaceae	Stem	Fresh young stem used as a garland to cure a special type of boil in children.
Clerodendrum infortunatum L. DD-105	Ghentu/Bhant	Verbenaceae	Leaf	Fresh leaf juice is given on an empty stomach to expel intestinal worm (tapeworm, guinea worm).
Clitoria ternatea L. DD-126	Aparajita	Papilionaceae	Root	Root paste used to prevent toothache.
Coccinia grandis (L.) Voigt DD-64	Telakucha	Cucurbitaceae	Leaf	Fresh leaf juice is given to control diabetes, cold and cough. The juice is also applied on head to reduce the body temperature.
Cocculus hirsutus (L.) W.Theob.* DD-81	Jaljamani/Fariboti	Menispermaceae	Leaf	Fresh leaf decoction mixed with pinch of sugar is given to cure gonorrhoea, physical weakness and also acts as sex stimulant and delays ejaculation.
Coix lacryma-jobi L. DD-65	Kanch	Poaceae	Root	Root decoction used for menstrual disorder.
Commelina benghalensis L. DD-106	Kanchire	Commelinaceae	Leaf	Leaves are crushed and applied on skin to reduce irritation especially in caterpillar strings.
Corchorus olitorius L. DD-37	Tita pat	Tiliaceae	Leaf	Dried leaf infusion mixed with pinch of salt and turmeric powder is given in morning on an empty stomach to expel intestinal worms.
Crateva religiosa G. Forst.* DD-82	Barun	Capparaceae	Stem bark	Stem bark paste used for rheumatism and burning sensation of body.
Croton bonplandianus Baill. DD-38	Churchuri	Euphorbiaceae	Latex	Latex of the plant used for stopping bleeding from cuts and wounds.
Cryptolepis dubia (Burm.f.) M.R.Almeida* DD-80	Kalmashna	Asclepiadaceae	Root	Root decoction mixed with sugar (100 g), garlic (50 g) and milk (250 g) is boiled, the preparation is taken twice daily for curing rheumatic pain.
Curculigo orchioides Gaertn. DD-85	Talmuli	Hypoxidaceae	Root	The root of the plant is used for increasing vitality and acts as sex stimulant. Root paste is used in arthritis and joint pains (gout). Root mixed with "Pan" (*Piper betel*) is given to cure piles.

Table 8.1 contd. ...

Table 8.1 contd. ...

Scientific Name and Voucher Specimen Number	Vernacular Name	Family	Parts used	Medicinal uses and preparation
Curcuma aromatica Salisb.* DD-43	Shut	Zingiberaceae	Rhizome	Fresh rhizome juice mixed with honey is given to children to cure cough and bronchitis.
Cuscuta reflexa Roxb DD-84	Swarnalata	Convolvulaceae	Whole plant	Whole plant decoction is to be taken on an empty stomach to cure gonorrhoea.
Cyanthillium cinereum (L.) H.Rob.* DD-02	Sahadebi	Asteraceae	Root	Fresh root juice used in piles, diarrhoea and stomachic.
Cynodon dactylon (L.) Pers. DD-31	Durba	Poaceae	Whole plant	Whole plant chewed and applied on cuts and wounds for stopping bleeding. Fresh plant decoction mixed with one slice of *Curcuma longa* is given to cure leucorrhoea and infertility.
Cyperus rotundus L. DD-56	MuthaGhas	Cyperaceae	Tuber	Juice of tuber with a pinch of table salt is given in morning on an empty stomach to cure chronic dysentery. Paste of the whole plant mixed with "Ada" (*Zingiber officinale*) is applied to cure boils on finger tips (Thosa).
Datura metel L. DD-127	Dhutura	Solanaceae	Leaf	Leaf paste applied as massage balm to get relief from rheumatic pain.
Deeringia amaranthoides (Lam.) Merr.* DD-79	Atmora/Atmutha	Amaranthaceae	Stem	Young mature stem combining with "Apang" (*Achyranthes aspera*) stem made into a chain and used to cure jaundice. Leaf also eaten as vegetable.
Desmodium triflorum (L.) DC. DD-122	Tin pata/Tepati	Papilionaceae	Leaf	Fresh leaves' juice is given on an empty stomach to cure flatulence.
Dioscorea alata L.* DD-42	Poraalu/Mach alu	Dioscoreaceae	Rhizome	Rhizome eaten as vegetable.
Dregea volubilis (L.f.) Benth. ex Hook.f.* DD-52	Jukti	Asclepiadaceae	Stem	Stem mixed with "Golmarich" (*Piper nigram*) and made into paste and externally applied for bone fracture.

Drynaria quercifolia (L.) J. Smith* DD-90	Pokhiraj	Polypodiaceae	Leaf	Dry leaf powder mixed with 12–18 "Golmarich" (*Piper nigrum*) and warmed in mustard oil and is applied locally to reduce muscular pain.
Eclipta prostrata (L.) L. DD-08	Kesut	Asteraceae	Whole plant	Whole plant juice is given to get relief from irritation and inflammation during urination. The whole plant paste is also applied on hair before 30 minutes of bath to reduce hair fall and promote hair growth.
Eleusine indica (L.) Gaertn DD-120	KanChulkani	Poaceae	Root	Fresh root decoction mixed with sugar is given in morning on an empty stomach to cure "Meho" (a sexual disease).
Euphorbia hirta L. DD-48	Dudhkushi	Euphorbiaceae	Whole plant	Whole plant paste used to treat rheumatism.
Euphorbia nerifolia L.* DD-88	ManasaSij	Euphorbiaceae	Leaf	Fresh leaves are crushed and fried in mustard oil and gently applied on the chest to cure bronchitis.
Flacourtia indica (Burm. f.) Merr. DD-45	Bainchi/Paniala	Flacourtiaceae	Stem bark	Stem bark mixed with stem bark of "Shimul" (*Bombax ceiba*) and boiled in water is given twice a day to cure physical weakness, Leucorrhoea and "Meho" (a sexual disease).
Geodorum densiflorum (Lam.) Schlt.* DD-70	Bon-ada	Orchidaceae	Tuber	Tuber paste applied on joint pain and arthritis.
Gloriosa superba L.* DD-09	Ulatchandal	Liliaceae	Root	Tuber paste mixed with "Ada" (*Zingiber officinale*) and "Tepati" (*Desmodium triflorum*) applied to cure joint pain.
Glycosmis pentaphylla (Retz.) DC. DD-110	Atiswar	Rutaceae	Root	Root paste applied for rupturing the boil or carbuncle and getting relief from joint pain. Mature stem used as tooth brush.
Heliotropium indicum L. DD-55	Hatisur	Boraginaceae	Root	Root paste warm with mustard oil and applied to cure rheumatism.
Helminthostachys zeylanica (L.) Hook.* DD-115	Akbir	Ophioglossaceae	Rhizome	Rhizome paste applied to cure arthritis.

Table 8.1 contd. ...

...*Table 8.1 contd.*

Scientific Name and Voucher Specimen Number	Vernacular Name	Family	Parts used	Medicinal uses and preparation
Hemidesmus indicus (L.) R. Br. ex Schult.* DD-73	Anantamul	Asclepiadaceae	Root	Root juice is given to promote sexual debilities and physical weakness.
Hibiscus rosa-sinensis L. DD-71	Jaba	Malvaceae	Leaf	Leaves' (5–7) juice mixed with "Gur" (Jaggery) taken on an empty stomach to reduce dysentery.
Holarrhena pubescens Wall. ex G. Don DD-86	Kurci/Indrajab	Apocyanaceae	Stem bark	Stem bark infusion is given early morning to cure chronic dysentery and seed powder mixed with water is used in diabetes and guinea worm.
Hygrophila auriculata (Schumach.) Heine DD-114	Kulekhara	Acanthaceae	Leaf	Fresh leaves' (5–7) juice is taken twice daily to get relief from allergy.
Ipomoea mauritiana Jacq.* DD-67	BhuiKumra	Convolvulaceae	Tuber	One teaspoonful tuber powder mixed with lukewarm milk is given during bed time as sexual stimulant.
Jatropha curcus L.* DD-29	Varenda	Euphorbiaceae	Latex	Latex used in toothache.
Jatropha gossypiifolia L. DD-47	LalVarenda	Euphorbiaceae	Latex	Latex used in pyorrhoea and applied on boil.
Justicia adhatoda L. DD-36	Harbashak	Acanthaceae	Leaf	Leaves boiled with water and sugar candy, mixture is given for better lactation.
Justicia gendarussa Burm. f.* DD-76	Bishtarak	Acanthaceae	Whole plant	Whole plant paste used in rheumatism.
Lawsonia inermis L.* DD-91	Mehendi	Lythraceae	Root	Root juice is used as sexual stimulant.
Leucas aspera (Willd.) Link.* DD-97	Dandakalash/Dulfi	Lamiaceae	Root	Root juice given on an empty stomach and the smell of the plant allowed to inhale for treatment of Tuberculosis.
Litsea glutinosa (Lour.) C. B. Rob. DD-104	Pipulti/Darodmaida	Lauraceae	Leaf	Leaf juice mixed with a pinch of salt and turmeric powder is given to cure dysentery and spermatorrhoea.
Ludwigia adscendens (L.) H. Hara DD-93	Keshra-dam	Onagraceae	Whole plant	Whole plant paste applied for bone fracture and rheumatism.

Botanical name	Local name	Family	Part used	Uses
Madhuca longifolia (Koen. ex L.) Macbride DD-53	Mahua	Sapotaceae	Stem bark	Stem bark boiled with water and the extract is used to cure physical weakness.
Mangifera indica L. DD-14	Aam	Anacardiaceae	Leaf	Bark decoction mixed with lime water is given to cure dysentery and diarrhoea. Tender leaf juice used for stomachic.
Marsilea quadrifolia L. DD-06	Susni	Marsileaceae	Leaf	Leaf juice mixed with a cup of lukewarm milk given at bed time to cure insomnia.
Meyna spinosa Roxb. ex Link.* DD-18	Moyenakanta	Rubiaceae	Spine	Spine fried with mustard oil is applied to cure piles.
Mimosa pudica L. DD-94	Lajjabati	Mimosaceae	Root	Fresh root boiled with water is used as gargle to cure toothache. Root decoction also used for leucorrhoea and blood dysentery.
Molineria capitulata (Lour.) Herb.* DD-17	Bansmora	Liliaceae	Root	Root juice is given early morning to cure piles.
Momordica charantia L. DD-19	Karola	Cucurbitaceae	Leaf	Leaf decoction and fruit juice is used to treat diabetes, allergy and guinea worms.
Momordica dioica Roxb. ex Willd. DD-109	Bhat-kalla	Cucurbitaceae	Fruit	Fruit eaten as vegetable to control blood sugar level.
Mucuna pruriens (L.) DC.* DD-15	Alkusi	Papilionaceae	Seed	The seeds are fried with "Ghee" and sugar and pills are made. Each pill taken during bed time as stimulant and for vitality.
Murraya koenigii (L.) Spreng. DD-28	Curry Pata	Rutaceae	Leaf	Leaf juice is given early morning to control blood sugar.
Nyctanthes arbor-tristis L. DD-30	Shiuly/Sephali	Oleaceae	Stem bark	Stem bark and leaf juice is given to control remittent fever and blood sugar.
Nymphaea rubra Roxb. ex Andrews* DD-13	Lalsaluk	Nymphaeaceae	Flower	Dried flowers' powder mixed with water is given to cure piles. Dried flowers mixed with root of *Eleusine indica* and a pinch of sugar is given on an empty stomach to cure leucorrhoea.

Table 8.1 contd. ...

...Table 8.1 contd.

Scientific Name and Voucher Specimen Number	Vernacular Name	Family	Parts used	Medicinal uses and preparation
Oxalis corniculata L. DD-92	Amrul	Oxalidaceae	Leaf	Leaf decoction used to cure smokers cough.
Ocimum americanum L. NBU-09797	Bon tulsi	Lamiaceae	Leaf/Seed	1. Infusion and mucilaginous seeds are given on an empty stomach to cure flatulence. 2. Half teaspoon of seed dust mixed with one cup of lukewarm milk is given continuously for 7 days during bed time to increase sexual potentiality. 3. The leaf used as mole repellents. 4. Dried burnt leaf smoke used for mosquito repellent.
Ocimum × africanum Lour.* NBU-09798	Lebu/bon tulsi	Lamiaceae	Leaf/Seed	1. Fresh leaves are made into paste and applied consecutively for 7 days before 30 minutes of bath to cure skin diseases. 2. 10–15 seeds mixed with drop of water and the mucilaginous seeds are applied on boils for quick rupture. 3. Leaf paste used on poisonous insect bites.
Ocimum basilicum L. NBU-09799	Babu/babuitulsi	Lamiaceae	Leaf	1. Fresh leaves' paste is applied on the forehead to get relief from the sinus and headache problem. 2. Leaves/whole plants worshipped during "Manasha puja" by "Polia tribe/Rajbangshi". 3. Infusion and mucilaginous seeds are given on an empty stomach to cure gonorrhoea and act as stimulant.
Ocimum basilicum L. NBU-09800	Maruatulsi	Lamiaceae	Leaf	1. Leaf juice is given on an empty stomach to reduce gastric problems. 2. Leaf used as chutney.

Ocimum gratissimum L.* NBU-09801	Ram tulsi	Lamiaceae	Leaf/Seed	1. Leaf juice mixed with a few drops of honey is given early morning for dry cough and gastrointestinal problems of child. 2. Leaf Juice is given to cure cold and cough. 3. Seed paste applied on the boils for suppuration.
Ocimum gratissimum L. NBU-09802	Ajowantulsi	Lamiaceae	Leaf/Seed	1. Leaf juice mixed with a few drops of honey is given early morning for dry cough and gastrointestinal problems of child. 2. Seed paste applied on the boils for suppuration. 3. Leaf paste used on poisonous insect stings.
Ocimum kilimandscharicum Gurke* NBU-09803	Karpurtulsi	Lamiaceae	Leaf	1. Leaf paste applied on forehead to reduce headache and sinus problem.
Ocimum tenuiflorum L. (Purple type) NBU-09795	Krishna tulsi	Lamiaceae	Leaf/Root	1. Leaf juice mixed with honey is used to reduce cold and cough. Fresh leaf decoction is taken twice a day for curing tuberculosis. 2. Leaf juice is used as drops for the ear to prevent earache. 3. One teaspoon of root extract is given with ½ cup of lukewarm milk during bed time for stimulant.
Ocimum tenuiflorum L. (Green type) NBU-09796	Radhatulsi	Lamiaceae	Leaf	1. Leaf juice mixed with honey is given for 3–7 days to cure cold and cough, bronchitis, whopping cough and fiver. 2. Leaf (10–15) paste mixed with lemon juice (1/4) is applied on face after bath for 7 consecutive days to cure fungal infection. 3. 5–6 leaves mixed with a pinch of salt and made into paste applied as poultice to cure rheumatic and gout pain. 4. Leaf paste used on poisonous insect bites.
Phyla nodiflora (L.) Greene DD-98	Koi Okhra	Verbenaceae	Whole plant	Whole plant paste used in bone fracture.

Table 8.1 contd. ...

...*Table 8.1 contd.*

Scientific Name and Voucher Specimen Number	Vernacular Name	Family	Parts used	Medicinal uses and preparation
Phyllanthus emblica L. DD-03	Amlaki	Phyllanthaceae	Fruit	Infusion of fruit is useful in acidity, constipation, insomnia and also in diabetes.
Physalis peruviana L.* DD-116	Fatki	Solanaceae	Root	Fresh root juice mixed with a few drops of honey is given to cure whooping cough.
Piper betel L.* DD-111	Pan	Piperaceae	Leaf	Leaf juice is given to cure indigestion and killing lice.
Piper longum L. DD-44	Pipul	Piperaceae	Whole plant	Whole plant cooked as vegetable is given after child birth for better lactation and relief from birth pain.
Plumbago zeylanica L. DD-113	Sadachita	Plumbaginaceae	Root	The root paste is used as emollient for gout.
Pongamia pinnata (L.) Pierre* DD-04	Karanja/Gokaranja	Papilionaceae	Seed	Seed oil is useful for hair growth and gout. Stem helpful in reducing toothache when used as tooth brush.
Portulaca oleracea L.* DD-75	Nunia sag	Portulacaceae	Whole plant	Whole plant cooked as vegetable to reduce constipation.
Premna serratifolia L.* DD-07	Ganiari	Verbenaceae	Leaf	Dry leaf infusion is given on an empty stomach to cure indigestion problem.
Psidium guajava L.* DD-100	Peyara	Myrtaceae	Leaf	Tender leaves (2–3) chewed to cure mouth ulcer, pyorrhoea, bad breath and sluggish fever. Bark infusion mixed with lime water used to cure diarrhoea and dysentery.
Pterospermum acerifolium (L.) Willd. DD-102	Muchkunda Chapa	Sterculiaceae	Flower	The dried flower's powder is mixed with jaggery and tablets are made. One tablet is given once in a day on an empty stomach to improve liver functions.
Rauvolfia serpentina (L.) Benth. ex Kurz DD-118	Sapagandha/Chandovado	Apocyanaceae	Leaf	Leaves (2–3) kept under the pillow to reduce insomnia.
Ricinus communis L. DD-128	Reri/Varenda	Euphorbiaceae	Seed	The leaves are warmed with coconut oil and applied to a woman's breast for improved lactation. Seed oil applied to cure chronic arthritis and gout.

Salvinia auriculata Aubl* DD-40	Khudipana/Muakarni	Salviniaceae	Whole plant	Whole plant mixed with 8–21 "Golmarich" (*Piper nigrum*) and applied to cure bone fracture.
Scoparia dulcis L. DD-105	Chinimichri/Jastimadhu	Scrophulariaceae	Leaf	Leaves and young shoots' decoction used in dysentery.
Senna sophera (L.) Roxb. DD-99	Chekenda/Kalkasunda	Caesalpiniaceae	Root	Root decoction mixed with "Ada" (*Zingiber officinale*) is given to cure stomachic and fever. Leaf paste applied for gout and body pain.
Sida cordifolia L. DD-123	SwetBerala	Malvaceae	Root	Root juice is given on an empty stomach to cure "Meho" (a sexual disease).
Smilax ovalifolia Roxb. ex D. Don DD-09	Bagnocha/Kumarilata	Smilaceae	Root	The root juice is used for dysentery and rheumatism. Tender leaves act as stimulant.
Solanum viarum Dunal* DD-07	Kantikari	Solanaceae	Fruit	Mature fruit mixed with "Golmarich" (*Piper nigrum*) and made into paste, is massaged to reduce pain and swelling of arthritis and paralysis.
Solena amplexicaulis (Lam.) Gandhi DD-89	RakhalSasa	Cucurbitaceae	Fruit	Fruit eaten as vegetable to control blood sugar level.
Sphagneticola calendulacea (L.) Pruskir* DD-51	Bhringaraj	Asteraceae	Leaf	Leaves' paste applied on hair before 30 minutes of bath for promoting hair growth.
Spondias pinnata (L.f.) Kurz.* DD-24	Amra	Anacardiaceae	Leaf	Fresh leaves decoction mixed with a pinch of salt is given to cure dysentery.
Stephania japonica (Thunb.) Miers DD-124	Aknadi/Takalati	Menispermaceae	Leaf	Fresh leaf applied on carbuncle or boil for rupture and relief from pain.
Syzygium cumini (L.) Skeels DD-27	Jam	Myrtaceae	Leaf	Fresh leaf juice mixed with a pinch of table salt is given on an empty stomach to control dysentery.
Terminalia arjuna (Roxb.ex DC.) Wt. &Arn. DD-46	Arjun	Combretaceae	Stem bark	Stem bark infusion is given in morning on an empty stomach to cure gastrointestinal troubles and heart problems.

Table 8.1 contd. ...

...*Table 8.1 contd.*

Scientific name and voucher specimen number	Vernacular name	Family	Parts used	Medicinal uses and preparation
Terminalia bellirica (Gaertn.) Roxb. DD-119	Bahera	Combretaceae	Fruit	Infusion of fruit is given in morning on an empty stomach to cure chronic dysentery, piles and constipation.
Terminalia chebula Retz. DD-121	Haritaki	Combretaceae	Fruit	Fruit infusion is given early morning on an empty stomach to promote liver functions and digestion.
Thunbergia laurifolia Lindl.* DD-60	SwetMahakal	Acanthaceae	Leaf	Leaf mixed with "Durba" (*Cynodon dactylon*) is given to cure leucorrhoea.
Tinospora sinensis (Lour.) Merr. DD-47	Gulancha	Menispermaceae	Stem	Stem (1.5 kg) boiled with water (3 lit) and made into 1 lit; from there half cup of mixture is given daily on an empty stomach to control blood sugar, cholesterol and physical weakness.
Trichosanthes dioica Roxb.* DD-96	Patol	Cucurbitaceae	Root	Fresh root juice is given early morning on an empty stomach to expel the intestinal worm of children.
Tylophora indica (Burm.f.) Merr.* DD-69	Antamul	Asclepiadaceae	Leaf	Leaf decoction is given to cure "Meho" (a sexual disease).
Vanda tessellata (Roxb.) G. Don DD-63	Rasna/Pargacha	Orchidaceae	Root	Root juice is given to cure physical weakness and promote sexual performances.
Vitex negundo L. DD-25	Nishinda	Verbenaceae	Leaf	Dry leaves boiled with mustard oil are applied to cure rheumatism. Leaf decoction used also in fever and diabetes.
Zingiber montanum (J. Koenig) Link ex A. Dietr.* DD-34	Bon Ada/Taraj	Zingiberaceae	Rhizome	Infusion of the fresh rhizome is given to cure leucorrhoea and physical weakness.
Zingiber zerumbet (L.) Roscoe ex Sm.* DD-62	Jabakusum	Zingiberaceae	Rhizome	Fresh rhizome juice mixed with "Aswagandha" (*Withania somnifera*) is given to cure leucorrhoea and physical weakness.

* New records/claims.

Dakshin Dinajpur. The study revealed that 39 plants were used as herbal medicine by different ethnic communities, namely Santhals and Mundas residing along the basin of the River Atreyee. Santhals are the dominant tribe in the study area, using the maximum number (67) of plants as indigenous health care system, and it was followed by Munda tribe (29). Further, they identified 62 plant species belonging to 34 families used by Santal, Oraon, Munda and other communities (Polia, Sabar, Lodha) for medicinal purposes in Hili Block of Dakshin Dinajpur district, West Bengal (Talukdar and Talukdar 2013). The study revealed that leaves of the plant species are frequently used by ethnic people for medicinal purposes. Among the tribals, Santals are the best in acquiring, maintaining and using traditional knowledge of herbal plants. Recently, Chowdhury et al. (2014a) documented some traditional uses of plants by the different communities of Dakshin Dinajpur district. Results revealed that the maximum number of plants was used in gastrointestinal problems followed by gynaecological problems. The study showed that among the various plant parts used, leaf was mostly used for medicinal preparation followed by root, whole plant, stem bark, etc. Total 132 plant species belonging to 65 families under 120 genera were recorded in the aforementioned study. Though all the 132 plants have their medicinal values, but these were unveiled before the study, and most importantly 56 plants are found to have medicinal importance, which was not known before the study (Table 8.1).

Ethnomedicines in Specific Ailments

To induce early stage pregnancy termination, some plants have been used by the various tribal communities from the different parts of West Bengal. Mitra and Mukherjee (2009) had documented the abortifacient plants used by the tribal communities of West Bengal. They documented total 22 Angiospermic plant species belonging to 21 genera under 18 families for abortifacient among the seven (Lodha, Lohar, Munda, Oraon, Polia, Sabar and Santal) native tribal communities of West Bengal. Gastro-intestinal problems like constipation, diarrhoea, dysentery, dyspepsia, indigestion and flatulence, inflammation of liver, stomachache, abdominal pain and intestinal worm mostly prevailed in the North Bengal area. Mitra and Mukherjee (2010) had identified 62 local plant species belonging to 60 genera under 36 families (33 dicots and 3 monocots) having medicinal importance to cure gastrointestinal problems. Leaf was the most commonly plant parts used for preparation of drug and it was also observed that diarrhoea and dysentery were the two commonest diseases which have been documented by 11 and 21 prescriptions, respectively. Diabetes is a very important disease because of its wide occurrence and severity. Chowdhury et al. (2011) documented the plants which were being used to treat diabetes by the ethnic tribal communities of Dinajpur (Uttar and Dakshin) and Malda District of West Bengal. The study recorded 31 plant species belonging to 21 families which were commonly used for the remedial of diabetes. Mitra and Mukherjee (2011) had listed 29 plant species, which have been used in 32 different ways for the treatment of diabetes by the 9 different tribal communities of the

North Bengal plains (Terai region of Darjeeling district, the districts of Jalpaiguri, Koch bihar, Uttar Dianjpur, Dakshin Dinajpur, and Malda) of West Bengal, India.

Ethnobotany of the Genus *Ocimum*

Ocimum tenuiflorum ("Tulsi") is considered as the most sacred and auspicious plant in India. The name 'Tulsi' is derived from the Sanskrit word which means "the matchless one" (Ghosh 1995). In Ayurveda, 'Tulsi' is known as the incomparable one, mother medicine of nature, the queen of herbs and elixir of life (Singh et al. 2002). Hindus worship this plant in the morning as well as in the evening by giving some "Prasad/Vog" with wet rice, flowers, vermillion and sweets. Many "Sadhus", particularly those belonging to the *Bairagi/Baishnab* sector, put on a garland of tulsi around their necks. They use 'Japmala' (Chanting beads) to count or chant the names of Ram from tulsi. The Baishnab devotee offers his daily meals to Vishnu by putting a leaf of tulsi in his food (Upadhyaya 1964). According to Hindu tradition, 'Tulsi' leaves are placed on the eyes of the dead body and planted at the funeral place and this plant is never burnt by Hindus (Kumar et al. 2006).

The traditional uses of *Ocimum* species are well documented in some Indian texts by different authors (Chopra 1953, Chopra et al. 1956, CSIR 1966, Kirtikar and Basu 1975, Nadkararni and Nadkarni 1976, Satyavati et al. 1976, Warrier et al. 1995). Tulsi is also a good home remedy for various diseases such as common cold and cough, asthma, headaches, bronchitis, liver diseases, fever, lumbago, hiccups, eye infections, ringworm, gastric disorders, diarrhoea, insomnia, arthritis, urinary disorders, skin diseases, sore throat, vomiting, antidote for snake bite and scorpion sting (Singh et al. 2002, Prajapati et al. 2003, Das and Vasudevan 2006, Ulbricht 2010, Cohen 2014). Another report showed the ethnoveterinary use of Lamiaceae *Ocimum* (Galav et al. 2013). Naghibi et al. (2005) documented the folk medicinal uses of Labiatae family from Iran. They documented total 410 species and subspecies of 46 genera. The 18% species of the family Lamiaceae were used as medicine. In this review, four *Ocimum* species and their ethnobotanical uses are well documented.

Prabhu et al. (2009) reviewed the chemical, pharmacological and ethnomedicinal properties of *O. gratissimum*. Folklore medicine claims its use in headache, fever, diarrhoea, pneumonia, etc. This review nicely represented the ethno-botanical, natural product chemistry, pharmacological, clinical and toxicological data of the plant. Kashyap et al. (2011) reviewed the ethnomedicinal, phytochemical and pharmacological survey of *O. kilimandscharicum*. It has been used generally in Kenya against cold and cough, measles, diarrhoea, abdominal pain and as a mosquito repellent. Singh and his associates (2011) reviewed the folk uses of *Ocimum sanctum*. The study clearly describes the symptoms of different ailments and various modes of administration of *Ocimum* for the management of healthcare system. Agarwal et al. (2013) studied the ethnomedicinal uses of *Ocimum* species from Rajasthan. The study showed that traditional healers of Rajasthan have commendable knowledge of the medicinal values of plants growing around them.

Mamun-Or-Rashid and his associates (2013) reviewed the ethno-medicobotanical study on *Ocimum sanctum*. This study revealed the enormous diversity of its medicinal uses and curing a wide range of common ailments like fever, malaria fever, asthma, bronchitis, colic pain, sore throat and hepatic diseases. Besides the ethnobotanical uses, they also listed the phytochemicals and various other important medicinal properties. Similar study has been conducted by Tiwari et al. (2014).

"Tulsi" (*Ocimum* sp.) is one of the most commonly used plant for curing ailments. During the period of ethnomedicinal study, it was found that nine different *Ocimum* species/varieties are present naturally in Dakshin Dinajpur district which were traditionally used by the different tribal people of this district (Chowdhury et al. 2016, 2017).

Uses Other Than Medicinal

Edible wild plants

In addition to medicinal uses, plant resources were also utilized as forage, manure, fishing, sheltering, vegetable and religious purposes. Talukdar and Talukdar (2012) nicely documented 19 plants used as vegetable by different ethnic communities, namely Santhals and Mundas, residing along the basin of the River Atreyee. Besides food value, some plants like *Hydrilla verticillata, Bambusa arundinacea, Coix lacryma-jobi* were used for fishing and *Echinochloa crus-galli* and *Hydrilla verticillata* were used as fish feed. Chowdhury et al. (2014b) documented a total of 91 plant species belonging to 51 families and 78 genera that have been used as wild edibles by the local communities as well as common people of Uttar Dinajpur district. Out of 91 species of wild edibles collected from this area, 89 species belong to angiosperms and 2 species are ferns. Of the angiosperms, 18 species belong to monocots and 71 species are dicotyledonous. Out of the 18 monocot species, herbs are predominant, followed by trees and climbers. On the other hand, out of 71 dicotyledonous species, herbs are predominant followed by trees, shrubs and climbers. The study also documented some interesting preparations of wild food plants, for example, rhizomes of *Alocasia macrorrhiza* are used as vegetable after cooking and are also cut into small juliennes and dried for future use as chips. Tubers of *Typhonium trilobatum* are consumed by the Santals after boiling. Rajbangsi people consume the spongy part of the petiole of *Musa balbisiana,* locally called "Kakna", after cooking with stem and leaf of *Piper longum,* especially after child birth for alleviating pain. Young sprouting shoot of *Bambusa tulda* is covered with earthen pot which makes the shoot grow into a cabbage like form. This cabbage shaped shoot is consumed as vegetable. Flowers of *Madhuca indica* are consumed raw and are also used as a flavoring agent of "Payasham"—a milk and rice containing dessert. Seeds of *Tamarindus indica* are fried in hot sand and consumed. Similar kind of uses has been observed by the author while doing ethnobotanical survey of Dakshin Dinajpur district.

Conclusion

The value of medicinal plants, herbs and spices as herbal remedies is getting lost due to lack of awareness, urbanization, deforestation and modernization. On the other hand, some traditional practitioners had a false apprehension that propagating the knowledge of medicinal plants by common people may reduce their efficacy and thus they kept it a secret. However, newer generations are not very keen to retain this tradition based knowledge from their predecessors. As a result, important knowledge based tradition is fading away. Therefore, in the present chapter, efforts have been made to accumulate traditional knowledge of the different tribal and non-tribal communities of Dakshin Dinajpur district before this huge wealth of traditional knowledge is lost forever and hence it must be documented properly. In conclusion, on the basis of the ethnobotanical review of Dakshin Dinajpur district, further phytochemical investigation is needed for clinical trials to test their efficacy and to develop a new herbal drug. Hopefully this documentation will serve for the future generations, researchers and common people as a whole.

References

Agarwal, S., Kumar, V.R. and Kumar, A. 2013. Ethnobotanical studies on *Ocimum* spp. in Rajasthan, India. Int. Res. J. Pharm. 4(4): 228–231.

Banerjee, R.N. and Ghora, C. 1996. On the domestic use of some unreported plants of West Dinajpur district (WB). J. Econ. Tax. Bot. Addl. Ser. 12: 325–328.

Bently, R. and Trimen, H. 1980. Medicinal Plants. I-IV, J & A. Publishers, Churchill, London.

Chaudhuri Rai, H.N. and Pal, D.C. 1976. Preliminary observation on ethnobotany of Medinipur district, West Bengal, India. Ind. Mus. Bull. 11(2): 51–53.

Chopra, R.N. 1953.Glossary of Indian medicinal plants, CSIR, New Delhi.

Chopra, R.N., Nayar, S.I. and Chopra, I.C. 1956. Glossary of Indian Medicinal Plants. CSIR, New Delhi.

Chowdhury, T., De Sarker, D. and Saha, M. 2011. Survey of plants used for the treatment of diabetes in Dinajpur (Uttar & Dakshin) and Malda districts of Paschimbanga. pp. 295–299. *In*: Ghosh, C. and Das, A.P. (eds.). Proceeding, Recent Studies in Biodiversity and Traditional Knowledge in India. Gour Mahavidyalaya, Malda.

Chowdhury, T., De Sarker, D. and Roy, S.C. 2014a. Local folk use of plants in Dakshin Dinajpur district of West Bengal, India. Int. Res. J. Biological. Sci. 3(5): 67–79.

Chowdhury, T., Roy, S.C. and De Sarker, D. 2014b. Wild edible plants of Uttar Dinajpur district, West Bengal. Life Sci. Leaflets 47: 20–36.

Chowdhury, T., Mandal, A., Jana, A.K., Roy, S.C. and De Sarker, D. 2016. Study of phytosociology and ecology of naturally growing *Ocimum* species with their conservational strategies in Dakshin Dinajpur district of West Bengal. Acta Ecol. Sin. 36: 483–491.

Chowdhury, T. 2017. Ethnobotany of Dakshin Dinajpur district with special reference to diversity and conservation of *Ocimum* species. Ph. D. Thesis Univ. of North Bengal.

Chowdhury, T., Mandal, A., Roy, S.C. and De Sarker, D. 2017. Diversity of the genus *Ocimum* (Lamiaceae) through morpho-molecular (RAPD) and chemical (GC-MS) analysis. J. Genet. Eng. Biotechnol. 15: 275–286.

Cohen, M.M. 2014. Tulsi—*Ocimum sanctum*: A herb for all reasons. J. Ayurveda. Integr. Med. 5(4): 251–259.

CSIR. 1966. The Wealth of India. A dictionary of Indian raw materials and industrial products. Vol. VII: N-Pe. Publications and Information Directorate, CSIR, New Delhi, India, pp. 79–89.

Das, S.K. and Vasudevan, D.M. 2006.Tulsi: The Indian holy power plant. Nat. Prod. Rad. 5(4): 279–83.

De, J.N. 1965. Some minor plant-fibres of aboriginal usage in the district of Purulia (West Bengal). Bull. Bot. Soc. Bengal. 19(2): 67–72.

De, J.N. 1969. Further observation on the ethnobotany of Purulia District in West Bengal. Indian Forester. 95(8): 551–559.

De Sarker, D., Chowdhury, T. and Saha, M. 2011. Biodiversity and medicinal plants of West Dinajpur and Malda, Vol. 1, Raiganj College (University College), Raiganj, Uttar Dinajpur, West Bengal.

Ernst, E. 2005. The efficacy of herbal medicine—an overview. Fund. Clin. Pharmacol. 19(4): 405–409.

Galav, P., Jain, A. and Katewa, S.S. 2013. Traditional veterinary medicines used by livestock owners of Rajasthan, India. Indian J. Tradit. Know. 12(1): 47–55.

Ghosh, A. 1986. A preliminary report on the ethnobotanical survey of Cooch Behar district, West Bengal. J. Bengal. Nat. Hist. Soc. 5(1): 69–73.

Ghosh, G.R. 1995. Tulasi (N.O. Labiatae, Genus-*Ocimum*). New Approaches to Medicine and Health (NAMAH) 3: 23–29.

Jain, S.K. and De, J.N. 1964. Some less known plants foods among the tribals of Purulia, West Bengal. Sci. Cult. 30: 285–286.

Jain, S.K. and Tarafder, C.R. 1970. Medicinal plant-lore of Santal (A revival of P. O. Boddings work). Econ. Bot. 24: 241–278.

Joshi, A.R. and Joshi, K. 2000. Indigenous knowledge and uses of medicinal plants by local communities of the Kali Gandaki Watershed area, Nepal. J. Ethnopharmacol. 73(12): 175–183.

Kamboj, V.P. 2000. Herbal medicine. Current. Sci. 78(1): 35–39.

Kashyap, C.P., Kaur, R., Vikrant, A. and Kumar, V. 2011. Therapeutic potency of *Ocimum Kilimandscharicum* Guerke—A review. Global J. Pharmacol. 5(3): 191–200.

Kirtikar, K.R. and Basu, B.D. 1975. Indian Medicinal Plants. Vol. 3 (2nd Eds.). Bishen Singh Mahendra Pal Singh, New Connaught Place, Dehradun, India, pp. 1965–1968.

Kumar, A., Tiwari, D.D. and Tiwari, J.P. 2006. Ethnomedicinal knowledge among Tharu tribe of Devipatan Division, UP, India. Indian J. Tradit Know. 5(3): 310–313.

Kundu, S. and Bag, A. 2012. Indigenous health care practices among Rajbanshi of Dakshin Dinajpur, West Bengal. Ethno. Med. 6(2): 117–120.

Mamun-Or-Rashid, A.N.M., Azam, M.M., Dash, B.K., Hafiz, F.B. and Sen, M.K. 2013. Ethnomedicobotanical study on *Ocimum sanctum* L. (Tulsi)—a review. Mintage. J. Pharm. Med. Sci. 2(2): 37–42.

Mitra, S. and Mukherjee, S.K. 2005a. Ethnobotanical usages of grasses by the tribals of West Dinajpur district, West Bengal. Indian J. Tradit. Know. 4(4): 396–402.

Mitra, S. and Mukherjee, S.K. 2005b. Root and rhizome drugs used by the tribals of West Dinajpur in Bengal. J. Trop. Med. Plants 6(2): 301–315.

Mitra, S. and Mukherjee, S.K. 2007. Plants used as ethnoveterinary medicine in Uttar and Dakshin Dinajpur districts of West Bengal, India. pp. 117–122. *In*: Das, A.P. and Pandey, A.K. (eds.). Advances in Ethnobotany. Bishen Singh Mahendra Pal Singh, Dehra Dun, India.

Mitra, S. and Mukherjee, S.K. 2009. Some abortifacient plants used by the tribal people of West Bengal. Nat. Prod. Rad. 8(2): 167–171.

Mitra, S. and Mukherjee, S.K. 2010. Ethnomedicinal uses of some wild plants of North Bengal plain for gastro-intestinal problems. Indian J. Tradit. Know. 9(4): 705–712.

Mitra, S. and Mukherjee, S.K. 2011. Plants used for the treatment of diabetes in West Bengal, India. J. Trop. Med. Plants 12(1): 99–105.

Mitra, S. and Mukherjee, S.K. 2013. Flora and ethnobotany of West Dinajpur district, West Bengal. Bishen Singh Mahendra Pal Singh, Dehra Dun.

Molla, H.A. and Roy, B. 1984. Folklore about some medicinal plants from the tribal areas of Jalpaiguri district, West Bengal. Bull. Bot. Surv. India 26(3-4): 160–163.

Molla, H.A. and Roy. B. 1985. Traditional uses of some medicinal plants by the Rabha tribals in Jalpaiguri district, West Bengal. J. Econ. Taxon. Bot. 7(3): 578–580.

Molla, H.A. and Roy, B. 1996. Some ethnobotanical claims from the Jalpaiguri district of West Bengal. J. Econ. Tax. Bot. Addl. Ser. 12: 322–324.

Nadkararni, A.K. and Nadkarni, K.M. 1976. Indian Materia Medica. Popular Prakashan Pvt. Ltd., Bombay.

Naghibi, F., Mosaddegh, M., Motamed, S.M. and Ghorbani, A. 2005. Labiatae family in folk medicine in Iran: from ethnobotany to pharmacology. Iran. J. Pharm. Res. 4(2): 63–79.

Pagan, J.A. and Pauly, M.V. 2005. Access to conventional medical care and the use of complementary and alternative medicine. Health Affairs 24(1): 225–262.

Pal, D.C. and Jain, S.K. 1989. Notes on Lodha medicine in Medinipur district, West Bengal, India. Econ. Bot. 43(4): 464–470.

Pant, N.C., Pandey, D.K., Banerjee, S.K. and Mishra, T.K. 1993. Some common ethnobotanical practices of Lodha community of Midnapore, West Bengal. J. Trop. Forestry 9(3): 215–218.

Prabhu, K.S., Lobo, R., Shirwaikar, A.A. and Shirwaikar, A. 2009. *Ocimumgratissimum*: a review of its chemical, pharmacological and ethnomedicinal properties. Open Complement Med. J. 1(1): 1–15.

Prajapati, N.D., Purohit, S.S., Sharma, A.K. and Kumar, T. 2003. A hand book of medicinal plant. Agrobios, India, pp. 367.

Ragupathy, S., Steven, N.G., Maruthakkutti, M., Velusamy, B. and Ul-Huda, M.M. 2008. Consensus of the 'Malasars' traditional aboriginal knowledge of medicinal plants in the Velliangiri holy hills, India. J. Ethnobiol. Ethnomed. 4(8): 1–14.

Rates, S.M.K. 2001. Plants as source of drugs. Toxicon 39(5): 603–613.

Saha, M.R. and De Sarker, D. 2013. Medicinal properties of ethnobotanically important plants of Malda and Uttar Dinajpur districts: a review. pp. 209–226. *In*: Sen, A. (ed.). Biology of Useful Plants and Microbes.

Saha, M.R., De Sarker, D., Kar, P., Sen Gupta, P. and Sen, A. 2014a. Indigenous knowledge of plants in local healthcare management practices by tribal people of Malda district, India. J. Intercult. Ethnopharmacol. 3(4): 179–185.

Saha, M.R., De Sarker, D. and Sen, A. 2014b. Ethnoveterinary practices among the tribal community of Malda district of West Bengal, India. Indian J. Tradit. Know. 13(2): 359–367.

Saha, M.R., Rai, R., Kar, P., Sen, A. and De Sarker, D. 2015. Ethnobotany, traditional knowledge and socioeconomic importance of native drink among the *Oraon* tribe of Malda district in India. J. Intercult. Ethnopharmacol. 4(1): 34–39.

Saha, M.R., Dey, P., Sarkar, I., Kar, P., De Sarker, D., Das, S., Haldar, B., Chaudhuri, T.K. and Sen, A. 2017. *Acacia nilotica* (L.) Delile could be a potential drug combating diabetes: an evidence based and *in-silico* approach. Diabetes Technol. Ther. 19(1): A-128.

Satyavati, G.V., Raina, M.K. and Sharma, M. 1976. Medicinal Plants of India. ICMR, New Delhi.

Schmidt, C., Fronza, M., Goettert, M., Geller, F., Luik, S., Flores, E.M., Bittencourt, C.F., Zanetti, G.D., Heinzmann, B.M., Laufer, S. and Merfort, I. 2009. Biological studies on Brazilian plants used in wound healing. J. Ethnopharmacol. 122(3): 523–532.

Siddiqui, H.H. 1993. Safety of herbal drugs—an overview. Drugs News & Views 1(2): 7–10.

Singh, N., Hoette, Y. and Miller, R. 2002. Tulsi: The mother medicine of nature. International Institute of Herbal Medicine, Lucknow, India.

Singh, V., Birendra, V. and Suvagiya, V. 2011. A review on ethnomedical uses of *Ocimum sanctum* (tulsi). Int. Res. J. Pharm. 2(10): 1–3.

Sirkar, N.N. 1989. Pharmacological basis of Ayurvedic therapeutics. *In*: Atal, C.K. and Kapoor, B.M. (eds.). Cultivation and Utilization of Medicinal Plants. PID CSIR.

Sur, P.R., Sen, R., Halder, A.C. and Bandyopadhyay, S. 1987. Observation on the ethnobotany of Malda-West Dinajpur districts, West Bengal-I. J. Econ. Tax. Bot. 10(2): 395–401.

Sur, P.R., Sen, R., Halder, A.C. and Bandyopadhyay, S. 1990. Observation on the ethnobotany of Malda-West Dinajpur districts, West Bengal-II. J. Econ. Tax. Bot. 14(2): 453–459.

Sur, P.R., Sen, R. and Halder, A.C. 1992. Ethnobotanical study of Purulia district, West Bengal, India. J. Econ. Taxon. Bot. Addl. Ser. 10: 259–264.

Talukdar, D. and Talukdar, T. 2012. Floral diversity and its indigenous use in old basin (Khari) of river Atreyee at Balurghat block of Dakshin Dinajpur district, West Bengal. NeBIO 3(2): 26–32.

Talukdar, T. and Talukdar, D. 2013. Ethno-medicinal uses of plants by tribal communities in Hili block of Dakshin Dinajpur district, West Bengal. Indian J. Nat. Prod. Resour. 4(1): 110–118.

Tiwari, A.K., Mishra, R. and Chaturvedi, A. 2014. Traditional uses of *Ocimum sanctum* (Tulsi). Int. J. Glob. Sci. Res. 1(2): 126–131.

Ulbricht, C.E. 2010. Natural standard: Herb and supplement guide—an evidence-based reference. St. Louis, MO: Elsevier Mosby.

Upadhyaya, K.D. 1964. Indian botanical folklore. Asian Folkl Stud. 23(2): 15–34.

Warrier, P.K., Nambiar, V.P.K. and Ramankutty, C. 1995. A compendium of 500 species. pp 157–168. *In*: Indian Medicinal Plants. Vol. 4. Orient Longman Publisher.

Xavier, T.F., Kannan, M., Lija, L., Auxillia, A., Rose, A.K.F. and Kumar, S.S. 2014. Ethnobotanical study of Kani tribes in Thoduhills of Kerala, South India. J. Ethnopharmacol. 152(1): 78–90.

Yamashita, H., Tsukayama, H. and Sugishita, C. 2002. Popularity of complementary and alternative medicine in Japan: A telephone survey. Complement Ther. Med. 10(2): 84–93.

CHAPTER 9

Antipsoriatic Medicinal Plants

From Traditional Use to Clinic

José Luis Ríos,[1,*] *Guillermo R. Schinella*[2] *and Isabel Andújar*[1,3,4]

Introduction

Psoriasis is a chronic autoimmune inflammatory disease of the skin with a worldwide prevalence of 2–3% (Nickoloff and Nestle 2004, Takeshita et al. 2017). The pathology is easy to recognize for the characteristic erythema, severe inflammation, excessive proliferation of keratinocytes, and scaly plaques (Lowes et al. 2007). In addition to the skin lesions, other associated pathologies could be present, mainly nail dystrophy accompanied by psoriatic arthritis (20–25%), which is similar to rheumatoid arthritis (Nickoloff and Nestle 2004). The pathogenesis of these comorbid diseases remains unknown, but common inflammatory pathways, cellular mediators, genetic susceptibility, and risk factors have been described as contributing elements (Takeshita et al. 2017).

Nickoloff (1991) proposed a hypothesis for the pathophysiology of psoriasis, in which a stimulus triggers a series of cellular events generating a cascade of cytokines, such as tumor necrosis factor (TNF)-α derived from dendritic antigen-presenting cells and keratinocytes, and interferon (IFN)-γ produced by activated Th1 lymphocytes (Nickoloff 1991, Boehncke 2007). The evolution of psoriasis treatment reflects the limitations in the knowledge of its pathogenesis. In the past, treatments were based on serendipity and chance because neither the specific target nor the mechanism of action for the treatment was known (Nickoloff and Nestle 2004). In this sense, arsenic (Fowler's solution), ammoniated mercury, crude

[1] Departament de Farmacologia, Facultat de Farmàcia, Universitat de València, Spain.
[2] Facultad de Ciencias Médicas, Universidad Nacional de La Plata, CIC-PBA, Argentina.
[3] FISABIO-Fundación Hospital Universitario Dr. Peset, Valencia, Spain.
[4] Departamento de Ciencias Biomédicas, Universidad Europea de Valencia, Valencia, Spain.
* Corresponding author: riosjl@uv.es

coal tar, anthralin, corticosteroids and ultraviolet (UV)-B radiation have been systematically used. The knowledge of the immunological components and some relevant mediators implicated allowed the use of selective immunosuppressors such as methotrexate, cyclosporine A, tacrolimus, psoralen + UV-A light therapy (PUBA), fumaric acid esters, 5-aminolevulinic acid, 6-thioguanine, salicilates, hydroxyurea and analogues of vitamins A (retinoids) and D. Advanced therapeutic options include the use of biologic drugs which specifically target cytokines that directly mediate the development of psoriasis, such as TNF-α inhibitors (etanercept, infliximab, and adalimumab), anti-interleukin (IL)-12/IL-23 (ustekinumab) or anti-IL-23 (secukinumab, ixekizumab) (Nickoloff and Nestle 2004, Herman and Herman 2016).

Due to the side effects and elevated cost of the treatment of psoriasis with these new biological drugs, patients often seek treatments outside the allopathic paradigm (Shawahna and Jarada 2017). In different parts of the world, patients with psoriasis employ medicinal plants and complementary and alternative medicine (CAM) as a potential solution. Some medicinal plants are used topically but others are administered orally for a systematic effect. Although the use of medicinal plants is based on the traditional employ and, consequently, is not an evidence-based practice, today different randomized clinical trials have been developed (Shawahna and Jarada 2017). The use of CAM among patients with psoriasis is quite common (43–69% prevalence), and the use of 1–6 herbs, special diets, or dietary supplements are the most common processes (Smith et al. 2009, Tirant et al. 2018). The number of randomized controlled clinical trials using different medicinal plants to treat signs/symptoms of psoriasis are considerable, and the results are varied due to cultural factors (Shawahna and Jarada 2017, Tirant et al. 2018).

The use of plant extracts or products as inhibitors of leukotriene synthesis is based on the previous reports on the increase of lipoxygenase activity, leukotrienes and other lipoxygenase products in the pathogenesis of psoriasis (Voorhees 1983). Consequently, 5- and 12-lipoxygenase inhibitors could have beneficial effect in this disease. Different pathologic events can be explained by the action of 12-hydroxyeicosatetraenoic acid (12-HETE): the normal epidermis synthesizes predominantly 12(S)-HETE, whereas the product derived from psoriatic skin is its enantiomer 12(R)-HETE (Schneider and Bucar 2005). However, Ford-Hutchinson (1993) had previously described doubts on the case of 5-lipoxygenase because there is no clear evidence of its presence in human skin, and selective leukotriene biosynthesis inhibitors have no therapeutic utility in psoriasis. Based on these features, many of the studied plants with potential as antipsoriatic agents are found in the arachidonate metabolism via lipoxygenase and in their antioxidant properties.

Traditional Use of Medicinal Plants as Antipsoriatic Agents

The treatment of skin diseases with plant extracts and natural products has been reported since ancient civilizations. Photochemotherapy was used in Egypt and India since 1200–2000 BC, as well as boiled extracts of seeds or leaves, which were applied to the skin or ingested before the patient was exposed to intense sunlight (Seyger et al. 1998), but also other complementary and alternative medicines have been used (Capella and Finzi 2003).

The first scientific report on the use of CAM in psoriatic patients was published by Jensen in 1990, who compiled data on 215 patients with psoriasis based on a questionnaire in Norway, which included different questions on the use of CAM. The conclusion drawn from the questionnaire was that the absence of a satisfactory effect of the standard therapy was the most decisive factor in using CAM, although, as he reported, patient showed no improvement, or psoriasis was even aggravated as a result of these alternative treatments (Jensen 1990a,b).

Amenta et al. (2000) compiled a series of 48 medicinal plants used around the world for treating psoriasis, and 9 used specifically in Sicily (Italy). These same authors indicated the extractive solvent and system for their application (Table 9.1). More recently, Talbott and Duffy (2015) reviewed the efficacy of herbal therapies against psoriasis and they observed the best efficacy in the case of *Berberis aquifolium* (syn.: *Mahonia aquifolium*) and indigo naturalis, while it was smaller for aloe (*Aloe vera*, syn: *Aloe barbadensis*), neem (*Azadirachta indica*), and extracts of sweet whey. Indigo naturalis is a blue powder obtained from the leaves of different plants, such as *Strobilanthes cusia* (syn.: *Baphicacanthus cusia*), *Persicaria tinctoria* (syn.: *Polygonum tinctorium*), *Isatis tinctoria* (syn.: *Isatis indigotica*) and *Strobilanthes formosanus* (McDermott et al. 2016). Herman and Herman (2016) also compiled a series of studies of plants used against psoriasis. All of them are included in Table 9.1. Some plant names have been modified and the present name was introduced.

Studies on Medicinal Plants as Potential Antipsoriatic Agents

Many of the cited plants in Table 9.1 have been studied in different experimental models of psoriasis. One of the hallmarks in the study of psoriasis is the lack of a model that reproduces appropriately the characteristic features of the disease. For that reason, the experimental models are not easily useful as antipsoriatic models and some of them are focused to evaluate concrete aspects of this disease, such as inflammation or cell proliferation. Of these potential protocols, different *in vitro* and *in vivo* experimental models have been developed, such as the *in vitro* effects on the proliferation of SVK-14 keratinocytes (Sampson et al. 2001), the anti-inflammatory and antiproliferative effects (Carrenho et al. 2015), the inhibition of mediators implicated in psoriasis, such as TNF-α (Sethi et al. 2009), the expression of relevant molecules in human psoriatic skin by immunohistochemistry (Augustin et al. 1999), and the *in vivo* orthokeratosis mouse-tail test (Herman and Herman 2016).

Epidermal keratinocytes provide a protective role against external stimuli forming a physical barrier, and its excessive proliferation is one of the most characteristic symptoms of psoriasis. Inhibiting its excessive proliferation could be a good target for the study of new antipsoriatic agents. In this sense, Sampson et al. (2001) carried out a rapid-throughput *in vitro* bioassay to look for plants

Table 9.1. Medicinal plants and fungi used in traditional medicine in treatment of psoriasis. All botanical names were updated using the recent taxonomical revision of 'The plant list. A working list of all plant species'.

Plant Name (updated)*	Family	Reference
Acanthus mollis L.	Acanthaceae	Amenta et al. 2000 Bader et al. 2015
Achillea ligustica All.	Compositae	Bader et al. 2015
Agave americana L.	Asparagaceae	Amenta et al. 2000
Aleurites moluccanus (L.) Willd.	Euphorbiaceae	Brown et al. 2005
Allium sativum L.	Amaryllidaceae	Shawahna and Jarada 2017
Aloe vera (L.) Burm.f. [1]	Xanthorrhoeaceae	Amenta et al. 2000 Smith et al. 2009 Singh and Tripathy 2014 Arora et al. 2016 Shawahna and Jarada 2017
Alpinia galanga (L.) Willd.	Zingiberaceae	Saelee et al. 2011 Singh and Tripathy 2014
Ammi visnaga (L.) Lam.	Apiaceae	Shawahna and Jarada 2017
Angelica dahurica (Hoffm.) Benth. & Hook.f. ex Franch. & Sav.	Apiaceae	Amenta et al. 2000 Prieto et al. 2003
Angelica dahurica var. *formosana* (Boissieu) Yen	Apiaceae	Shan et al. 2006
Angelica pubescens Maxim.	Apiaceae	Prieto et al. 2003
Angelica sinensis (Oliv.) Diels	Apiaceae	Amenta et al. 2000 Singh and Tripathy 2014
Annona squamosa L.	Annonaceae	Saelee et al. 2011 Singh and Tripathy 2014

Table 9.1 contd. ...

...*Table 9.1 contd.*

Plant Name (updated)*	Family	Reference
Anthemis cotula L.	Compositae	Shawahna and Jarada 2017
Arctium lappa L.	Compositae	Amenta et al. 2000
Arnebia euchroma (Royle) I.M.Johnst.	Boraginaceae	Yao et al. 2016 Dai et al. 2014
Arnebia guttata Bunge	Boraginaceae	Yao et al. 2016 Dai et al. 2014
Artemisia arborescens (Vaill.) L.	Compositae	Amenta et al. 2000 Bader et al. 2015
Artemisia capillaris Thunb.	Compositae	Lee et al. 2018
Astragalus propinquus Schischkin [2]	Leguminosae	Prieto et al. 2003
Avena barbata Pott ex Link	Poaceae	Shawahna and Jarada 2017
Azadirachta indica A. Juss	Meliaceae	Amenta et al. 2000 Smith et al. 2009
Berberis aquifolium Pursh [3]	Berberidaceae	Amenta et al. 2000 Singh and Tripathy 2014 Smith et al. 2009
Betula alleghaniensis Britton	Betulaceae	Kaur and Kumar 2012
Borago officinalis L.	Boraginaceae	Amenta et al. 2000
Caesalpinia bonduc (L.) Roxb.	Leguminosae	Muruganantham et al. 2011 Kaur and Kumar 2012
Camptotheca accuminata Decne.	Cornaceae	Singh and Tripathy 2014
Capsicum annuum L. [4]	Solanaceae	Shawahna and Jarada 2017 Singh and Tripathy 2014

Species	Family	References
Carica papaya L.	Caricaceae	Amenta et al. 2000
Catharanthus roseus (L.) G. Don	Apocynaceae	Shawahna and Jarada 2017
Celastrus paniculatus Willd.	Celastraceae	Arora et al. 2016
Celastrus orbiculatus Thunb.	Celastraceae	Kaur and Kumar 2012
Centella asiatica (L.) Urb.	Apiaceae	Kaur and Kumar 2012 Sampson et al. 2001
Celastrus paniculatus Willd.	Celastraceae	Arora et al. 2016
Chelidonium majus L.	Papaveraceae	Amenta et al. 2000
Citrus limon (L.) Osbeck	Rutaceae	Shawahna and Jarada 2017
Coptis chinensis Franch.	Ranunculaceae	Kaur and Kumar 2012
Crotalaria juncea L.	Leguminosae	Amenta et al. 2000 Singh et al. 2015
Cryptostegia grandifolia Roxb. ex R.Br.	Apocynaceae	Amenta et al. 2000
Cullen corylifolium (L.) Medik [5]	Leguminosae	Amenta et al. 2000 Sampson et al. 2001
Curcuma aromatica Salisb. [6]	Zingiberaceae	Dai et al. 2014
Curcuma kwangsiensis S.G.Lee & C.F.Liang	Zingiberaceae	Dai et al. 2014
Curcuma longa L. [7]	Zingiberaceae	Saelee et al. 2011 Singh and Tripathy 2014 Arora et al. 2016 Shawahna and Jarada 2017
Curcuma phaeocaulis Valeton	Zingiberaceae	Dai et al. 2014
Dictamnus albus L.	Rutaceae	Amenta et al. 2000
Dioscorea composita Hemsl.	Dioscoreaceae	Amenta et al. 2000

Table 9.1 contd. ....

...*Table 9.1 contd.*

Plant Name (updated)*	Family	Reference
Dittrichia viscosa (L.) Greuter [8]	Compositae	Amenta et al. 2000 Bader et al. 2015 Shawahna and Jarada 2017
Dodonaea polyandra Merr. & L.M.Perry	Sapindaceae	Simpson et al. 2014
Duchesnea indica (Jacks.) Focke	Rosaceae	Song et al. 2010
Ecballium elaterium (L.) A.Rich.	Cucurbitaceae	Amenta et al. 2000
Echinacea sp.	Compositae	Amenta et al. 2000
Eruca vesicaria (L.) Cav. [9]	Brassicaceae	Kaur and Kumar 2012
Eupatorium cannabinum L.	Compositae	Amenta et al. 2000
Falconeria insignis Royle [10]	Euphorbiaceae	Amenta et al. 2000
Ficus carica L.	Moraceae	Shawahna and Jarada 2017
Forsythia suspensa (Thunb.) Vahl	Oleaceae	Prieto et al. 2003
Fucus vesiculosus L.	Fucaceae	Amenta et al. 2000
Fumaria officinalis L.	Papaveraceae	Amenta et al. 2000
Gardenia gummifera L.f.	Rubiaceae	Nagar et al. 2016
Gaultheria procumbens L.	Ericaceae	Singh and Tripathy 2014
Givotia moluccana (L.) Sreem [11]	Euphorbiaceae	Amenta et al. 2000
Glycyrrhiza glabra L.	Leguminosae	Amenta et al. 2000 Shawahna and Jarada 2017
Glycyrrhiza inflata Batalin	Leguminosae	Yao et al. 2016
Glycyrrhiza uralensis Fisch.	Leguminosae	Yao et al. 2016
Handroanthus impetiginosus (Mart. ex DC.) Mattos [12]	Bignoniaceae	Kaur and Kumar 2012

Helichrysum italicum (Roth) G.Don	Compositae	Amenta et al. 2000
Helicteres isora L.	Malvaceae	Amenta et al. 2000
Hydnocarpus anthelmintica Pierre ex Laness.	Achariaceae	Amenta et al. 2000
Hydrastis canadensis L.	Ranunculaceae	Amenta et al. 2000
Hypericum perforatum L.	Hypericaceae	Amenta et al. 2000
Iris versicolor L.	Iridaceae	Amenta et al. 2000
Isatis tinctoria L. [13]	Brassicaceae	McDermott et al. 2016
Juglans regia L.	Juglandaceae	Shawahna and Jarada 2017
Kigelia africana (Lam.) Benth.	Bignoniaceae	Herman and Herman 2016
Lawsonia inermis L.	Lythraceae	Shawahna and Jarada 2017
Leucas aspera (Willd.) Link	Lamiaceae	Amenta et al. 2000 Singh et al. 2015
Ligusticum striatum DC. [14]	Apiaceae	Liu et al. 2001
Linum usitatissimum L.	Linaceae	Shawahna and Jarada 2017
Lithospermum erythrorhizon Siebold & Zucc.	Boraginaceae	Yao et al. 2016 Dai et al. 2014
Malva sylvestris L.	Malvaceae	Shawahna and Jarada 2017
Matricaria chamomilla L. [15]	Compositae	Singh and Tripathy 2014
Melaleuca alternifolia (Maiden & Betche) Cheel	Myrtaceae	Amenta et al. 2000 Singh and Tripathy 2014
Melaleuca leucadendra (L.) L.	Myrtaceae	Amenta et al. 2000
Menispermum dauricum DC.	Menispermaceae	Song et al. 2010
Musa × *paradisiaca* L.	Musaceae	Shawahna and Jarada 2017
Nerium oleander L.	Apocynaceae	Amenta et al. 2000

Table 9.1 contd. ...

...Table 9.1 contd.

Plant Name (updated)*	Family	Reference
Nepeta tenuifolia Benth. [16]	Lamiaceae	Amenta et al. 2000
Nigella arvensis L.	Ranunculaceae	Shawahna and Jarada 2017
Nigella sativa L.	Ranunculaceae	Herman and Herman 2016 Singh and Tripathy 2014
Oenothera biennis Scop.	Onagraceae	Amenta et al. 2000
Olax scandens Roxb.	Olacaceae	Singh and Tripathy 2014
Oldenlandia diffusa (Willd.) Roxb.	Rubiaceae	Song et al. 2010
Olea europaea L.	Oleaceae	Shawahna and Jarada 2017
Origanum jordanicum Danin & Kunne	Lamiaceae	Shawahna and Jarada 2017
Paeonia anomala subsp. *veitchii* (Lynch) D.Y.Hong & K.Y.Pan [17]	Paeoniaceae	Yao et al. 2016
Paeonia lactiflora Pallas	Paeoniaceae	Choi et al. 2015
Panax ginseng C.A. Mey.	Araeliaceae	Kaur and Kumar 2012
Parietaria officinalis L.	Urticaceae	Amenta et al. 2000
Paronychia argentea Lam.	Caryophyllaceae	Shawahna and Jarada 2017
Pedalium murex L.	Pedaliaceae	Singh and Tripathy 2014
Persea americana Mill.	Lauraceae	Shawahna and Jarada 2017
Persicaria tinctoria (Aiton) H. Gross [18]	Polygonaceae	McDermott et al. 2016
Phellodendron amurense Rupr.	Rutaceae	Li et al. 2017
Phellodendron chinense C.K.Schneid.	Rutaceae	Li et al. 2017
Phlebodium decumanum (Willd.) J. Sm. [19]	Polypodiaceae	Kaur and Kumar 2012
Phyllanthus reticulatus Poir.	Phyllanthaceae	Singh and Tripathy 2014 Singh et al. 2015
Phyllanthus virgatus G.Forst. [20]	Phyllanthaceae	Singh and Tripathy 2014
Picea mariana (Mill.) Britton, Sterns & Poggenb.	Pinaceae	Kaur and Kumar 2012
Pilocarpus jaborandi Holmes	Rutaceae	Amenta et al. 2000

Pinus halepensis Mill. [21]	Pinaceae	Kaur and Kumar 2012 / Shawahna and Jarada 2017
Platycladus orientalis (L.) Franco	Cupressaceae	Song et al. 2010
Podophyllum peltatum L.	Berberidaceae	Amenta et al. 2000
Pongamia pinnata (L.) Pierre	Leguminosae	Kaur and Kumar 2012
Prunus dulcis (Mill.) D.A.Webb [22]	Rosaceae	Shawahna and Jarada 2017
Prunus mume (Sieb.) Sieb. et Zucc.	Rosaceae	Yao et al. 2016
Pseudolarix amabilis (J. Nelson) Rehder	Pinaceae	Amenta et al. 2000
Rauwolfia vomitoria Afzel.	Apocynaceae	Amenta et al. 2000
Rehmannia glutinosa (Gaertn.) DC.	Plantaginaceae	Amenta et al. 2000
Reynoutria japonica Houtt. [23]	Polygonaceae	Song et al. 2010
Rhagadiolus stellatus (L.) Gaertn.	Compositae	Amenta et al. 2000
Rhinacanthus nasutus (L.) Kurz	Acanthaceae	Singh and Tripathy 2014
Rhus chinensis Mill.	Anacardiaceae	Song et al. 2010
Rhus mysorensis G.Don	Anacardiaceae	Singh and Tripathy 2014
Ricinus communis L.	Euphorbiaceae	Shawahna and Jarada 2017
Rubia cordifolia L.	Rubiaceae	Kaur and Kumar 2012 / Herman and Herman 2016
Rumex nepalensis Spreng.	Polygonaceae	Song et al. 2010
Ruta graveolens L.	Rutaceae	Amenta et al. 2000
Saccharum officinarum L.	Poaceae	Kaur and Kumar 2012
Salvia fruticosa Mill.	Lamiaceae	Shawahna and Jarada 2017
Salvia miltiorriza Bunge	Lamiaceae	Deng et al. 2014
Sarcandra glabra (Thunb.) Nakai	Chloranthaceae	Yao et al. 2016
Scrophularia striata Boiss	Scrophulariaceae	Monsef-Esfahani et al. 2014
Scutellaria baicalensis Georgi	Lamiaceae	Herman and Herman 2016

Table 9.1 contd. ...

...Table 9.1 contd.

Plant Name (updated)*	Family	Reference
Securidaca longipedunculata Fresen.	Polygalaceae	Amenta et al. 2000
Silybum marianum (L.) Gaertn.	Compositae	Amenta et al. 2000
Simmondsia chinensis (Link) C.K. Schneid.	Simmondsiaceae	Shawahna and Jarada 2017
Senna alexandrina Mill.	Leguminosae	Shawahna and Jarada 2017
Senna tora (L.) Roxb. [24]	Leguminosae	Herman and Herman 2016
Smilax china L.	Smilacaceae	Herman and Herman 2016 Singh and Tripathy 2014
Smilax glabra Roxb.	Smilacaceae	Song et al. 2010 Yao et al. 2016
Smilax officinalis Kunth	Smilacaceae	Amenta et al. 2000
Solanum dulcamara L.	Solanaceae	Amenta et al. 2000
Solanum lyratum Thunb.	Solanaceae	Song et al. 2010
Solanum pubescens Willd.	Solanaceae	Singh and Tripathy 2014
Sophora tonkinensis Gagnep. [25]	Leguminosae	Amenta et al. 2000
Stellaria media (L.) Vill.	Caryophyllaceae	Amenta et al. 2000
Stellera chamaejasme L.	Thymelaeaceae	Tian et al. 2005
Strobilanthes cusia (Nees) Kuntze [26]	Acanthaceae	McDermott et al. 2016
Strobilanthes formosanus S. Moore	Acanthaceae	McDermott et al. 2016
Silybum marianum (L.) Gaertn.	Compositae	Singh and Tripathy 2014
Teucrium capitatum L.	Lamiaceae	Shawahna and Jarada 2017
Thespesia populnea (L.) Sol. ex Corrêa	Malvaceae	Kaur and Kumar 2012 Herman and Herman 2016
Tinospora sinensis (Lour.) Merr. [27]	Menispermaceae	Arora et al. 2016
Trigonella arabica Delile	Leguminosae	Shawahna and Jarada 2017
Ulmus rubra Muhl.	Ulmaceae	Singh and Tripathy 2014
Urtica urens L.	Urticaceae	Shawahna and Jarada 2017

Vataireopsis araroba (Aguiar) Ducke [28]	Leguminosae	Amenta et al. 2000
Verbascum sinuatum L.	Scrophulariaceae	Amenta et al. 2000
Verbena officinalis L.	Verbenaceae	Amenta et al. 2000
Viola tricolor L.	Violaceae	Amenta et al. 2000
Vitis vinifera L.	Vitaceae	Shawahna and Jarada 2017
Vitex glabrata R.Br.	Lamiaceae	Singh et al. 2015
Wolfiporia extensa (Peck) Ginns [29]	Polyporaceae	Prieto et al. 2003
Woodfordia fructicosa (L.) Kurz	Lythraceae	Nagar et al. 2016
Wrightia tinctoria R.Br.	Apocynaceae	Herman and Herman 2016 Singh and Tripathy 2014
Zea mays L.	Poaceae	Amenta et al. 2000
Zingiber officinale Roscoe	Zingiberaceae	Shawahna and Jarada 2017

Table 9.1 contd. ...

* All the plant names have been actualized according to the new taxonomic review cited in 'The plant list. A working list of all plant species': http://www.the-plantlist.org/.

1 *Aloe barbadensis* Miller is a synonym of *Aloe vera* (L.) Burm.f.
2 *Astragalus membranaceus* (Fisch.) Bunge is a synonym of *Astragalus propinquus* Schischkin
3 *Mahonia aquifolium* (Pursh) Nutt. is a synonym of *Berberis aquifolium* Pursh.
4 *Capsicum frutescens* L. is a synonym of *Capsicum annuum* L.
5 *Psoralea corylifolia* L. is a synonym of *Cullen corylifolium* (L.) Medik
6 *Curcuma wenyujin* Y.H.Chen & C.Ling is a synonym of *Curcuma aromatica* Salisb.
7 *Curcuma domestica* Valeton is a synonym of *Curcuma longa* L.
8 *Inula viscosa* (L.) Aiton is a synonym of *Dittrichia viscosa* (L.) Greuter
9 *Eruca sativa* Mill. is a synonym of *Eruca vesicaria* (L.) Cav.
10 *Sapium insigne* (Royle) Trimen is a synonym of *Falconeria insignis* Royle
11 *Givotia rottleriformis* Griff. ex Wight is a synonym of *Givotia moluccana* (L.) Sreem
12 *Tabebuia avellanedae* Lorentz ex Griseb. is a synonym of *Handroanthus impetiginosus* (Mart. ex DC.) Mattos
13 *Isatis indigotica* Fortune ex Lindl. is a synonym of *Isatis tinctoria* L.

...Table 9.1 contd.

14   *Ligusticum wallichii* Franch. is a synonym of *Ligusticum striatum* DC.

15   *Matricaria recutita* L. is a synonym of *Matricaria chamomilla* L.

16   *Schizonepeta tenuifolia* (Benth.) Briq. is a synonym of *Nepeta tenuifolia* Benth

17   *Paeonia veitchii* Lynch is a synonym of *Paeonia anomala* subsp. *veitchii* (Lynch) D.Y.Hong & K.Y.Pan

18   *Polygonum tinctorium* Aiton. is a synonym of *Persicaria tinctoria* (Aiton) H. Gross

19   *Polypodium decumanum* Willd. is a synonym of *Phlebodium decumanum* (Willd.) J. Sm.

20   *Phyllanthus simplex* Retz. is a synonym of *Phyllanthus virgatus* G.Forst.

21   *Pinus maritima* Mill. is a synonym of *Pinus halepensis* Mill.

22   *Prunus amygdalus* var. *amara* (DC.) Focke is a synonym of *Prunus dulcis* (Mill.) D.A.Webb

23   *Polygonum cuspidatum* Siebold & Zucc. is a synonym of *Reynoutria japonica* Houtt.

24   *Cassia tora* L. is a synonym of *Senna tora* (L.) Roxb.

25   *Sophora subprostrata* Chun & T.Chen is a synonym of *Sophora tonkinensis* Gagnep.

26   *Baphicacanthus cusia* (Nees) Bremek is a synonym of *Strobilanthes cusia* (Nees) Kuntze

27   *Tinospora cordifolia* (Willd.) Miers is a synonym of *Tinospora sinensis* (Lour.) Merr.

28   *Andira araroba* Aguiar is a synonym of *Vataireopsis araroba* (Aguiar) Ducke

29   *Poria cocos* F.A.Wolf is a synonym of *Wolfiporia extensa* (Peck) Ginns

with inhibitory effects on the growth of SVK-14 keratinocytes and established the effects of Gotu kola (*Centella asiatica*) and Babchi (*Cullen corylifolium*, syn: *Psoralea corylifolia*) as inhibitors of keratinocyte replication, with IC_{50} values of 18.4 and 209.9 mg/mL, respectively. The effect produced by gotu kola was due to madecassoside and asiaticoside (Fig. 9.1), which had IC_{50} values of 8.6 and 8.4 µM, respectively, similar to the standard dithranol (IC_{50} 5.1 µM). Other extracts such as of the bark of *Berberis aquifolium* also inhibited keratinocyte growth (IC_{50} 35 µM), the alkaloids berberine, berbamine and oxyacanthine being the potential active compounds (Müller et al. 1995). Singh et al. (2015) studied the skin keratinocyte antiproliferative activity of the petroleum ether and ethanol extracts obtained from four medicinal plants, of *Phyllanthus virgatus* (syn: *Phyllanthus simplex*), *Crotalaria juncea*, *Leucas aspera* and *Vitex glabrata*. Of them, the petroleum ether extract from *C. juncea* and the ethanol extract of *L. aspera* showed significant activity, with IC_{50} values of 45.45 and 55.36 µg/mL, respectively. The antiproliferative and antipsoriatic activities were correlated with the action against nitric oxide production and lipid peroxidation.

Other authors focused their studies on mediators and transcription factors from keratinocytes implicated in psoriasis. It is the case of *Paeonia lactiflora*, which reduced the production of crucial psoriatic cytokines on polyinosinic:polycytidylic acid-stimulated human epidermal keratinocytes (SV-HEKs), such as IL-6, IL-8, chemokine (C-C motif) ligand 20 (CCL20) and TNFα, via down-regulation of NF-κB signaling pathway. In addition, the extract also inhibited the induction of the inflammasome, in terms of IL-1β and caspase-1 secretion. These results justify the employ of *Paeonia lactiflora* for the treatment of psoriasis (Choi et al. 2015).

Other studies were focused on the suppression of NF-κB signaling and its consequences. In this sense, Saelee et al. (2011) studied the effects of *Alpinia galanga* (Thai ginger), *Curcuma longa* (turmeric) and *Annona squamosa* (sweetsop)

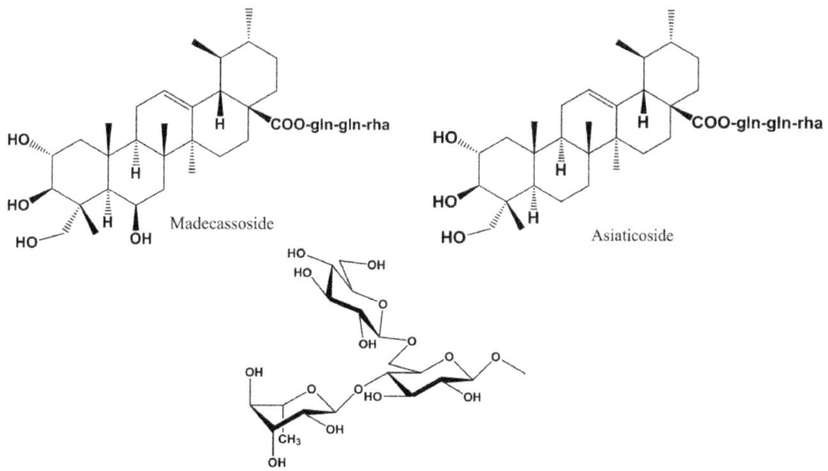

Fig. 9.1. Chemical structures of madecassoside and asiaticoside. Gln = glucose; rha = rhamnose.

extracts on ten different genes of the NF-κB signaling network in HaCaT cells, and observed that Thai ginger extract reduced the expression of NF-κB2, turmeric extract significantly decreased the expression of both NF-κB2 and NF-κB1, while *Annona squamosa* extract significantly lowered the expression of NF-κB1. So, this *in vitro* study suggested that these medicinal plants might exert their antipsoriatic activity by controlling the expression of NF-κB signaling biomarkers.

Other relevant mediators in psoriasis are matrix metalloproteinases (MMP), and its inhibition by *Scrophularia striata* extract and some isolated compounds were studied by Monsef-Esfahani et al. (2014). Among these isolated compounds, the inhibitory effects of nepitrin at 20 μg/mL (56%) and acteoside at 80 μg/mL (73%) on MMP-2 and MMP-9 were remarkable. Lipoxygenase and elastase could also be potential therapeutic targets in psoriasis. In this sense, Prieto et al. (2003) screened 15 extracts from traditional Chinese medicinal plants/fungi used to treat topical inflammations such as psoriasis. They were screened for their inhibitory effect on lipoxygenase, cyclooxygenase and elastase activity in intact leukocytes and platelets. *Astragalus propinquus* (syn: *Astragalus membranaceus*), *Forsythia suspensa* and *Wolfiporia extensa* (syn: *Poria cocos*) inhibited 5-lipoxygenase (IC$_{50}$ values of 141, 80 and 141 μg/mL, respectively). *Angelica dahurica, Angelica pubescens, F. suspensa* and *W. extensa* also inhibited elastase (IC$_{50}$ values of 80, 123, 68 and 93 μg/mL, respectively). Previously, Cuéllar et al. (1996) had demonstrated the inhibitory effect of *W. extensa* extract and its metabolites dehydrotumulosic and pachymic acids on phospholipase A$_2$ activity as well as experimental dermatitis (Cuéllar et al. 1997).

The metabolism of arachidonic acid via lipoxygenase enzymes, 5-, 12- and 15-lipoxygenases has been highlighted as the potential target for antipsoriatic treatments. The 5-lipoxygenase pathway is considered important since it participates in pro-inflammatory regulations, but the 12- and 15-lipoxygenase pathway may also play an important role in the progression of psoriasis (Schneider and Bucar 2005). Bader et al. (2015) tested four species used in Southern Italy for the treatment of psoriasis against 5-, 12-, 15-lipoxygenase and NF-κB activation: *Acanthus mollis, Achillea ligustica, Artemisia arborescens* and *Dittrichia viscosa* (syn: *Inula viscosa*). According to their results, the effect of *A. ligustica* was the most relevant as it had the highest anti-5-lipoxygenase activity (IC$_{50}$ = 49.5 μg/mL) and also enhanced the biosynthesis of the anti-inflammatory eicosanoid 15(*S*)-HETE. These species also reduced the activation of NF-κB, having IC$_{50}$ values of 16.7, 19.2 and 30.4 μg/mL for *A. ligustica, A. arborescens* and *D. viscosa*, respectively.

Nagar et al. (2016) tested the *in vivo* activity of *Woodfordia fructicosa* and *Gardenia gummifera* on a psoriasis model induced in Wistar rats: 10% of total body area was exposed to UV radiations after topical (0.1% gel of the extract) and oral (dose of 100 mg/kg) administration. The antipsoriatic activity (severity index), histological analysis and biochemical estimation suggest positive antipsoriatic effects of both plant extracts. Other interesting species against psoriasis could be *Tinospora sinensis, Curcuma longa, Celastrus paniculatus,* and *Aloe vera*, which were tested in a model of psoriasis-like dermatitis using topical application of 5%

imiquimod in mice, and oral/topical administration to mice. The results suggested that these plants can act as preventive agents against the disease (Arora et al. 2016).

Calaguala is an extract obtained from the fern *Phlebodium decumanum* (syn: *Polypodium decumanum*). It is used to treat psoriasis in the counties of South America as well as Spain. Several *in vitro* studies have tried to shed light on the mechanism responsible for its antipsoriatic properties (Tuominen et al. 1992, Vasänge-Tuominen et al. 1994, Vasänge et al. 1997). According to their results, calaguala extract has a double antipsoriatic mechanism: on the one hand it exerts an inhibitory effect on the inflammation induced by platelet activating factor (PAF). This effect is due to the PAF receptor antagonism exerted by sulphoquinovosyl diacylglycerol 1,2-di-*O*-palmitoyl-3-*O*-(6-sulpho-α-D-quinovopyranosyl)-glycerol (IC_{50} = 2 µM) (Vasänge et al. 1997), and the presence of adenosine in the extract (Tuominen et al. 1992). On the other hand, the polyunsaturated fatty acids (linoleic, linolenic and arachidonic acid) found in the calaguala extract inhibit the LTB_4 formation, with IC_{50} values mostly between 20–60 µM.

The ethanol extract of *Artemisia capillaris* was tested in human keratinocyte cells (HaCaT) and imiquimod-induced psoriasis-like mouse models. The extract had an IC^{50} of 37.5 µg/mL (72 h) in the antiproliferative test, the percentage of apoptotic population in the treated group being higher than that of the control. In the *in vivo* test, the 'Psoriasis Area and Severity Index' (PASI) score in treated group (50 mg/mL) was significantly lower than that of the negative control group (day 4). The topical application of the extract on psoriasis-like lesion during 4 days improved the skin damage, with lower values than the negative control (not treated) in epidermal thickness, expression levels of Ki-67 and intracellular adhesion molecule-1 (ICAM-1). After the evaluation of these results, the authors remark the high interest of this extract as an antipsoriatic agent (Lee et al. 2018). The HPLC analysis of this extract showed a content with prevalence of flavonoids, caffeoyl acid-quinic acid derivatives and coumarins, with chlorogenic acid and 3,5-dicaffeoylquinic acid as main compounds (Lee et al. 2018).

Natural Products with Potential Antipsoriatic Properties

The different compounds from natural origin with antipsoriatic properties can be classified into three different chemical structures: phenolics and terpenoids, which have been shown to inhibit angiogenesis and psoriasis-associated inflammation through the inhibition of NF-κB pathway (Simpson et al. 2014, Wen et al. 2014a, 2015, Venkatesha et al. 2016), and alkaloids, which mostly act by inhibiting lipid peroxidation (Müller and Ziereis 1993, Misik et al. 1995, Bezáková et al. 1996).

Among the phenolics, gambogic acid, the main active compound isolated from the resin of the *Garcinia hanburyi* Hook.f. (Clusiaceae) tree, and honokiol, a biphenolic neolignan isolated from *Magnolia officinalis* Rehder & E.H.Wilson (Magnoliaceae), have been reported to have anti-angiogenic and anti-inflammatory properties through the inhibition of the transcription factor NF-κB (Wen et al. 2014a, 2015). Gambogic acid and honokiol (Fig. 9.2) were tested *in vitro* and *in vivo*:

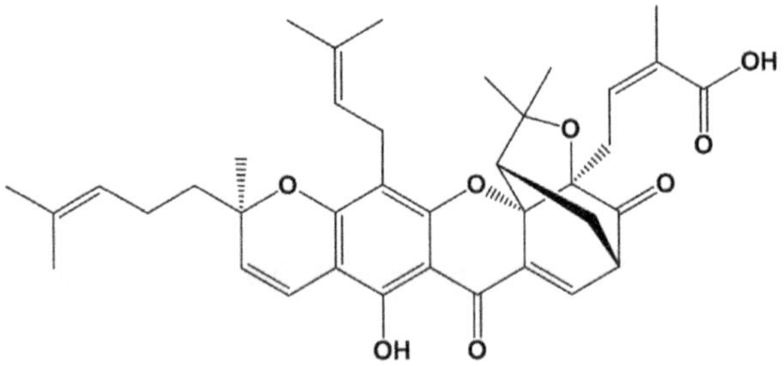

Fig. 9.2. Chemical structures of gambogic acid and honokiol.

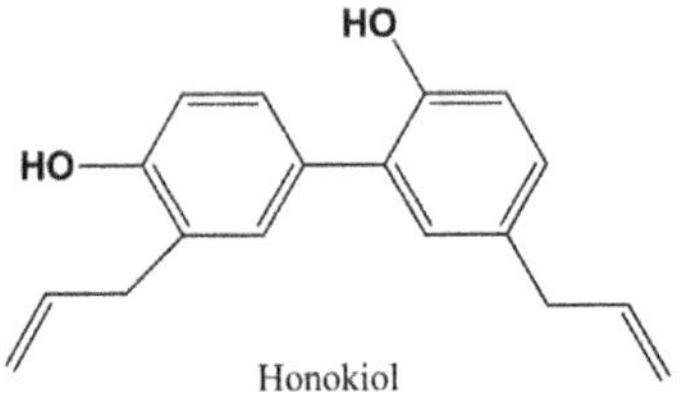

in vitro, gambogic acid inhibited the proliferation of the human keratinocyte cell line HaCaT (IC_{50} = 0.09 μM) and it was also demonstrated to be active in human umbilical vein endothelial cells (HUVEC) by inhibiting TNF-α-induced activation of NF-κB, both processes being highly active in psoriatic patients. Honokiol also inhibited TNF-α-induced NF-κB activation in HUVEC and decreased the ratio of Th_1/Th_2-expression CD4$^+$ T cells.

Gambogic acid and honokiol were also demonstrated to be effective *in vivo*. K14-VEGF transgenic mice with moderate psoriasis treated with gambogic acid showed a reduction in the erythema, resolution of the epidermal hyperplasia and acanthosis, decreased parakeratosis with reduced inflammatory infiltrate and reduced vascular hyperplasia and inflammation (evidenced by reduced expression of adhesion molecules such as E-selectin and ICAM-1). In the case of honokiol, it also normalized the psoriatic phenotype in K14-VEGF mice, producing macroscopic and histologic improvement, and dose-dependently decreased TNF-α and IFN-γ levels, this reduction being associated with suppression of p65-NF-κB expression in the ear tissues analyzed. Both gambogic acid and honokiol inhibited angiogenesis and the expression of vascular endothelial growth factor-2 (VEGF2) and p-VEGFR2 and, in the case of honokiol, this inhibition was accompanied by suppression of phosphorylation of extracellular signal–regulated kinase1/2 (ERK1/2), protein kinase B (AKT) and p38 mitogen-activated protein kinase (p38 MAPK). Gambogic acid was tested in a second animal model: a psoriasis-like model of guinea-pig, where improvements in epidermis and dermic could be detected (Wen et al. 2014a).

Paeoniflorin (Fig. 9.3), a monoterpene glycoside isolated from paeony root (*Paeonia lactiflora*), has also been tested in this psoriasis-like model of guinea-pig, where it relieved the lesions improving parakeratosis and hyperkeratinization.

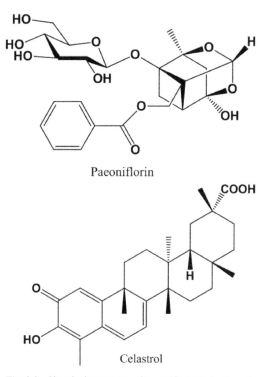

Paeoniflorin

Celastrol

Fig. 9.3. Chemical structures of paeoniflorin and celastrol.

In vitro, in HaCaT cells, it inhibits the expression of IL-17A, and IL-22 (2.08 µM) through a mechanism involving p38 MAPK (Yu et al. 2017).

Celastrol (Fig. 9.3) is also a terpenoid isolated from *Celastrus orbiculatus.* This herb has been traditionally used in Chinese medicine for anti-inflammatory, anti-cancer and antioxidant activities, especially in the treatment of rheumatoid arthritis and skin diseases. Zhou et al. (2011) demonstrated that this triterpene is able to inhibit HaCaT growth (IC_{50} = 1.1 µM) and induce their apoptosis through the inhibition of NF-κB pathway.

Alkaloids with described antipsoriatic properties can be found in *Berberis aquifolium,* whose bark has been traditionally used in North America for the treatment of several skin diseases, including psoriasis, and are mostly active in the lipoxygenase metabolism (Müller and Ziereis 1993, Misik et al. 1995, Bezáková et al. 1996). *Berberis aquifolium* extract inhibits the production of LTB_4 and 5-HETE (IC_{50} = 50 µM) in isolated bovine polymorphonuclear cells (Müller and Ziereis 1993), although the authors could not explain this effect by the action of the studied alkaloids berberine, oxyacanthine and berbamine. None of the three were as effective in inhibiting lipid peroxidation as the whole extract, although these alkaloids did show some antioxidant effect (much less than the whole extract). In this sense, protoberberine alkaloids (berberine, oxyberberine, jatrorrhizine, and columbamine) and aporphine alkaloids (magnoflorine, and corytuberine) were tested for lipoxygenase inhibition and lipid antioxidant properties (Misik et al. 1995). Oxyberberine, corytuberine, and columbamine (Fig. 9.4) showed the highest lipoxygenase inhibition potency, whereas berberine and magnoflorine had lower effects. The authors suggested that these differences in effects could be, at least partially, explained by the presence of a free electron pair in the nitrogen atom, which would play a key role in this activity. Moreover, they established a strong linear correlation (r = 0.866) between lipid antioxidant properties and inhibition of lipoxygenase, indicating that the mechanism of action of these alkaloids may be related to the prevention of the accumulation of lipid hydroperoxide substrate. This same conclusion was drawn after studying the lipid antioxidant properties of six bisbenzylisoquinolines alkaloids (oxyacanthine, armoline, baluchistine, berbamine, obamegine, and aquifoline) also found in *B. aquifolium* (Bezáková et al. 1996). In this study, oxyacanthine and berbamine (Fig. 9.5) were among the most active compounds, which is consistent with the fact that they have a phenolic domain that can act as free radical scavengers. Also, a linear correlation between lipoxygenase inhibition and lipid peroxidation was found (r = 0.9533).

Since it was first described in 2009 (van der Fits et al. 2009), the imiquimod model of psoriasis is being increasingly used due to its convenience and because it reproduces psoriasis morphologically and immunologically. The most recent studies dealing with natural products active against psoriasis are being carried out using this model. In this sense, it has been described that topically applied curcumin or orally administered resveratrol or quercetin significantly ameliorate psoriasis (Sun et al. 2013, Kjær et al. 2015, Chen et al. 2017, Di Nardo et al. 2018). Curcumin (Fig. 9.6) reduces epidermal hyperplasia and cytokine production (TNF-α, IL-1β,

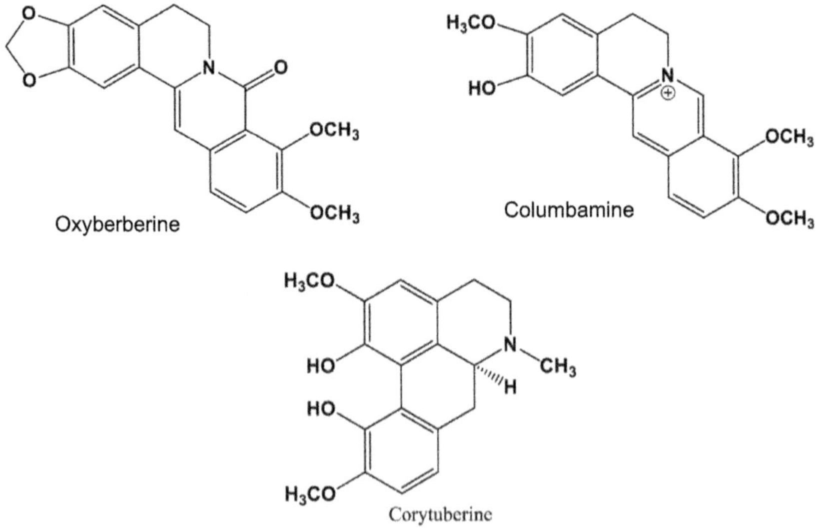

Oxyberberine

Columbamine

Corytuberine

Fig. 9.4. Chemical structures of berberine, magnoflorine and columbamine.

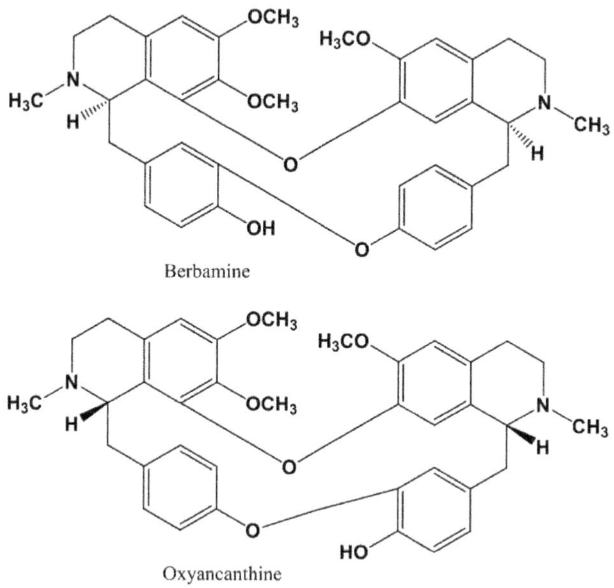

Fig. 9.5. Chemical structures of oxyacanthine and berbamine.

IL-6, IL17A, IL-17F, and IL-22) through a mechanism involving NF-κB inhibition (Sun et al. 2013), whilst resveratrol (Fig. 9.6) inhibits the expression of IL-17A, IL-19, and IL-23p19 (Kjær et al. 2015). Quercetin (Fig. 9.6) caused a reduction in the levels of TNF-α, IL-6 and IL-17 in serum, increased the activities of glutathione, catalase and superoxide dismutase, and decreased the skin accumulation of malonaldehyde through a mechanism which involved the downregulation of the NF-κB pathway (Chen et al. 2017).

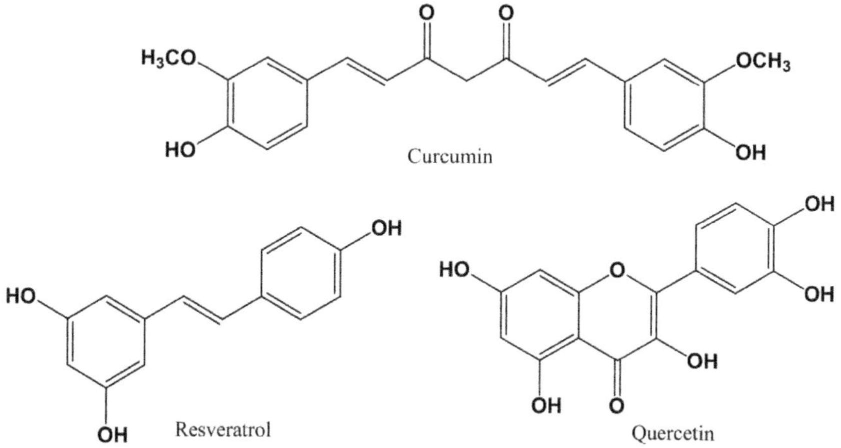

Fig. 9.6. Chemical structures of curcumin, resveratrol and quercetin.

Dodonaea polyandra, a medicinal plant containing polyandric acid A and traditionally used by the Kuuku I'yu (Northern Kaanju) indigenous people of Cape York Peninsula, Australia (Fig. 9.7), was described as a potent inhibitor of pro-inflammatory cytokine production (TNF-α, IL-1α, IL-1β, IL-6, and IL-8) and other inflammatory mediators, such as nitric oxide. Polyandric acid A may be useful in applications for skin inflammatory conditions including psoriasis and dermatitis (Simpson et al. 2014).

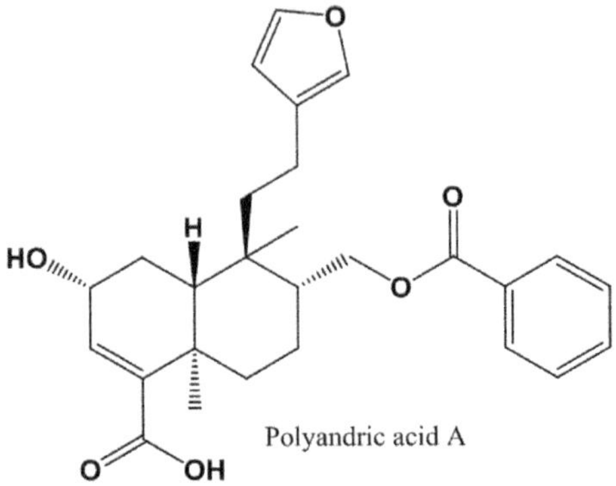

Polyandric acid A

Fig. 9.7. Chemical structure of polyandric acid A.

Clinical Trials

In traditional Chinese medicine (TCM), the application of medicinal plants has been used for centuries in the management of psoriasis and is believed to be effective, with few long-term side-effects (Koo and Arain 1998). Psoriasis is commonly classified into three main syndromes in TCM: 'blood heat', 'blood dryness' and 'blood stasis', and different medicinal plants are used to fight these conditions and restore the balanced state of the body (Tse 2003). However, the number of clinical trial dealing with medicinal plants in the treatment of psoriasis is quite limited (Table 9.2) and show big differences in their design, number of participants, parameters evaluated and form of treatment. It is worth noting that clinical trials evaluating 'indigo naturalis' are the most numerous and results conclude that its topical application is a safe and effective therapy for plaque-type psoriasis and nail psoriasis.

Regarding the species contained in the clinical trials revised, *Paeonia lactiflora* and *Smilax glabra* are the most commonly studied. Paeoniae radix and smilacis glabrae rhizoma are two of the most well-known herbs in China and have been used for over 1200 years. Both species have been tested in different experimental models of inflammation and immunomodulation with excellent results (Spelman et al. 2006, He and Dai 2011). Other species of interest for treating psoriasis are *Aloe vera* (Miroddi et al. 2015) and *Berberis aquifolium* (Jong et al. 2013). The quality of the trials and methodological approaches vary considerably, and conclusions on the effectiveness of aloe in psoriasis is still uncertain. Cutaneous application seems to be very safe as serious side effects have not been reported.

Several open-label or placebo-controlled clinical studies have shown beneficial effects of similar *Berberis aquifolium* extracts in patients with psoriasis, possibly explained by its anti-inflammatory and antibacterial properties. The main selected trials are compiled in Table 9.2.

Regarding *Aloe vera*, different clinical trials have been performed. For example, in 1996, Syed et al. tested the clinical efficacy and tolerability of a 0.5% concentrated extract in a hydrophilic cream to treat patients with psoriasis vulgaris. In this randomized, double-blind, placebo-controlled and parallel study, sixty patients (36 males and 24 females, mean 25.6 years) with slight to moderate chronic plaque-type psoriasis, *Aloe vera* cream improved the psoriatic plaques (82.8%) vs. placebo (7.7%) and decreased PASI score in patients (83.3%) vs. placebo (6.6%). These results demonstrated that topical application of *Aloe vera* extract 0.5% in a hydrophilic cream is more effective than placebo and did not present relevant side-effects. These effects were confirmed by Vogler and Ernst (1999), who reviewed all controlled clinical trials until date, and ratified the positive effect of topical application of *Aloe vera* for genital herpes and psoriasis, but not in preventive radiation-induced injuries.

Deng et al. (2013a,b, 2017) reviewed and evaluated the efficacy and safety of topically used plant extract preparations by psoriasis patients. A total of twelve studies were investigated, including three with *Aloe vera*, five with *Berberis aquifolium*, two with indigo naturalis, one with kukui nut oil (*Aleurites moluccanus*) and one with *Camptotheca acuminata* nut. Some of these plant extracts have partial effects on psoriasis symptoms, such as inflammation and cell proliferation. However, the clinical trials analyzed did not provide clear evidences that would support their topical use on psoriatic plaques, probably due to the small size of most studies and methodological weaknesses (Deng et al. 2013a). Similar results were obtained by the same authors after reviewing the evidence for the efficacy and safety of herbal medicines used topically in conjunction with antipsoriatic pharmacotherapy (Deng et al. 2013b). In a third study, they analyzed the efficacy and safety of oral forms of phytotherapy in psoriasis management and discussed the pharmacological actions of the plants used in clinical trials. Their results demonstrated that the most commonly used plants were *Oldenlandia diffusa*, *Rehmannia glutinosa* and *Salvia miltiorrhiza*, which also have anti-inflammatory, anti-proliferative and anti-oxidative properties, which are of relevance to psoriasis management (Deng et al. 2014). In a recent study, these same authors analyzed a preparation formed by different medicinal plants (PSORI-CM01, a modified form of Yinxieling) comparing the results with the original Chinese herbal formula for psoriasis Yinxieling (in tablet) and placebo, and demonstrated in a randomized, double-blinded and multicentral clinical trial, that this formula improved PASI scores and relapse rates in psoriasis vulgaris (Deng et al. 2017). The formula was then modified (PSORI-CM02) eliminating two herbs (liquorice and lithospermum), keeping the remaining five from PSORI-CM01 (Curcumae rhizoma, Radix paeoniae rubra, Rhizoma smilacis glabrae, Mume fructus, and Sarcandrae herba). PSORI-CM02 was tested as a possible suppressor of alloimmunity; results show that it

Table 9.2. Clinical trials of traditional Chinese medicines used in the treatment of psoriasis.

Chinese Herbal Components	Posible Botanical Species	Treatment	Reference
Rhizoma Smilacis Glabrae	*Smilax glabra*	Systemic Topical	Song et al. 2010
Folium Isatidis	*Isatis tinctoria*		
Rhizoma Menispermi	*Menispermum dauricum*		
Oldenlandia (Bai Hua She She Cao)	*Oldenlandia diffusa*		
Rhizoma Curcumae	*Curcuma longa*		
Rhizoma Polygoni Cuspidati	*Polygonum cuspidatum*		
Herba Solani Lyrati	*Solanum lyratum*		
Herba Duchesneae Indicae	*Duchesnea indica*		
Cacumen Platycladi	*Platycladus orientalis*		
Nepal dock root	*Rumex nepalensis*		
Wubeizi	*Rhus chinensis*		
Radix Paeoniae Rubra	*Paeonia lactiflora* or *Paeonia veitchii*	Systemic combined with topical (calcipotriol betamethasone ointment)	Yao et al. 2016
Rhizoma Curcumae	*Curcuma longa*		
Sarcandra	*Sarcandra glabra*		
Radix Glycyrrhizae	*Glycyrrhiza uralensis Glycyrrhiza inflata Glycyrrhiza glabra*		
Fructus Mume	*Prunus mume*		
Radix Arnebiae	*Arnebia euchroma Lithospermum erythrorhizon* or *Arnebia guttata*		
Rhizoma Smilacis Glabrae	*Smilax glabra*		
Indigo Naturalis	*Strobilanthes cusia Persicaria tinctoria Isatis indigotica*	Topical	Lin et al. 2015 Lin et al. 2006 Cheng et al. 2017 Lin et al. 2008 Lin et al. 2012 Lin et al. 2014
Cortex Phellodendri	*Phellodendron amurense* or *Phellodendron chinense*	Topical	Li et al. 2017
Radix Scutellariae	*Scutellaria baicalensis*		
Radix Rehmanniae	*Rehmannia glutinosa*	Systemic	Shan et al. 2006
Radix Angelicae Dahuricae	*Angelica dahurica* or *A. dahurica* var. *formosana*		
Radix Paeoniae Rubra	*Paeonia lactiflora* or *Paeonia veitchii*		
Herba Schizonepetae Tenuifoliae	*Schizonepeta tenuifolia*		
Radix Paeoniae Rubra	*Paeonia lactiflora* or *Paeonia veitchii*	Systemic	Parker et al. 2014 Wen et al. 2014b
Sarcandra	*Sarcandra glabra*		
Rhizoma Smilacis Glabrae	*Smilax glabra*		

Table 9.2 contd. ...

Table 9.2 contd. ...

Chinese Herbal Components	Posible Botanical Species	Treatment	Reference
Radix Rehmanniae	*Rehmannia glutinosa*	Systemic	Dai et al. 2014
Radix Angelicae Sinensis	*Angelica sinensis*		
Radix Paeoniae Rubra	*Paeonia lactiflora* or *Paeonia veitchii*		
Rhizoma Chuanxiong	*Ligusticum wallichii*		
Radix Lithospermi	*Arnebia euchroma Lithospermum erythrorhizon* or *Arnebia guttata*		
Curcuma Zedoary	*Curcuma phaeocaulis* or *Curcuma kwangsiensis* or *Curcuma wenyujin*		

inhibited murine skin allograft rejection and reduced graft-infiltration of CD3[+] T cells, which led the authors to hypothetize a potential interest of this new formula against autoimmune psoriasis (Lu et al. 2018).

Finally, *Hypericum perforatum* has also been reported to have a reasonable interest in psoriasis treatment as it has both anti-inflammatory and anti-proliferative properties, and was recently reported to be clinically helpful for the improvement of psoriatic lesions. For this purpose, Mansouri et al. (2017) performed a double-blind, placebo-controlled, pilot study with intra-individual comparison on twenty patients with mild to moderate plaque-type psoriasis. They observed that TNF-α concentrations in dermis, endothelial cells, and dendrite cells were reduced in the lesions of the treated patients vs. placebo, and the PASI scores, erythema, scaling and thickness also decreased vs. placebo. Because high concentrations of TNF-α are present in the skin lesions and plasma of patients with psoriasis, *Hypericum perforatum* could be a promising treatment for this disease.

Conclusions

The majority of medicinal plants used in the treatment of psoriasis are based on their traditional use and folk medicine, without relevant studies on humans. The number of clinical trials is limited and, those available, are not clearly defined, the number of patients is limited and the evaluation of results is variable. Moreover, some of these trials involve mixtures of plants (Deng et al. 2017, Na Takuathung et al. 2017) or use them simultaneously with standard drugs, such as calcipotriol and betamethasone (Wen et al. 2014b), which hinders their evaluation. Nevertheless, many of these plants still have potential as antipsoriatic drugs. Among these, the species studied with a high level of interest include *Aloe vera*, *Paeonia lactiflora*, *Smilax glabra*—which improves the psoriatic plaques by 83% and PASI score by 83% vs. placebo—and *Berberis aquifolium*. *Phlebodium decumanum*, *Camptotheca acuminata*, *Oldenlandia diffusa*, *Rehmannia glutinosa* and *Salvia miltiorrhiza* are also remarkable since they have the three pharmacological activities which are useful in the treatment of psoriasis: anti-inflammatory, anti-proliferative and anti-oxidative.

References

Amenta, R., Camarda, L., Di Stefano, V., Lentini, F. and Venza, F. 2000. Traditional medicine as a source of new therapeutic agents against psoriasis. Fitoterapia. 71(Suppl. 1): S13–S20.

Arora, N., Shah, K. and Pandey-Rai, S. 2016. Inhibition of imiquimod-induced psoriasis-like dermatitis in mice by herbal extracts from some Indian medicinal plants. Protoplasma. 253: 503–515.

Augustin, M., Andrees, U., Grimme, H., Schöpf, E. and Simon, J. 1999. Effects of *Mahonia aquifolium* ointment on the expression of adhesion, proliferation, and activation markers in the skin of patients with psoriasis. Forsch. Komplementarmed. 6(Suppl. 2): 19–21.

Bader, A., Martini, F., Schinella, G.R., Ríos, J.L. and Prieto, J.M. 2015. Modulation of COX-1, 5-, 12- and 15-LOX by popular herbal remedies used in southern Italy against psoriasis and other skin diseases. Phytother. Res. 29: 108–113.

Bezáková, L., Misik, V., Máleková, L., Svajdlenka, E. and Kostálová, D. 1996. Lipoxygenase inhibition and antioxidant properties of bisbenzylisoquinoline alkaloids isolated from *Mahonia aquifolium*. Pharmazie. 51: 758–761.

Boehncke, W.H. 2007. Efalizumab in the treatment of psoriasis. Biologics. 1: 301–309.

Brown, A.C., Koett, J., Johnson, D.W., Semaskvich, N.M., Holck, P., Lally, D., Cruz, L., Young, R., Higa, B. and Lo, S. 2005. Effectiveness of kukui nut oil as a topical treatment for psoriasis. Int. J. Dermatol. 44: 684–687.

Capella, G.L. and Finzi, A.F. 2003. Complementary therapy for psoriasis. Dermatol. Ther. 16: 164–174.

Carrenho, L.Z., Moreira, C.G., Vandresen, C.C., Gomes Junior, R., Gonçalves, A.G., Barreira, S.M., Noseda, M.D., Duarte, M.E., Ducatti, D.R., Dietrich, M., Paludo, K., Cabrini, D.A. and Otuki, M.F. 2015. Investigation of anti-inflammatory and anti-proliferative activities promoted by photoactivated cationic porphyrin. Photodiagnosis Photodyn. Ther. 12: 444–458.

Chen, H., Lu, C., Liu, H., Wang, M., Zhao, H., Yan, Y. and Han, L. 2017. Quercetin ameliorates imiquimod-induced psoriasis-like skin inflammation in mice via the NF-κB pathway. Int. Immunopharmacol. 48: 110–117.

Cheng, H.M., Wu, Y.C., Wang, Q., Song, M., Wu, J., Chen, D., Li, K., Wadman, E., Kao, S.T., Li, T.C., Leon, F., Hayden, K., Brodmerkel, C. and Chris Huang, C. 2017. Clinical efficacy and IL-17 targeting mechanism of Indigo naturalis as a topical agent in moderate psoriasis. BMC Complement. Altern. Med. 17: 439.

Choi, M.R., Choi, D.K., Sohn, K.C., Lim, S.K., Kim, D.I., Lee, Y.H., Im, M., Lee, Y., Seo, Y.J., Kim, C.D. and Lee, J.H. 2015. Inhibitory effect of *Paeonia lactiflora* Pallas extract (PE) on poly (I:C)-induced immune response of epidermal keratinocytes. Int. J. Clin. Exp. Pathol. 8: 5236–5241.

Cuéllar, M.J., Giner, R.M., Recio, M.C., Just, M.J., Máñez, S. and Ríos, J.L. 1996. Two fungal lanostane derivatives as phospholipase A₂ inhibitors. J. Nat. Prod. 59: 977–979.

Cuéllar, M.J., Giner, R.M., Recio, M.C., Just, M.J., Máñez, S. and Ríos, J.L. 1997. Effect of the basidiomycete *Poria cocos* on experimental dermatitis and other inflammatory conditions. Chem. Pharm. Bull. (Tokyo) 45: 492–494.

Dai, Y.J., Li, Y.Y., Zeng, H.M., Liang, X.A., Xie, Z.J., Zheng, Z.A., Pan, Q.H. and Xing, Y.X. 2014. Effect of Yinxieling decoction on PASI, TNF-α and IL-8 in patients with psoriasis vulgaris. Asian Pac. J. Trop. Med. 7: 668–670.

Deng, S., May, B.H., Zhang, A.L., Lu, C. and Xue, C.C. 2013a. Plant extracts for the topical management of psoriasis: a systematic review and meta-analysis. Br. J. Dermatol. 169: 769–782.

Deng, S., May, B.H., Zhang, A.L., Lu, C. and Xue, C.C. 2013b. Topical herbal medicine combined with pharmacotherapy for psoriasis: a systematic review and meta-analysis. Arch. Dermatol. Res. 305: 179–189.

Deng, S., May, B.H., Zhang, A.L., Lu, C. and Xue, C.C. 2014. Phytotherapy in the management of psoriasis: a review of the efficacy and safety of oral interventions and the pharmacological actions of the main plants. Arch. Dermatol. Res. 306: 211–229.

Deng, J., Yao, D., Lu, C., Wen, Z., Yan, Y., He, Z., Wu, H. and Deng, H. 2017. Oral Chinese herbal medicine for psoriasis vulgaris: protocol for a randomised, double-blind, double-dummy, multicentre clinical trial. BMJ Open 7: e014475.

Di Nardo, V., Gianfaldoni, S., Tchernev, G., Wollina, U., Barygina, V., Lotti, J., Daaboul, F. and Lotti, T. 2018. Use of curcumin in psoriasis. Open Access Maced. J. Med. Sci. 6: 218–220.

Ford-Hutchinson, A.W. 1993. 5-Lipoxygenase activation in psoriasis: a dead issue? Skin. Pharmacol. 6: 292–297.

He, D.Y. and Dai, S.M. 2011. Anti-inflammatory and immunomodulatory effects of *Paeonia lactiflora* Pall., a traditional Chinese herbal medicine. Front. Pharmacol. 2: 10.

Herman, A. and Herman, A.P. 2016. Topically used herbal products for the treatment of psoriasis— Mechanism of action, drug delivery, clinical studies. Planta Med. 82: 1447–1455.

Jensen, P. 1990a. Use of alternative medicine by patients with atopic dermatitis and psoriasis. Acta Derm. Venereol. 70: 421–424.

Jensen, P. 1990b. Alternative therapy for atopic dermatitis and psoriasis: patient-reported motivation, information source and effect. Acta Derm. Venereol. 70: 425–428.

Jong, M.C., Ermuth, U. and Augustin, M. 2013. Plant-based ointments versus usual care in the management of chronic skin diseases: a comparative analysis on outcome and safety. Complement. Ther. Med. 21: 453–459.

Kaur, A. and Kumar, S. 2012. Plants and plant products with potential antipsoriatic activity—a review. Pharm. Biol. 50: 1573–1591.

Kjær, T.N., Thorsen, K., Jessen, N., Stenderup, K. and Pedersen, S.B. 2015. Resveratrol ameliorates imiquimod-induced psoriasis-like skin inflammation in mice. PLoS One 10: e0126599.

Koo, J. and Arain, S. 1998. Traditional Chinese medicine for the treatment of dermatologic disorders. Arch. Dermatol. 134: 1388–1393.

Lee, S.Y., Nam, S., Hong, I.K., Kim, H., Yang, H. and Cho, H.J. 2018. Antiproliferation of keratinocytes and alleviation of psoriasis by the ethanol extract of *Artemisia capillaris*. Phytother Res. doi: 10.1002/ptr.6032.

Li, N., Zhao, W., Xing, J., Liu, J., Zhang, G., Zhang, Y., Li, Y., Liu, W., Shi, F. and Bai, Y. 2017. Chinese herbal Pulian ointment in treating psoriasis vulgaris of blood-heat syndrome: a multi-center, double-blind, randomized, placebo-controlled trial. BMC Complement. Altern. Med. 17: 264.

Liu, H.P., Liu, H.C. and Li, G.Y. 2001. Integrated traditional Chinese and western medicine in treating 37 psoriatic patients. Chin. J. Integr. Med. 7: 53.

Lin, Y.K., Yen, H.R., Wong, W.R., Yang, S.H. and Pang, J.H. 2006. Successful treatment of pediatric psoriasis with indigo naturalis composite ointment. Pediatr. Dermatol. 23: 507–510.

Lin, Y.K., Chang, C.J., Chang, Y.C., Wong, W.R., Chang, S.C. and Pang, J.H. 2008. Clinical assessment of patients with recalcitrant psoriasis in a randomized, observer-blind, vehicle-controlled trial using indigo naturalis. Arch. Dermatol. 144: 1457–1464.

Lin, Y.K., See, L.C., Huang, Y.H., Chang, Y.C., Tsou, T.C., Leu, Y.L. and Shen, Y.M. 2012. Comparison of refined and crude indigo naturalis ointment in treating psoriasis: Randomized, observer-blind, controlled, intrapatient trial. Arch. Dermatol. 148: 397–400.

Lin, Y.K., See, L.C., Huang, Y.H., Chang, Y.C., Tsou, T.C., Lin, T.Y. and Lin, N.L. 2014. Efficacy and safety of indigo naturalis extract in oil (Lindioil) in treating nail psoriasis: a randomized, observer-blind, vehicle-controlled trial. Phytomedicine 21: 1015–1020.

Lin, Y.K., Chang, Y.C., Hui, R.C., See, L.C., Chang, C.J., Yang, C.H. and Huang, Y.H. 2015. A Chinese herb, indigo naturalis, extracted in oil (Lindioil) used topically to treat psoriatic nails: a randomized clinical trial. JAMA Dermatol. 151: 672–674.

Lowes, M.A., Bowcock, A.M. and Krueger, J.G. 2007. Pathogenesis and therapy of psoriasis. Nature. 445: 866–873.

Lu, C., Liu, H., Jin, X., Chen, Y., Liang, C.L., Qiu, F. and Dai, Z. 2018. Herbal components of a novel formula PSORI-CM02 interdependently suppress allograft rejection and induce CD8+CD122+PD-1+ regulatory T cells. Front. Pharmacol. 9: 88.

Mansouri, P., Mirafzal, S., Najafizadeh, P., Safaei-Naraghi, Z., Salehi-Surmaghi, M.H. and Hashemian, F. 2017. The impact of topical Saint John's Wort (*Hypericum perforatum*) treatment on tissue tumor necrosis factor-alpha levels in plaque-type psoriasis: A pilot study. J. Postgrad. Med. 63: 215–220.

McDermott, L., Madan, R., Rupani, R. and Siegel, D. 2016. A review of indigo naturalis as an alternative treatment for nail psoriasis. J. Drugs Dermatol. 15: 319–323.

Miroddi, M., Navarra, M., Calapai, F., Mancari, F., Giofrè, S.V., Gangemi, S. and Calapai, G. 2015. Review of clinical pharmacology of *Aloe vera* L. in the treatment of psoriasis. Phytother. Res. 29: 648–655.

Misík, V., Bezáková, L., Máleková, L. and Kostálová, D. 1995. Lipoxygenase inhibition and antioxidant properties of protoberberine and aporphine alkaloids isolated from *Mahonia aquifolium*. Planta Med. 61: 372–373.

Monsef-Esfahani, H.R., Shahverdi, A.R., Khorramizadeh, M.R., Amini, M. and Hajiaghaee, R. 2014. Two matrix metalloproteinase inhibitors from *Scrophularia striata* Boiss. Iran. J. Pharm. Res. 13: 149–155.

Müller, K. and Ziereis, K. 1993. Effects of oxygen radicals, hydrogen peroxide and water-soluble singlet oxygen carriers on 5- and 12-lipoxygenase. Arch. Pharm. (Weinheim). 326: 819–821.

Müller, K., Ziereis, K. and Gawlik, I. 1995. The antipsoriatic *Mahonia aquifolium* and its active constituents; II. Antiproliferative activity against cell growth of human keratinocytes. Planta Med. 61: 7475.

Muruganantham, N., Basavaraj, K.H., Dhanabal, S.P., Praveen, T.K., Shamasundar, N.M. and Rao, K.S. 2011. Screening of *Caesalpinia bonduc* leaves for antipsoriatic activity. J. Ethnopharmacol. 133: 897–901.

Na Takuathung, M., Wongnoppavich, A., Pitchakarn, P., Panthong, A., Khonsung, P., Chiranthanut, N., Soonthornchareonnon, N. and Sireeratawong, S. 2017. Effects of Wannachawee Recipe with antipsoriatic activity on suppressing inflammatory cytokine production in HaCaT human keratinocytes. Evid. Based Complement. Alternat. Med. 2017: 5906539.

Nagar, H.K., Srivastava, A.K., Srivastava, R. and Ranawat, M.S. 2016. Evaluation of potent phytomedicine for treatment of psoriasis using UV radiation induced psoriasis in rats. Biomed. Pharmacother. 84: 1156–1162.

Nickoloff, B.J. 1991. The cytokine network in psoriasis. Arch. Dermatol. 127: 871–884.

Nickoloff, B.J. and Nestle, F.O. 2004. Recent insights into the immunopathogenesis of psoriasis provide new therapeutic opportunities. J. Clin. Invest. 113: 1664–1675.

Parker, S., Zhang, A.L., Zhang, C.S., Goodman, G., Wen, Z., Lu, C. and Xue, C.C. 2014. Oral granulated Chinese herbal medicine (YXBCM01) plus topical calcipotriol for psoriasis vulgaris: study protocol for a double-blind, randomized placebo controlled trial. Trials 15: 495.

Prieto, J.M., Recio, M.C., Giner, R.M., Máñez, S., Giner-Larza, E.M. and Ríos, J.L. 2003. Influence of traditional Chinese anti-inflammatory medicinal plants on leukocyte and platelet functions. J. Pharm. Pharmacol. 55: 1275–1282.

Saelee, C., Thongrakard, V. and Tencomnao, T. 2011. Effects of Thai medicinal herb extracts with anti-psoriatic activity on the expression on NF-κB signaling biomarkers in HaCaT keratinocytes. Molecules 16: 3908–3932.

Sampson, J.H., Raman, A., Karlsen, G., Navsaria, H. and Leigh, I.M. 2001. *In vitro* keratinocyte antiproliferant effect of *Centella asiatica* extract and triterpenoid saponins. Phytomedicine 8: 230–235.

Schneider, I. and Bucar, F. 2005. Lipoxygenase inhibitors from natural plant sources. Part 2: medicinal plants with inhibitory activity on arachidonate 12-lipoxygenase, 15-lipoxygenase and leukotriene receptor antagonists. Phytother. Res. 19: 263–272.

Sethi, G., Sung, B., Kunnumakkara, A.B. and Aggarwal, B.B. 2009. Targeting TNF for treatment of cancer and autoimmunity. Adv. Exp. Med. Biol. 647: 37–51.

Seyger, M.M., van de Kerkhof, P.C., van Vlijmen-Willems, I.M., de Bakker, E.S., Zwiers, F. and de Jong, E.M. 1998. The efficacy of a new topical treatment for psoriasis: Mirak. J. Eur. Acad. Dermatol. Venereol. 11: 13–18.

Shan, C., Yuan, L., Xiuzhen, B. and Aiju, Q. 2006. Treatment of psoriasis vulgaris by oral administration of yin xie ping granules—a clinical report of 60 cases. J. Tradit. Chin. Med. 26: 198–201.

Shawahna, R. and Jaradat, N.A. 2017. Ethnopharmacological survey of medicinal plants used by patients with psoriasis in the West Bank of Palestine. BMC Complement. Altern. Med. 17: 4.

Simpson, B.S., Luo, X., Costabile, M., Caughey, G.E., Wang, J., Claudie, D.J., McKinnon, R.A. and Semple, S.J. 2014. Polyandric acid A, a clerodane diterpenoid from the Australian medicinal plant *Dodonaea polyandra*, attenuates pro-inflammatory cytokine secretion *in vitro* and *in vivo*. J. Nat. Prod. 77: 85–91.

Singh, K.K. and Tripathy, S. 2014. Natural treatment alternative for psoriasis: a review on herbal resources. J. Appl. Pharm. Sci. 4: 114–121.

Singh, S.K., Chouhan, H.S., Sahu, A.N. and Narayan, G. 2015. Assessment of *in vitro* antipsoriatic activity of selected Indian medicinal plants. Pharm. Biol. 53: 1295–1301.

Smith, N., Weymann, A., Tausk, F.A. and Gelfand, J.M. 2009. Complementary and alternative medicine for psoriasis: a qualitative review of the clinical trial literature. J. Am. Acad. Dermatol. 61: 841–856.

Song, P., Lysvand, H., Yuhe, Y., Liu, W. and Iversen, O.J. 2010. Expression of the psoriasis-associated antigen, Pso p27, is inhibited by traditional Chinese medicine. J. Ethnopharmacol. 127: 171–174.

Spelman, K., Burns, J., Nichols, D., Winters, N., Ottersberg, S. and Tenborg, M. 2006. Modulation of cytokine expression by traditional medicines: a review of herbal immunomodulators. Altern. Med. Rev. 11: 128–150.

Sun, J., Zhao, Y. and Hu, J. 2013. Curcumin inhibits imiquimod-induced psoriasis-like inflammation by inhibiting IL-1beta and IL-6 production in mice. PLoS One 8: e67078.

Syed, T.A., Ahmad, S.A., Holt, A.H., Ahmad, S.A., Ahmad, S.H. and Afzal, M. 1996. Management of psoriasis with *Aloe vera* extract in a hydrophilic cream: a placebo-controlled, double-blind study. Trop. Med. Int. Health. 1: 505–509.

Takeshita, J., Grewal, S., Langan, S.M., Mehta, N.N., Ogdie, A., Van Voorhees, A.S. and Gelfand, J.M. 2017. Psoriasis and comorbid diseases: Epidemiology. J. Am. Acad. Dermatol. 76: 377–390.

Talbott, W. and Duffy, N. 2015. Complementary and alternative medicine for psoriasis: what the dermatologist needs to know. Am. J. Clin. Dermatol. 16: 147–165.

Tian, Q., Li, J., Xie, X., Sun, M., Sang, H., Zhou, C., An, T., Hu, L., Ye, R.D. and Wang, M.W. 2005. Stereospecific induction of nuclear factor-κB activation by isochamaejasmin. Mol. Pharmacol. 68: 1534–1342.

Tirant, M., Lotti, T., Gianfaldoni, S., Tchernev, G., Wollina, U. and Bayer, P. 2018. Integrative dermatology—the use of herbals and nutritional supplements to treat dermatological conditions. Open Access Maced. J. Med. Sci. 6: 185–202.

Tse, T.W. 2003. Use of common Chinese herbs in the treatment of psoriasis. Clin. Exp. Dermatol. 28: 469–475.

Tuominen, M., Bohlin, L. and Rolfsen, W. 1992. Effects of calaguala and an active principle, adenosine, on platelet activating factor. Planta Med. 58: 306–310.

van der Fits, L., Mourits, S., Voerman, J.S., Kant, M., Boon, L., Laman, J.D., Cornelissen, F., Mus, A.M., Florencia, E., Prens, E.P. and Lubberts, E. 2009. Imiquimod-induced psoriasis-like skin inflammation in mice is mediated via the IL-23/IL-17 axis. J. Immunol. 182: 5836–5845.

Vasänge, M., Rolfsen, W. and Bohlin, L. 1997. A sulphonoglycolipid from the fern *Polypodium decumanum* and its effect on the platelet activating-factor receptor in human neutrophils. J. Pharm. Pharmacol. 49: 562–566.

Vasänge-Tuominen, M., Perera-Ivarsson, P., Shen, J., Bohlin, L. and Rolfsen, W. 1994. The fern *Polypodium decumanum*, used in the treatment of psoriasis, and its fatty acid constituents as inhibitors of leukotriene B$_4$ formation. Prostaglandins Leukot. Essent. Fatty Acids 50: 279–284.

Venkatesha, S.H. and Moudgil, K.D. 2016. Celastrol and its role in controlling chronic diseases. Adv. Exp. Med. Biol. 928: 267–289.

Vogler, B.K. and Ernst, E. 1999. *Aloe vera*: a systematic review of its clinical effectiveness. Br. J. Gen. Pract. 49: 823–828.

Voorhees, J.J. 1983. Leukotrienes and other lipoxygenase products in the pathogenesis and therapy of psoriasis and other dermatoses. Arch. Dermatol. 119: 541–547.

Wen, J., Pei, H., Wang, X., Xie, C., Li, S., Huang, L., Qiu, N., Wang, W., Cheng, X. and Chen, L. 2014a. Gambogic acid exhibits anti-psoriatic efficacy through inhibition of angiogenesis and inflammation. J. Dermatol. Sci. 74: 242–250.

Wen, Z.H., Xuan, M.L., Yan, Y.H., Li, X.Y., Yao, D.N., Li, G., Guo, X.F., Ou, A.H. and Lu, C.J. 2014b. Chinese medicine combined with calcipotriol betamethasone and calcipotriol ointment for psoriasis vulgaris (CMCBCOP): study protocol for a randomized controlled trial. Trials 15: 294.

Wen, J., Wang, X., Pei, H., Xie, C., Qiu, N., Li, S., Wang, W., Cheng, X. and Chen, L. 2015. Anti-psoriatic effects of honokiol through the inhibition of NF-κB and VEGFR-2 in animal model of K14-VEGF transgenic mouse. J. Pharmacol. Sci. 128: 116124.

Yao, D.N., Lu, C.J., Wen, Z.H., Yan, Y.H., Xuan, M.L., Li, X.Y., Li, G., He, Z.H., Xie, X.L., Deng, J.W., Guo, X.F. and Ou, A.H. 2016. Oral PSORI-CM01, a Chinese herbal formula, plus topical sequential therapy for moderate-to-severe psoriasis vulgaris: pilot study for a double-blind, randomized, placebo-controlled trial. Trials 17: 140.

Yu, J., Xiao, Z., Zhao, R., Lu, C. and Zhang, Y. 2017. Paeoniflorin suppressed IL-22 via p38 MAPK pathway and exerts anti-psoriatic effect. Life Sci. 180: 17–22.

Zhou, L.L., Lin, Z.X., Fung, K.P., Cheng, C.H.K., Che, C.T., Zhao, M., Wu, S.H. and Zuo, Z. 2011. Celastrol-induced apoptosis in human HaCaT keratinocytes involves the inhibition of NF-κB activity. Eur. J. Pharmacol. 670: 399–408.

CHAPTER 10

Knowledge of the Ethnomedicinal Plants Used by Tobas and Mocovíes Tribes in the Central-North of Argentina

María I. Stegmayer,[1] Norma H. Alvarez,[1] Melina G. Di Liberto,[2] Lucas D. Daurelio[1] and Marcos G. Derita[1,2,]*

Introduction

Among the flora of different regions of the world, Latin America represents one of the wealthiest sources of material with pharmacological activities due to its biodiversity (Brandão et al. 2008). It possesses a very high number of vascular plants (85,000) (Grornbridge 1992) and there is a recent evidence that neotropical forests located in Latin America possess the highest diversity of plants in the world (Berry 2002). In addition, some factors critically distinguish the medicinal plants of Latin America: (1) this region possesses a huge unexplored biodiversity (Cruz et al. 2007); (2) there is a rich traditional use of medicinal plants (Gupta 1995, 2008); (3) the ethnopharmacological knowledge has been tightly kept or transmitted by the many indigenous populations still living in this region (Murillo 1889, Rosenblat 1954, Morton 1981, Correia 1984, Cleaves 2001, Portillo et al. 2001, Coelho de

[1] CONICET, Universidad Nacional del Litoral/Facultad de Ciencias Agrarias/Laboratorio de Investigaciones en Fisiología y Biología Molecular Vegetal, Kreder 2805, Esperanza, Santa Fe, Argentina.
[2] CONICET, Universidad Nacional de Rosario/Facultad de Ciencias Bioquímicas y Farmacéuticas/ Cátedra de Farmacognosia, Suipacha 531, Rosario, Santa Fe, Argentina.
* Corresponding author: mgderita@hotmail.com

Souza et al. 2004, Scarpa 2004a, Goleniowski et al. 2006, Cruz et al. 2007, Estévez et al. 2007); and (4) these resources have been poorly studied.

In Argentina, there is a large number of indigenous people living in different communities that are present in several geographical areas, each one with its own cultural characteristics. During the 200 years of history of the nation, these populations were victims of all kinds of physical and symbolic violence by the ruling classes, overwhelming their customs and lifestyles. They were exiled from their territories and forced to assume a religion and social norms, which were completely alien to them. However, some tribes maintained their identity and still endure, although they had to adapt to new living conditions in large cities.

Many studies on the medical folklore of northwestern Argentina (Di Lullo 1929, 1946, Sosa Verón and Vivante 1951, Daoud 1954, Ávila 1960, Carrizo 1960, Torres 1975) are of great ethnographical value as they unravel ancient therapeutic practices, many of which have been lost. However, these studies are incomplete in their review of plants and the medicinal uses related to them. Later studies carried out by Sturzenegger (1987, 1989, 1999) focusing on medical anthropology do not provide enough information on medicinal plants either. Finally, Scarpa (2000) points out 81 uses of 61 plant species in veterinary medicine by the Hispano-Quechua community called "*Criollo*" groups. He also showed that their veterinary health practices involve different approaches to treatment: the ancient Hippocratic medicine, magical procedures and Christian religious practices. Healing with plant remedies is not a specialized activity within this society. As with most folk knowledge, it is mainly well spread within the population. When *Criollos* fall ill, they resort to the experience of an elderly member of the community with a good knowledge about home remedies. For unknown diseases or when home remedies fail, they resort to the nearest village hospital for treatment or, alternatively, to somebody who "cures by secret". The latter, called "*curanderos*" (medicine-man), are endowed with a gift known as "the secret"; they carry out a special therapy by invoking supernatural forces. These therapies are performed by rituals called "*cura de palabra*" (healing by words) and "*cura por el rastro*" (healing through traces) which involve magical and/or religious elements such as spells, prayers, or invocations to God and/or saints.

It is difficult to approach the ethnobotanical knowledge of all the indigenous communities of Argentina. Many authors studied and compiled data related to the inhabitants of different geographic regions of our territory (Scarpa 2004a, Svetaz et al. 2010). In this chapter, we will discuss the survey information of the medicinal plants that grows in the states of Santa Fe, Chaco and Formosa, mainly inhabited by the original towns Tobas and Mocovíes. The phytogeographic characteristics of the region as well as the use and properties of the medicinal plants used by these tribes will also be presented. Finally, we will analyze statistically the main plant families reported for medicinal purpose and the more common uses of them.

Phytogeographic Description of the Region Inhabited by the Toba and Mocoví Tribes

Nowadays, there are 18 folk communities among the vast Argentinean geographical land: 3 in the Patagonian region, 3 in the Central region, 1 in Cuyana region, 6 in the Northwest and 5 in the Northeast (Fig. 10.1). Two of the main communities that inhabit the last region mentioned above, which includes the provinces of Santa Fe, Chaco and Formosa, are the Tobas and Mocovíes tribes. Tobas used to be one of the greatest communities of this Argentine region and they currently maintain one of the highest numbers of inhabitants, with almost 70,000 people. Its strong cultural imprint and its ability to adapt made these people (also called "*quom*") to keep on their customs over time and nowadays they have a powerful legal representation. Mocovíes also used to be one of the majority groups of this zone, but the advance of the civilized society destroyed its customs and according to the last census, they are about 15,000 inhabitants of this town.

Regarding phytogeographic zones, both tribes belong to Chaqueño domain. This is the largest one in the Argentinean national territory, and is presented with

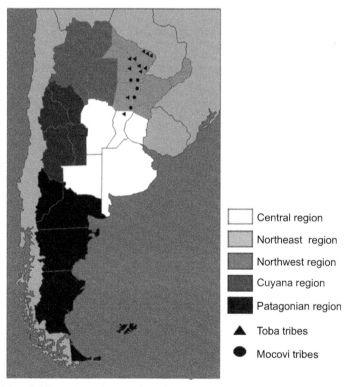

Fig. 10.1. Map of different Argentinean regions showing the zones inhabited by Tobas and Mocovíes tribes.

diverse physiognomies such as savanna, xerophilous forest and bushes, among others. The clime is subtropical with mild and dry winters but warm and rainy summers. The boundaries of this domain, in which five phytogeographic provinces can be distinguished, are generally transitional with broad zones of ecotone.

The Chaqueña phytogeographic province (Fig. 10.2) extends around Argentinean states of Chaco, Formosa, Santiago del Estero; eastern parts of Jujuy, Salta, Tucumán, La Rioja and Catamarca; northern regions of San Luis, Córdoba and Santa Fe; west of Corrientes and the southeast end of San Juan and bordering sectors of Mendoza around the Desaguadero River. Due to its vast extension, this province is affected by rainfall gradients from more than 1000 mm per year in the northeastern sector to around 400 mm per year in the southwest. It is possible to distinguish four phytogeographic districts in this province: Chaqueño Occidental, Chaqueño Serrano, Chaqueño Oriental and Savanna district. Tobas tribes inhabit the last two districts that include north of Santa Fe, Chaco and Formosa (Fig. 10.1). Chaqueño Oriental district has been characterized mainly by the

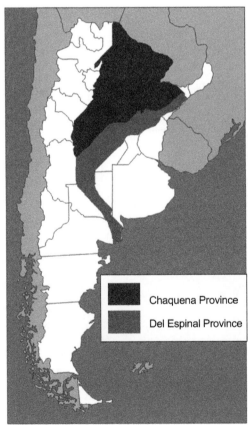

Chaquena Province

Del Espinal Province

Fig. 10.2. Argentinean phytogeographic regions according to Cabrera (1976), showing Chaqueña and the Espinal provinces where Toba and Mocoví tribes live nowadays.

exploitation throughout the history of the tree "*quebracho colorado chaqueño*" (*Schinopsis balansae*), for the use of its wood and extraction of tannin. Savanna district is an area dominated by meadows and isolated forests, some of them made up of palm trees. The main species of this phytogeographic district are: "*quebracho blanco*" (*Aspidosperma quebracho-blanco*), "*itín*" (*Prosopis kuntzei*), "*lapacho negro*" (*Handroanthus heptaphyllus*), "*mistol*" (*Ziziphus mistol*), "*chañar*" (*Geoffroea decorticans*), and "*horco-quebracho*" (*Schinopsis haenkeana*) among others.

The Espinal phytogeographic province (Fig. 10.2), also known as the Pampeano forest, extends as an arch from the states of Corrientes, the north of Entre Ríos, and the central regions of Santa Fe, Córdoba and La Pampa; south-central of San Luis, southeast of Buenos Aires and a small extension of Río Negro. A fraction of this province is constituted by the xerophytic coastal forests of the rivers Paraná, de la Plata and the Argentinean Sea. It is worth mentioning that since 1971, the Argentinean botanic expert Angel Lulio Cabrera stopped representing Espinal province in the maps, so it usually does not appear in the phytogeographic maps of Argentina. At present, it is very difficult to observe native vegetation as a result of the transformation occurred due to the agricultural expansion. This region exhibits important climatic variations from north to south, such as rainfall that varies from about 1100 millimeters per year in Corrientes and Entre Ríos to a minimum of 350 millimeters per year in the west of La Pampa. The predominant vegetation of this province is the open xerophilous deciduous forest and shrub steppes, the carob tree (*Prosopis alba*) being the characteristic species. This phytogeographic province is subdivided into three districts: "Ñandubay", "del Caldén" and "del Algarrobo". Tobas and Mocovíes tribes inhabit the last districts that include the central region of Santa Fe (Fig. 10.1) (Cabrera 1971, Adámoli et al. 1972, Cabrera and Willink 1980).

The Plants in the Subsistence of the Original Communities

The use of plants is a reflex of the cultural, ideological and technological peculiarities that have existed at a given time throughout the course of human history (Bates 1985). For this reason, it is considered that botanical reports devoid of their cultural context do not constitute what is understood by ethnobotany (Arenas 1987). Conversely, ethnobotanical data look upon a wide range of information regarding the ways in which a cultural group or community interacts with its plant environment (Martin 1995). In most cases, a knowledge or application is associated to a plant species. It also describes the parts of the species that is used as well as the method of obtaining, preparing and administration of the plant products.

The vegetation constitutes a direct source of appropriation of vital products through the collection of foodstuffs, medicines, fuels and materials for construction. The plant environment of these communities is also an indirect source to access other types of practices such as animal breeding and wood production. The products derived from these activities are used, in turn, for self-consumption or they are sold in the markets through barter or cash, thus allowing the acquisition of manufactured products. Although at first sight the strategies of appropriation of vegetal resources from the surroundings can be varied, in quantitative terms the animal production destined to self-consumption and commercialization predominates widely and has accentuated in the last decades. Tobas and Mocovíes reside in the rural area and are mainly engaged in the extensive grazing of beef and goats cattle on fiscal lands (Scarpa 2000).

The Idea of Disease and Diagnosis

On another level of connotations, the disease is mostly conceptualized by these indigenous people, both relationally and ontologically. Within the first group, it is necessary to distinguish between those representations derived from a disharmony produced in relation to the person itself, from those derived due to a disharmony with the social environment. The ontological idea of illness, on the other hand, is manifested by an intrusion process of an external object (a bad air, a wind, a cold, a heat, a thing), as well as by a process of loss of a vital component (loss of the soul). The symptomatology is expressed mainly through the topography of the pain or the appearance of alterations in the skin surface, the latter interpreted as coming from the bad state of the blood.

The fresh-warm characterization of diseases, medicines, foods, environments and actions alludes both to a thermal and metaphorical character. That is, it refers to a sensation that causes cold or heat in the human body; for example: fatty foods, fevers, burning, refreshing, sweat, winds, airs, hard work as well as to an intrinsic nature of the disease, food, remedy or environment, not related to a thermal character.

The fresh-warm categories also apply to medicinal plants. The plants are the most important remedies of these communities and their techniques of preparation and administration are quite varied, although the main one is to prepare a decoction that is administered orally. It was also found that, in most cases, warm plants are taken "by tea" and fresh ones "by water". Finally, the excessive environmental heat, a prolonged exposure to the sun, the north wind, the hard work and the rough woods, received the characterization of warm. The south wind, a prolonged exposure to the dew, the rest and the soft woods, are characterized as fresh (Scarpa 2004b).

Main Ways to Prepare Remedies Using Plants

According to these tribes, each plant has its particular way of preparing. Most of the times they are drunk in the form of infusion (tea), but in some cases they present a strange form of preparation such as:

- "*Quemadillo*" (burnt) is a traditional and popular home remedy used to remove cough and colds, which is made by burning. This recipe was made in the past, with charcoal embers, sugar, wine and medicinal plants. Currently, the original recipe has undergone some modifications but it is still as fast and effective. In the northwest of Chaco, the "*quemadillo*" is prepared from toasted leaves and sprinkled with sugar, to accompany different remedies, for example, those presented by the genus *Prosopis* sp. In this case, its stem embers are used to prepare the "*quemadillo*" and this element accompanies many vegetable remedies referred to as "warm" (Pen 2013).

- "*Mate*" has its origins in the Guaraní town and is currently consumed mostly in southern Brazil and northern Argentina. It is a drink that is prepared using the leaves of the "*yerba mate*" plant (*Ilex paraguariensis*), which is filtered in hot water. Drinking *mate* is a very important social practice, similar to coffee or tea. In fact, this infusion is so popular that it is the national drink of Argentina and Paraguay. The correct preparation and consumption of the drink is essential and specific utensils are required. The container used to drink *mate* is a hollow pumpkin that is left to dry in the sun. In addition to the pumpkin, a small bulb tube is needed, traditionally made of silver and includes ornamental decorations although the most modern varieties can be made of nickel or stainless steel, which are cheaper (Fig. 10.3). The *mate* is prepared in the pumpkin, filling it with dried leaves of *Ilex paraguariensis*. Then, a strong stir is necessary, so that the dusty residue is separated from the leaves and hot water is poured into the pumpkin. It is important that the water temperature oscillates between 70 and 80°C but not boiling (Burtnik 2006, Dellacassa 2007).

- "*Tereré*" is a refreshing drink and also originated in the Guarani town. Its preparation is similar to that of "*mate*" with the difference that "*tereré*" employs water cooled with pieces of ice, and fruit juices such as orange, grapefruit or lemon may be added (Oberti 1960).

A   **B**

Fig. 10.3. (A) "*Yerba mate*" plant (*Ilex paraguariensis*). (B) Hollow pumpkin and small bulb tube used to prepare the "*mate*".

Most Important Plants Used by Toba and Mocoví Communities

Table 10.1 describes the particular uses and knowledge of 156 plant species belonging to the monocotyledonous and dicotyledonous classes, linked to the subsistence of Toba and Mocoví communities. In each group, the taxa are arranged according to the botanical family in alphabetical order. Within each family, species

Table 10.1. Most important plants used by Toba and Mocoví communities.

Species	Vernacular Name	Medicinal Uses	Utilized Parts
Acanthaceae			
Ruellia hygrophila Mart.	"Reventador"	antidiarrheic, diuretic, antiacid, vulnerary	leaves
Amaranthaceae			
Alternanthera pungens Kunth	Yerba 'el pollo, yerba 'e pollo	febrifuge,digestive, diuretic, against renal diseases, preventive, stomach refreshing, depurative of blood, vulnerary, oral antiseptic	whole plant
Alternanthera sp.	Novalgina, Coramina	antiacid, against headache, vulnerary, febrifuge, depurative of blood, hepatic affections	aerial parts
Amaranthus muricatus (Moq.) Hieron.	Yerba meona, Cola de gama	intestinal antiinflammatory	whole plant
Amaranthus viridis L.	Ataco	general preventive, diuretic, hematuria	roots
Gomphrena tomentosa (Griseb.) Fries var. tomentosa	Yerba meona	diuretic, digestive, hepatic affections, depurative	whole plant
Anacardiaceae			
Schinus lorentzii (Griseb.) Engl.	Quebracho colorao, Quebracho	emenagogue, vulnerary, against bronchitis, odontalgic	bark and heartwood
Schinus fasciculata (Griseb.) I.M. Johnst.	Molle pispito, Molle fragante, Molle	curative of cough, antidiarrheic, contraceptive, against flue	stems and leaves
Apiaceae			
Petroselinum crispum (Mill.) Nyman ex A.W. Hill	Perejil	strengthen gums, curative of cough, contraceptive, abortive	aerial parts
Pimpinella anisum L.	Anis	anthelmintic, against pains, abortive	seeds
Apocynaceae			
Asclepias mellodora A.St.-Hil	Matatodo	antimicrobial	leaves

Species	Common name	Uses	Parts used
Aspidosperma quebracho-blanco Schltdl	Quebracho blanco	purgant, expectorant, against cooling and flu, antidysenteric, abortive, depurative of blood, curative of cough, against malaria, for ripen pimples, antimicrobial	bark and seeds
Funastrum gracile (Decne.) Schltr	Tramontana, Enredadera	diuretic, against renal affections, antihypertensive, hepatic anti-inflammatory, antiperistaltic, oral antiseptic, against Chagas disease, depurative of blood, preventive, antidandruff, antimicrobial, vulnerary	aerial parts
Morrenia odorata (Hook. & Arn.) Lindl.	Doca	against palpitations, against bronchitis, against snake bites, antimicrobial	Stems and flowers
Vallesia glabra (Cav.) Link	Ancoche, Ancochi	vulnerary, for ripen pimples, against rheumatic pain, against cooling, childbirth stimulative	aerial parts
Aquifoliaceae			
Ilex paraguariensis A.St.-Hil.	Yerba	purgant, diuretic, to cure warts, childbirth stimulative	aerial parts
Asteraceae			
Acanthospermum hispidum DC.	Guasdrilla, Guarilla, Guasdiya	diuretic, against malaria, febrifuge	roots and branches
Ambrosia tenuifolia Spreng.	Altamisa, Altamisa de las islas	anthelmintic, hepatic digestive, against sunstroke, febrifuge	aerial parts
Artemisia absinthium L.	Ajenjo	abortive, hepatic digestive, against flu and cooling	leaves
Baccharis salicifolia (Ruiz & Pav.) Pers.	Suncho	against cooling and flu, against bones and kidneys pain, diuretic	aerial parts
Cyclolepis genistoides D.Don	Palo azul	diuretic, against renal affections, febrifuge, hepatic digestive, depurative of blood, antihypertensive	aerial parts
Eupatorium christieanum Baker	Eupatorio	against cooling and flu	aerial parts
Flaveria bidentis (L.) Kuntze	Balda, Balta	against sunstroke	branches
Parthenium hysterophorus L.	Altamisa, Altamisa de ajuera, Altamisa del bordo	anthelmintic, against rheumatic pain, against cooling and flu, depurative of blood	aerial parts

Table 10.1 contd. ...

...*Table 10.1 contd.*

Species	Vernacular Name	Medicinal Uses	Utilized Parts
Pectis odorata Griseb.	Manzanilla ´el monte	against childs gastritis, infant purgant, antidiarrheic, oxytocic, emenagogue, abortive	aerial parts
Pluchea sagittalis Less.; Pluchea microcephala R.K.Godfrey	Cuatro cantos	hepatic digestive, against the "cold stomach", general preventive, antihypertensive, antidisenteric, vulnerary	leaves
Tessaria integrifolia Ruiz et Pav	Palo bobo, Bobo	against insect bites	barks
Verbesina encelioides (Cav.) Benth. & Hook.f. ex A.Gray	Quellusisa, Margarita	for infected wounds	leaves
Xanthium spinosum L.	Cepacaballo	to cure warts, ophthalmic, stomach refreshing, against renal affections, depurative of blood, hepatic digestive, oral antiseptic	whole plant
Basellaceae			
Anredera cordifolia (Ten.) Steenis	Zarzaparrilla	depurative of blood, emenagogue	aerial parts
Bignonaceae			
Amphilophium cynanchoides (DC.)	Lengua de vaca	antipruritus	fruits
Tabebuia nodosa (Griseb.) Griseb	Palo cruz	emenagogue, abortive	branches and leaves
Bombacaceae			
Ceiba chodatii (Hassl.) Ravenna	Yuchán, Lluchán	against sunstroke, curative of cough, for infected wounds, diuretic, sedative, hypocholesterolemic, against snake bites, hepatic digestive	leaves and flowers
Borraginaceae			
Borago officinalis L.	Borraja, Borraja de Castilla	to treat measles	stems and leaves
Heliotropium elongatum (Lehm.) Gürke	Borraja e´campo, Borraja	to treat measles, febrifuge, stomach digestive	stems and leaves
Heliotropium procumbens Mill.	Cola de gama	vulnerary, antiarthritic	stems and leaves

Brassicaceae			
Lepidium didymum L.	Quimpe, Quimpi	against gingivitis, curative of cough, expectorant	aerial parts
Buddlejaceae			
Buddleja sp	Salvia	against sunstroke	aerial parts
Cactaceae			
Gymnocalycium mihanovichii (Fric ex Gürke) Britton & Rose		against bone pains	stems
Opuntia elata Link & Otto ex Salm-Dyck		febrifuge	fruits
Opuntia ficus-indica (L.) Mill.		against rheumatic pain, against kidney stones	stems
Opuntia quimilo K. Schum.	Quimil	against sunstroke, diuretic, for infected wounds, febrifuge, hepatic antiinflammatory, antiinflammatory, for ripen pimples	stems
Quiabentia verticillata (Vaupel) Borg	Sacha rosa	for infected wounds	leaves
Stetsonia coryne (Salm-Dyck) Britton et Rose	Cardón	against cutaneous eruptions, hepatic digestive	fruits
Capparaceae			
Anisocapparis speciosa (Griseb.) X. Cornejo & H.H.Iltis	Bola verde, Palo verde	odontalgic, against the "cold stomach", antidysenteric	aerial parts
Capparicordis tweedieana (Eichler) Iltis & Cornejo	Hoja redonda, Sacha mamita, Mataburro, Comida de burro	antidisenteric, against the "cold stomach", for ripening pimples	aerial parts
Capparis salicifolia Griseb.	Sacha sandia	diuretic	fruits
Caricaceae			
Carica papaya L.	Mamón, Papaya	anthelmintic	seeds

Table 10.1 contd. ...

...Table 10.1 contd.

Species	Vernacular Name	Medicinal Uses	Utilized Parts
Celastraceae			
Maytenus spinosa (Griseb.) Lourteig & O'Donell; *Maytenus scutioides* (Griseb.) Lourteig & O'Donell	Abriboca	antidiarreic, antiacid	aerial parts
Maytenus vitis-idaea Griseb.	Coike yuyo, Coiki yuyo, Con'ko yuyo	febrifuge, oral antiseptic	leaves
Celtidaceae			
Celtis chischape (Wedd.) Miq.	Tala pispita, Tala	against the "cold stomach", against child gastritis, febrifuge, curative of cough, antidisenteric	aerial parts
Chenopodiaceae			
Dysphania ambrosioides (L.) Mosyakin & Clemants	Paico	against the "cold of the stomach", against cooling and flu, antidisenteric, anthelmintic	aerial parts
Convolvulaceae			
Ipomoea carnea Jacq.	Mandiyona	vulnerary	leaves
Cucurbitaceae			
Citrullus lanatus (Thunb.) Matsum. & Nakai	Sándia	against measles, against general cooling	fruits and seeds
Cucumis melo L.	Melón	emenagogue	seeds
Cucurbita maxima Duchesne	Zapallo	anthelmintic	seeds
Cucurbita moschata Duchesne	Anco	against burns	mucilage of the fruit
Cuscutaceae			
Cuscuta indecora Choisy	Cabello de ángel, Rayo de sol	antialopecic	stems

Erythroxylaceae			
Erythroxylum coca Lam.	Coca	against the "side pain", stomach digestive, oxytocic	leaves
Phyllantaceae			
Phyllanthus niruri L.	Rompepiedras	vesicular stones, against renal affections	aerial parts
Euphorbiaceae			
Croton bonplandianus Baill.	Escoba negra, Tinajero, Paloma yuyo, Comida de paloma	hepatic digestive, antidiarreic, gastric pains	aerial parts
Croton hieronymi Griseb.	Poleo'el monte, Poleo	stomach digestive	aerial parts
Euphorbia serpens Kunth	Yerba 'e la golondrina, Yerba golondrina, Golondrina	stomach refreshing, oral antiseptic, hepatic digestive, against urinary tract irritations, diuretic, febrifuge	roots and aerial parts
Jatropha hieronymi Kuntze; *Jatropha excisa* Griseb.	Piñón, Manchador, Higuera 'el monte, Higuerilla	for infected wounds	latex
Manihot esculante Crantz	Mandioca	antidiarrheic, gut depurative, against sunstroke	roots
Ricinus communis L.	Tártago, Ricino	against sunstroke and headache, against the "cold stomach"	leaves
Sapium haematospermum Müll. Arg.	Lecherón	vulnerary, cicatrizant, dermic affections	bark and leaves
Tragia hieronymi Pax & K. Hoffm.	Yuyo quemador, Quemador, Ortiguilla	against insect bites	aerial parts
Fabaceae			
Acacia albicorticata Burkart	Espinillo	vulnerary, against throat pain	bark and leaves

Table 10.1 contd. ...

...Table 10.1 contd.

Species	Vernacular name	Medicinal uses	Utilized parts
Acacia aroma Hook. & Arn.	Tusca	curative of cough, against throat pain, hepatic digestive, conception stimulant, against snake bites, vulnerary, antihypertensive, febrifuge, for ripen pimples, against irritation of the urinary tract	stems, leaves and bark
Bauhinia argentinensis Burkart	Pata´e buey	diuretic, hepatic digestive, against kidney pains	roots and leaves
Caesalpinia paraguariensis (Parodi) Burkart	Guayacán	stomach and liver digestive, febrifuge, abortive, antidisenteric, against throat pain, curative of cough	bark
Caesalpinia stuckertii Hassl.	Guaycurú	diuretic, for kidney pains, intestinal refreshing	bark
Cercidium praecox (Ruiz & Pavon) Harmas	Brea, Brea ´el bordo	vulnerary, antiacid, against the "cold stomach", curative of cough	bark and roots
Desmanthus virgatus (L.) Willd.	Rompepiedras	vesicular stones	stems and leaves
Enterolobium contortisiliquum (Vell.) Morong	Pacará	antidandruff	fruits
Geoffroea decorticans (Hook. & Arn.) Burkart	Chañar	oxytocic, curative of cough, abortive, for infected wounds, against snakes bites	bark, leaves and flowers
Prosopis alba Griseb	Árbol, Algarrobo blanco, Algarrobo	antiacid, ophthalmic, the embers of the stems are used to prepare the "quemadillo"	leaves and stems
Prosopis nigra (Griseb.) Hieron		the embers of its wood are used to prepare the "quemadillo"	stems
Prosopis ruscifolia Griseb; *Prosopis vinalillo* Stuck.	Vinal	hepatic digestive, for ophthalmic infections, depurative of blood, against rheumatic pain, against bone pain, antihypertensive, diuretic, antiacid, vulnerary	leaves
Prosopis sericantha Hook	Guaschín, Palo mataco, Barba ´i tigre	abortive	seeds
Senna morongii (Britton) Irwin et Barneby	Pitacanuto	vulnerary, against sunstroke and snakes bites, against rheumatic pain and vesicular stones, febrifuge, for infected wounds, hepatic digestive, against gastric pains	aerial parts

Senna occidentalis (L.) Link	Café del monte	anthelmintic, abortive	whole plant
Senna pendula (Willd.) Irwin & Barneby var. *paludicola* Irwin & Barneby	Café del agua, Pitacanuto del agua	febrifuge, hepatic digestive	whole plant
Lamiaceae			
Melissa officinalis L.	Torongil	anthelmintic	leaves
Mentha spicata L.	Menta	against the "cold stomach", against palpitations, purgant, ophthalmic, against throat pain, general preventive	leaves
Ocimum basilicum L.	Albahaca	against ophthalmic infections, general preventive	stems and seeds
Rosmarinus officinalis L.	Romero	oxytocic, antimocotic	branches
Loranthaceae			
Struthanthus uraguensis G. Don	Liga del mistol, Liga	oxytocic	aerial parts
Tripodanthus acutifolius (Ruiz & Pav.) Tiegh.	Corpo del palo santo, Corpo, Liga del palo santo	abortive, against kidney stones, against hypothermia, oxytocic, emenagogue	aerial parts
Lythraceae			
Heimia salicifolia (Kunth) Link	Quiebra 'arao, Enchullador	antidisenteric, against the "cold stomach"	whole plant
Malvaceae			
Sida cordifolia L.	Malva, Malvisco	oral antiseptic, gut depurative, general preventive, stomach refreshing, depurative of blood, against childhood gastritis, against irritation of the urinary tract, diuretic, infantile febrifuge	roots and leaves
Sphaeralcea bonariensis (Cav.) Griseb	Malvavisco, Malva	gut depurative, febrifuge, stomach refreshing, oral antiseptic, antidiarrheic, curative of cough	roots and leaves
Byttneria filipes Mart. ex K.Schum.	Garabato 'el agua	febrifuge	stem, leaves

Table 10.1 contd. ...

...Table 10.1 contd.

Species	Vernacular Name	Medicinal Uses	Utilized Parts
Meliaceae			
Cedrela sp.	Cedro	for the "internal blows", against the "side pain"	stems
Melia azederach L.	Paraiso	for cutaneous eruptions, headache, abortive, anthelmintic, emenagogue	leaves and roots
Menispermaceae			
Cissampelos pareira L.	Mil hombres	against the "cold stomach", depurative of blood, hepatic digestive, diuretic, vulnerary, anti inflammatory, abortive	stems
Menyanthaceae			
Nymphoides indica (L.) Kuntze		febrifuge	leaves
Myrthaceae			
Eucalyptus tereticornis Sm.	Ucalito	expectorant, curative of cough, against throat pain, against the "air"	leaves and stems
Nyctaginaceae			
Boerhavia diffusa var. *leiocarpa* (Heimerl) C.D. Adams	Batata e´cuchi, Batata de chancho	stomach refreshing, diuretic, depurative of blood, febrifuge	roots
Pisonia zapallo Griseb.	Caspi zapallo, Caspi	Against testicle inflammation	bark
Olacaceae			
Ximenia americana L. var. *argentinensis* DeFilipps	Pata, Pata pata	antidisenteric	seeds
Passifloraceae			
Passiflora mooreana Hook.	Granadilla, Granada	antidiarrheic, antidisenteric, against venereal disease, sedative, antihypertensive, purgant, vermifuge	aerial parts

Phytolacaceae			
Petiveria alliaceae L.	Calaj'ch'n, Calauchin	against the "bone pain", against sunstroke, against malaria, general preventive, against palpitations, febrifuge, vulnerary, against the "cold stomach"	aerial parts
Plantaginaceae			
Plantago myosuros Lam.	Llanten	diuretic, febrifuge, hepatic antiinflammatory, ophthalmic, antisyphilitic, depurative	leaves
Polygonaceae			
Polygonum punctatum Elliot.	Vinagrillo, Yuyo picante	Anti inflammatory, vulnerary, antimicrobial	leaves
Salta triflora Griseb.	Duraznillo, Duraznillo del bordo, Duraznillo del monte	antidiarreic, antidisenteric, antihemorrhoidal, abortive, vulnerary	whole plant
Portulacaceae			
Portulaca oleraceae L.; *Portulaca* sp.	Verdolaga	against sunstroke, febrifuge, anthelmintic	aerial parts
Talinum paniculatum (Jacq.) Gaertn.	Carne gorda	diuretic	roots
Ranunculaceae			
Clematis campestris A.St.-Hil.	Barba'e chivato, Barba'e chivo	hemostatic, against venereal diseases, antifungal, vulnerary	roots and fruits
Rhamnaceae			
Ziziphus mistol Griseb.	Mistol, Mistol cuaresmillo	antidandruff, emetic, curative of cough, vulnerary, antidisenteric	leaves and bark
Rutaceae			
Citrus limon (L.) Osbeck	Limonero, Limón	antihypertensive, digestive, febrifuge, against cooling and flu, sedative	fruits

Table 10.1 contd. ...

...Table 10.1 contd.

Species	Vernacular Name	Medicinal Uses	Utilized Parts
Citrus sinensis (L.) Osbeck	Naranjo	against spider bites, febrifuge, digestive, against rheumatic pain, sedative, cardiotonic, antihypertensive	seeds and leaves
Ruta chapalensis L.	Ruda	against spider bites, for the "bones pain", against palpitations, digestive, carminative, febrifuge, emenagogue, abortive, anthelmintic	leaves
Salicaceae			
Salix humboldtiana Willd.	Sauce	vulnerary, depurative of blood, emenagogue, oxytocic, preventive, purgant, febrifuge, analgesic, sedative, tonic, against rheumatic pain, astringent, digestive	bark and leaves
Sapotaceae			
Sideroxylon obtusifolium (Roem. & Schult.) T.D.Penn.	Molle	against throat pain, curative of cough, against rheumatic pain, oxytocic, hypertensive	bark, stems and leaves
Scrophulariaceae			
Scoparia dulcis L.	Flor de casamiento del monte	antidiarreic	aerial parts
Simaroubaceae			
Castela coccinea Griseb.	Meloncillo, Melonciyo	antidisenteric, against hypothermia, against the "astonishment"	roots
Solanaceae			
Capsicum chacoense Hunz.	Ají del monte, Ají	stomach digestive, to wean	fruits
Cestrum parqui (Lam.) L'Hér.	Hediondilla	against headache and sunstroke, for the "Inner fever", febrifuge, stomach refreshing, antihemorrhoidal, antipruritus	roots and leaves
Datura ferox L.	Chamico	for odontological pains, vulnerary, against asthma, for ripen pimples, against headache	leaves
Nicotiana glauca Graham	Palán, Palancho	Anti inflammatory, against headache, for ripening pimples, febrifuge, against sunstroke, vulnerary	leaves

Nicotiana tabacum L.	Tabaco	against the "air", against hypothermia	leaves
Solanum aridum Morong	Pocote, Pocotillo, Pocote 'e perro	antifungal	fruits
Solanum argentinum Bitter & Lillo	Cabrayuyo	general preventive, against "bone pain", against rheumatic pain, against sunstroke, febrifuge, against hypothermia, against cooling and flu, for ripen pimples	aerial parts
Solanum glaucophyllum Desf	Corcho 'el agua, Sunho 'el agua	emetic	aerial parts
Solanum hieronymi Kuntze	Pocote	antifungal	fruits
Solanum sisymbrifolium Lam.	Vila vila	intestinal refreshing, hepatic digestive, febrifuge, oral antiseptic, diuretic, against irritation of the urinary tract, depurative of blood, curative of cough, general preventive, contraceptive	roots
Solanum tuberosum L.	Papa	Anti inflammatory	tubers
Theaceae			
Camellia sinensis (L.) Kuntze	Té	for ophthalmic infections	leaves
Urticaceae			
Parietaria debilis G. Forst.	Paletaria	against irritation of the urinary tract, hepatic and intestinal refreshing, febrifuge	aerial parts
Verbenaceae			
Aloysia gratissima (Gillies & Hook.) Tronc.	Poleo del campo	digestive, carminative, tonic, against palpitations, nervous diseases, antifungal	leaves
Aloysia polystachya (Griseb.) Moldenke	Burrito, Poleo, Poleo de la casa, Cedrón	stomach and hepatic digestive, hypertensive, against palpitations, for nervous diseases	aerial parts
Glandularia tweedieana (Niven ex Hook.) P. Peralta	Margarita, Sangre de Cristo	febrifuge, for ripen pimples, emenagogue	aerial parts

Table 10.1 contd. ...

...Table 10.1 contd.

Species	Vernacular name	Medicinal uses	Utilized parts
Lantana trifolia L.	Salvia 'e monte	stomach digestive	leaves
Lippia alba (Mill.) N.E.Br. ex Britton & P.Wilson	Salvia de castilla, Salvia de la casa, Salvia	against cooling and flu, against throat pain, against gastric pains and headache, curative of cough	aerial parts
Phyla reptans (Kunth) Greene	Mosko yuyo, Mosko yuyo del bajo	for infected wounds and gangrene, against skin rashes, against sunstroke	aerial parts
Viscaceae			
Phoradendron liga (Gillies ex. Hook. & Arn.) Eichler; *Phoradendron bathyoryctum* Eichler	Liga, Corpo	oxytocic, antihypertensive, against hypothermia	aerial parts
Vitaceae			
Cissus palmata Poir	Zarzaparrilla, Zarza, Enredadera, Bejuco´el agua	blood depurative, hepatic refreshing	leaves
Zygophylaceae			
Bulnesia sarmientoi Lorentz ex Griseb.	Palo santo	hepatic digestive, against the "internal blows" and the "side pain", febrifuge, blood depurative, against hypothermia, against bronchitis, cooling and flu, against rheumatic pain, against cutaneous eruptions	aerial parts
Arecaceae			
Copernicia alba Morong	Palma, Palma negra, Palma blanca	Childbirth stimulator	roots
Bromeliaceae			
Tillandsia loliacea Mart. ex Schult. & Schult.f.; *Tillandsia recurvifolia* Hook.	Chasca	against traumatisms	leaves
Commelinaceae			
Commelina erecta L.	Santa lucia	against ophthalmic pains and infections	leaves

Species	Common name	Uses	Parts used
Hydrocharitaceae			
Limnobium laevigatum (Humb. & Bonpl. ex Willd.) Heine		febrifuge	aerial parts
Lilaceae			
Allium cepa L.	Cebolla	stomach antiacid	bulb
Allium sativum L.	Ajo	against snake bites, antihypertensive, against teeth pains	bulb
Aloe vera L.	Penca e 'sábila, Sábila	Vulnerary, for ripen pimples, antifungal, antiinflammatory, for infected wounds, hemostatic, anticancerigen, against insect bites, preventive, depurative of blood, vesicular stones, against the "inner fever", antidandruff and antialopecic, against sunstroke, headache, febrifuge, for stomach ulcers, against snake bites	leaves
Orchidaceae			
Cyrtopodium punctatum (L.) Lindl.	Chacra 'el monte	antihypertensive, diuretic, febrifuge, against renal pains	leaves and bulbs
Poaceae			
Cenchrus myosuroides Kunth	Cadillo	abortive	roots
Cymbopogon citratus (DC.) Stapf	Cedrón, Cedrón pasto	stomachal digestive, antihypertensive, against teeth pains	leaves
Elionurus muticus (Spreng.) Kuntze	Aibe	antialopecic	roots
Oryza sativa L.	Arroz	antidisenteric	seeds
Triticum aestivum L.	Trigo	antidisenteric, emetic	seeds
Zea mays L.	Chacra, Máiz	oxytocic, against flu and cooling, febrifuge, diuretic, stomach antacid	seeds
Typhaceae			
Typha dominguensis Pers.	Totora	diuretic, purgant, antidisenteric, anti-allergic	rhizomes and leaves

are sorted according to the same criteria, by their scientific name including their vernacular names, medicinal applications and used parts. All the data related to the uses are briefly presented, since all the information is much broader and is related to the respective cultural context (Crovetto 1968, Marzocca 1997, Lahitte 1998, Hurrell 2011, Scarpa 2013).

From the information summarized in Table 10.1, it can be observed that plant families most used by these communities belong to dicotyledonous class (90%), while monocots are scarcely used for medicinal purposes, representing only 10% of the total species reported in this work. Within dicotyledons, the most prominent used families are: Fabaceae (11%), Asteraceae (9%), Solanaceae (8%), Euphorbiaceae (6%), Cactaceae (4%), Verbenaceae (4%), Amaranthaceae (4%), Apocynaceae (4%), Cucurbitaceae (3%), Lamiaceae (3%), Borraginaceae (2%), Capparaceae (2%), Malvaceae (2%), Rutaceae (2%), Anacardiaceae (2%) and another 34 families that each one constitutes 1% of the total dicotyledons surveyed (Fig. 10.4).

Regarding the most important medicinal uses of the 156 plants surveyed, they were grouped into 17 different categories. It should be noted that the description of the uses of plants by the inhabitants of these communities is much more detailed and descriptive, but this grouping into categories was carried out in order to shorten the information and draw some broad conclusions about the use of their plants. It is important to mention that a plant species can be used for one or several ailments categories.

The percentages of plants used for the different categories are the following (Fig. 10.5): odontalgic disease (10%), ophthalmic disease (4%), antiparasitic (17%), antiinflammatory (4%), kidney dysfunctions (21%), female health problems, childbirth & abortive (21%), digestive tract ailments (45%), antimicrobials (11%), febrifuge (24%), respiratory disease (20%), depurative and purgant (15%), rheumatic and general pains (12%), sunstroke (10%), anti-hypertensive and heart problem (12%), headache (5%), skin diseases and wounds (35%), others (21%).

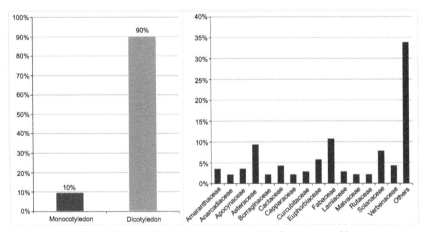

Fig. 10.4. Plant families most commonly used by these communities.

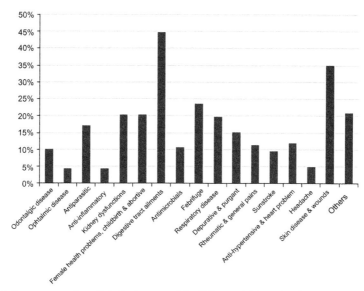

Fig. 10.5. Percentages of medicinal plants used by Tobas and Mocovíes for different ailments categories.

Botanical Features of the Seven Most Important Medicinal Species of the "Chaco" and the "Espinal" Regions of Argentina

Figure 10.6 shows seven most important medicinal species of this Argentinean region and then a slight botanical description of each one are informed (Lahitte 1998).

- *Schinus molle*: It is native to the warm, temperate and tropical regions of South America but it was introduced in some regions of Europe as a shade and ornamental tree. It is a perennial, aromatic, and fast growing tree commonly known as "*aguaribay*" or red pepper tree that usually measures between 6 to 8 m height, although in optimal conditions it reaches 25 m. The leaves are divided into numerous narrow and lanceolate leaflets of intense green color. Its flowers are of small size, hermaphrodite or unisexual, grouped in hanging terminal and axillary panicles, yellowish or yellow-greenish. Its fruit is a bright pink drupe with the size of a peppercorn.
- *Salix humboldtiana*: This species is found from Mexico to Argentina. It flowers in spring and fructifies in summer. They are deciduous trees from 5 to 18 m high and have alternate, linear or lanceolate leaves from 4 to 15 cm long and 1 cm wide presenting finely serrated edge. Its flowers are yellowish or greenish, without

Fig. 10.6. Pictures of the seven most important medicinal species of this Argentinean region: (A) *Schinus molle,* (B) *Salix humboldtiana,* (C) *Gomphrena tormentosa,* (D) *Commelina erecta,* (E) *Clematis montevidiensis,* (F) *Cestrum parqui,* (G) *Baccharis salicifolia.*

perianth, grouped in 3 to 6 cm long amentums. The fruit is an ovoid capsule of 0.5 cm long and its seeds have a large tuft of characteristic white hair.

- *Gomphrena tomentosa*: It is widely distributed in warm and temperate zones of Argentina. It is a perennial and decumbent herb with a thick pivoting root. Its leaves, which are 2 to 3 cm long by 0.3 to 0.5 cm, wide are subsessile, linear or lanceolate and hairy. The tepals are linear, pink-purple or rarely white and grouped in long-capped pedunculated inflorescences. Its fruit is usually dry, monosperm, indehiscent or dehiscent and with floury endosperm.

- *Commelina erecta*: Species distributed from the United States to Uruguay and north-center of Argentina. In the region, it grows in sandy soils and flowers in spring and summer. It is a perennial herb, with erect or decumbent stems. Its leaves have reddish pods with oval sheets of 3 to 10 cm long by 2 to 3.5 cm wide presenting slightly wavy edge. Its zygomorphic flowers are found in terminal tops having 3 petals, two of them of 1.8 to 2 cm long and showing blue or light blue colors; the third one is reduced to a scale. Its fruit is a bivalve capsule.

- *Clematis montevidiensis*: This species is found from subtropical and temperate South America to Patagonia Argentina. It is common in forests, hygrophilous forests, roadsides and over fences. It flowers in spring and fructifies in summer. They are rhizomatous lianas with striated and glabrous stems. Its leaves are opposite and petiolate, with 3 to 5 whole leaflets, dentate, split or sectarian, asymmetric, ovate or elliptical from 2 to 8 cm long and 1 to 5 cm wide. Its flowers have 10 mm long petals grouped in multifloral inflorescences. The fruit is a polyachene with fluffy stylus of 10 cm long.

- *Cestrum parqui*: Species distributed in Peru, Bolivia, Chile, Brazil, Paraguay and Argentina land. It is common in forests near the rivers, modified soils and disturbed areas. It flowers in spring and fructifies in summer. It is a shrub from 1 to 4 m tall, highly branched and glabrous. Its leaves are alternate, lanceolate, from 4 to 14 cm long and 1 to 4 cm wide. Its numerous flowers, from 2 to 2.5 cm long and with tubular-infundibuliform corolla, are yellow or green arranged in glomeruli or axillary and terminal tops. Its fruit is a globose and black berry from 7 to 10 mm long, with persistent calyx.

- *Baccharis salicifolia*: It is a common polymorphic species in America distributed from California to southern Argentina. It grows in humid to slightly dry soils, in sandy ones or near watercourses, fences and modified sites. It flowers in summer and autumn and fruits in autumn and winter. They are shrubs, which grow from 1 to 3 m tall, evergreen, dioecious, erect, with striated stems. Its leaves are alternate, simple, briefly petiolate and elliptical to linear from 4 to 9 cm long by 0.7 to 1.5 cm wide, with acute apex and whole or paucidentate border in the upper half. Its flowers are grouped into pedunculated segments arranged in composite corimbiform tops and its fruits are 1 mm long and reddish achenes.

Conclusion

Native peoples of Argentina represent a very important cultural value for the country. Thermal and climatic amplitudes of the region offer a biodiversity of plants that constitutes an amazing natural resource. Two of the main communities that inhabit the "Chaqueña" and the "Espinal" phytogeographic provinces are the Tobas and Mocovíes tribes. Ethnobotanical data look upon a wide range of information regarding the ways in which these cultural communities interact with its plant environment and the knowledge associated to a plant species. It was possible to compile the ethnobotanical information of 156 vegetal species existing in different texts and specialized reports, concluding that the main families used by these tribes are: Fabaceae, Asteraceae, Solanaceae, Euphorbiaceae, Cactaceae, Verbenaceae, Amaranthaceae, Apocynaceae, Cucurbitaceae, Lamiaceae, Borraginaceae, Capparaceae, Malvaceae, Rutaceae and Anacardiaceae. The most important medicinal uses of the 156 plants surveyed correspond to the categories of: odontalgic disease, ophthalmic disease, antiparasitic, anti-inflammatory, kidney dysfunctions, female health problems, digestive tract ailments, antimicrobials, febrifuge, respiratory disease, depurative and purgant, rheumatic and general pains, sunstroke, anti-hypertensive and heart problem, headache and skin diseases and wounds. The forms of the remedies' preparation using medicinal plants is quite varied and these tribes use three particular ways such as "*quemadillo*", "*mate*" and "*tereré*". Finally, we can conclude that most important medicinal species of this Argentinean region are: *Schinus molle, Salix humboldtiana, Gomphrena tormentosa, Commelina erecta, Clematis montevidiensis, Cestrum parqui* and *Baccharis salicifolia*, among others.

References

Adámoli, J., Neumann, R., Ratier de Colina, A. and Morello, J. 1972. El Chaco aluvional salteño. Revista de Investigaciones Agropecuarias 9: 165–238.

Arenas, P. 1987. La etnobotánica en el Gran Chaco. VI Congreso Latinoamericano de Botánica, Simposio de Etnobotánica, Serie Memorias de Eventos Científicos Colombianos, Medellín ICFES, 35–42.

Ávila, M. 1960. Flora y Fauna en el folklore de Santiago del Estero, Ed. M. Violetto, San Miguel de Tucumán, p. 254.

Bates, D. 1985. Plant utilization: Patterns and prospects. Economic Botany 39: 241–265.

Berry, P.E. 2002. Diversidad y endemismo en los bosques neotropicales de bajura. pp. 83–96. *In*: Guariguata, M. and Catan, G. (eds.). Ecología y Conservación de Bosques. Libro Universitario Regional (Eulac/GTZ), Cartago.

Brandão, M., Zanetti, N., Oliveira, P., Grael, C., Santos, A. and Monte-Mór, R. 2008. Brazilian medicinal plants described by 19th century European naturalists in the Official Pharmacopoeia. Journal of Ethnopharmacology 120: 141–148.

Burtnik, O. 2006. Yerba Mate: Manual de Producción. INTA, AER Santo Tomé, Corrientes, Argentina, p. 52.

Cabrera, A. 1971. Fitogeografía de la República Argentina. Boletín de la Sociedad Argentina de Botánica 14: 1–42.

Cabrera, A. 1976. Enciclopedia Argentina de Agricultura y Jardinería: Regiones Fitogeográficas Argentinas. Acme.

Cabrera, A. and Willink, A. 1980. Biogeografía de América Latina. Monografía 13. Serie de Biología, Secretaría General de la Organización de los Estados Americanos, Washington DC, EEUU, p. 120.

Carrizo, J. 1960. Algunas supersticiones medicinales del Norte argentino. Cuadernos del Instituto Nacional de Investigaciones Folklóricas 1: 160–175.

Cleaves, C. 2001. Etnobotánica médica participativa en el Parque Nacional Lachua (Thesis). Universidad de San Carlos de Guatemala, Guatemala.

Coelho de Souza, G., Haas, A., von Poser, G., Schapoval, E. and Elisabetsky, E. 2004. Ethnopharmacological studies of antimicrobial remedies in the south of Brazil. Journal of Ethnopharmacology 90: 135–143.

Correia, P. 1984. Dicionário de Plantas Úteis do Brasil e Das Exóticas Cultivadas. Instituto Brasileiro de Desenvolvimento Forestal, Imprenta Nacional, Río de janeiro, p. 490.

Crovetto, R.M. 1968. Introducción a la Etnobotánica aborigen del nordeste argentino. Etnobiológica 11: 1–10.

Cruz, M., Santos, P., Barbosa Jr., A., de Mélo, D., Alviano, C., Antoniolli, A., Alviano, D. and Trindade, R. 2007. Antifungal activity of Brazilian medicinal plants involved in popular treatments of mycoses. Journal of Ethnopharmacology 111: 409–412.

Daoud, D. 1954. Veterinaria y medicina popular en Tucumán. Boletín de la Asociación Tucumana de Folklore 6(3): 30–87.

Dellacassa, E., Cesio, V., Vázquez, A., Echeverry, S., Soule, S., Ferreira, F. and Heinzen, H. 2007. Yerba mate. Historia, uso y propiedades. Revista de la Asociación de Química y Farmacia del Uruguay 51: 16–20.

Di Lullo, O. 1929. La medicina popular de Santiago del Estero, Ed. El Liberal, Santiago del Estero, p. 171.

Di Lullo, O. 1946. Contribución al estudio de las voces santiagueñas. Ed. Gobierno de la Provincia de Santiago del Estero, Santiago del Estero, p. 371.

Estévez, Y., Castillo, D., Tangoa Pisango, M., Arévalo, J., Rojas, R., Alban, J., Deharo, E., Bourdy, G. and Sauvain, M. 2007. Evaluation of the leishmanicidal activity of plants used by Peruvian Chayahuita ethnic group. Journal of Ethnopharmacology 114: 254–259.

Goleniowski, M., Bongiovanni, G., Palacio, L., Núñez, C. and Cantero, J. 2006. Medicinal plants from the "Sierra de Comechingones", Argentina. Journal of Ethnopharmacology 107: 324–341.

Grombridge, E. 1992. Global diversity Status of the Earth's Living Resources.

Gupta, M. 1995. 270 Plantas Medicinales Latinoamericanas, CYTED-SECAB, Bogotá, p. 617.

Gupta, M. 2008. Plantas medicinales Iberoamericanas, Convenio Andrés Bello Ed., Bogotá, p. 520.

Hurrell, J., Ulibarri, E., Arenas, P. and Pochettino, M. 2011. Plantas de herboristería, Ed. LOLA, Buenos Aires, Argentina, p. 242.

Lahitte, H., Hurrell, J., Belgrano, J., Jankowski, M., Haloua, L. and Mehltreter, M. 1998. Plantas medicinales rioplatenses: plantas nativas y naturalizadas utilizadas en medicina popular en la región del Delta del Paraná, Isla Martín García y Ribera Platense. Ed. LOLA, Buenos Aires, República Argentina, p. 240.

Martin, G. 1995. Ethnobotany. Chapman & Hall, London, p. 268.

Marzocca, A., Fernández, M. and Fernández, A. 1997. Vademécum de malezas medicinales de la Argentina: indígenas y exóticas. Orientación Gráfica, Buenos Aires, Argentina, p. 352.

Morton, J. 1981. Atlas of medicinal plants of Middle America: Bahamas to Yucatan. Charles C. Thomas.

Murillo, A. 1889. Plantes médicinales du Chili. Exposition Universelle de Paris, section chilienne, Lagny, París.

Oberti, F. 1960. Disquisiciones sobre el origen de la bombilla. Cuadernos del Instituto Nacional de Antropología y Pensamiento Latinoamericano 1: 151–158.

Pen, C., Romero, C., Deza, M., Durando, P. and Barioglio, C. 2013. Las Prácticas Culturales de Pequeñas Productoras Caprinas de los Departamentos Cruz del Eje e Ischilín de la Provincia de Córdoba, Argentina. VII Jornadas Santiago Wallace de Investigación en Antropología Social. Sección de Antropología Social. Instituto de Ciencias Antropológicas. Facultad de Filosofía y Letras, UBA, p. 11.

Portillo, A., Vila, R., Freixa, B., Adzet, T. and Cañigueral, S. 2001. Antifungal activity of Paraguayan plants used in traditional medicine. Journal of Ethnopharmacology 76: 93–98.

Rosenblat, A. 1954. La población indígena y el mestizaje en América: La población indígena, 1492–1950. Instituto de Filología de la Universidad de Buenos Aires, Biblioteca Americanista, Ed. Nova, p. 79.

Scarpa, G. 2000. Plants employed in traditional veterinary medicine by the Criollos of the Northwestern Argentine Chaco. Darwiniana 38: 253–265.

Scarpa, G. 2004a. Medicinal plants used by the criollos of northwestern Argentine Chaco. Journal of Ethnopharmacology 91: 115–135.

Scarpa, G. 2004b. El síndrome cálido-fresco en la medicina popular criolla del Chaco argentino. Revista de Dialectología y Tradiciones Populares 59(2): 5–29.

Scarpa, G. 2013. Las plantas en la vida de los criollos del oeste formoseño: medicina, ganadería, alimentación y viviendas tradicionales. Ed. Rumbo Sur Asociación Civil, Buenos Aires, República Argentina, p. 237.

Sosa Verón, H. and Vivante, A. 1951. Algunas recetas supersticiosas de Río Hondo (Santiago del Estero). Revista del Instituto de Antropología de la Universidad Nacional de Tucumán 5: 89–102.

Sturzenegger, O. 1987. Medecine traditionnelle et pluralisme medical dans une culture creole du Chaco Argentin. Mémoire pour le D.E.A. Anthropologie: Biologie Humaine et Société, Universite de Droit, D'Economie et des Sciences D'Aix-Marseille, France, p. 72.

Sturzenegger, O. 1989. Maladie et environnement culturel: à propos des "Culture-bond syndromes". Ecologie Humaine 7: 53–62.

Sturzenegger, O. 1999. Le mauvais oeil de la lune: Ethnomédecine créole en Amérique du Sud. Ed. Karthala, Paris, p. 302.

Svetaz, L., Zuljan, F., Derita, M., Petenatti, E., Tamayo, G., Cáceres, A., Cechinel Filho, V., Giménez, A., Pinzón, R., Zacchino, S. and Gupta, M. 2010. Value of the ethnomedical information for the discovery of plants with antifungal properties. A survey among seven Latin American countries. Journal of Ethnopharmacology 127(1): 137–158.

Torres, M. 1975. Ingeniero Guillermo Nicasio Juárez y los parajes del oeste de Formosa, Tiempo de hoy, Buenos Aires, p. 167.

CHAPTER 11

Ethnobotany of *Teucrium* Species

Milan S. Stanković* and Nenad M. Zlatić

Introduction

At the dawn of 20th century, fresh plants were more widely used for medicinal purposes. Increased investments into cultivation of medicinal plants and their more organised and effective usage have marked the decades of the century and opened novel paths in the study of plants and its potentially beneficial properties. Despite many advantages of modern medicine, traditional medical treatment is of great importance. Medicinal plants are widely applied in traditional medicine and have been an irreplaceable aspect of cultural heritage of diverse nations, both in the past and the present. The knowledge of beneficial active substances of plants has lately influenced the global increase in the medicinal application of various herbs. Their popularity may have been due to their status of inexpensive raw materials which contain products with increased biological activity particularly useful in prevention of various metabolism disorders (Petrovska 2012).

Medicinal plants are applied either independently or in combination with synthetic drugs. The efficiency of phytotherapy largely depends on the active components identified in the herbs used as well as on the appropriately established diagnosis and dose. Previous extensive studies and technological developments have contributed to the recognition of significance medicinal plants have in medicine and pharmacy. Plant origin preparatus has become widely applicable in various domains. Pharmacies all over the world sell teas, extracts or pills made of medicinal plants. Plant products in the medicines show multilevel biological activity and therefore have diuretic, sedative, antiseptic, antioxidant, antidiabetic

Department of Biology and Ecology, Faculty of Science, University of Kragujevac, Radoja Domanovića 12, 34000 Kragujevac, Republic of Serbia.
* Corresponding author: mstankovic@kg.ac.rs

and anti-inflammatory effects. The research on medicinal effects of plants improves understanding of both plant toxicity and protection from natural toxins (Petrovska 2012, Hosseinzadeh et al. 2015).

The aim of this chapter is to present the value of certain medicinal plants of the genus *Teucrium*, used in both traditional and modern medical studies as natural bioactive compounds. The species of the genus *Teucrium* are particularly popular in folk medicine. The first records of the medicinal properties of these species date back to Greek mythology from ten centuries BC. The origins of the name of the genus can be traced to the name of Trojan king Teucer (Τεῦκρος). He is believed to have been the first to observe the medicinal properties of the species. The species of this genus have been used in the treatment of tuberculosis, scurvy, jaundice, rheumatism and disorders of digestive system (Petrović 1883). Therapeutic products made of above ground parts of these plants are of prominently bitter taste. This particularity is important in terms of its beneficial effect on digestive organs. High quantity of aromatic compounds with antimicrobial activity contributes to their efficiency in the treatment of numerous infectious diseases (Gostuški 1979). As the species of the genus *Teucrium* contain bitter aromatic compounds, these are used as culinary spices, food supplements and beverage flavour enhancers (Keršek 2006, Maccioni et al. 2007).

In order to prepare tinctures and other medicinal preparatus green, above ground plant parts are picked and subsequently dried in a cool, dark and airy place. Some species of the genus are important in commercial terms and hence are used in industry, medicine and pharmacy (Table 11.1).

The Main Characteristics of the Species of the Genus *Teucrium* L.

The genus *Teucrium* belongs to the family Lamiaceae. The species of this genus are for the most part pernennial, bushy or semi-bushy, rarely annual herbaceous plants. The leaves are oppositely arranged, with entire or dentate margins and if existent, rather short petiole. Flowers are zygomorphic, with longer or shorter stalk. Inflorescence is determinate and realized as dichasium. Bracteoles are practically underdeveloped. Flower calyx is bent downwards and consists of 10 nerves and 5 triangular dents. Crown tube is of the same length as calyx tube. There are 4 stamens longer than corolla. The pistil with two part stigma is set high. Fruits are of ovoid shape (Diklić 1974, Stanković 2012a).

The species of the genus *Teucrium* grow in moderate climate zones, particularly in the Mediterranean and Central Asia. As these species can be found in southern, southwestern and southeastern part of Europe, the continent is regarded as one with the greatest recorded variety of the species. The significant number of the species have been observed in southwestern Asia, northernmost parts of Africa, southern North America and southwestern South America. As for Australia, the species

Table 11.1. General ethnobotanical properties of the selected *Teucrium* species.

Botanical Name	Common Names	Part Used	Therapeutic Use	Manipulation	Biological Activity	Notes
Teucrium chamaedrys L.	Wall germander, podubica, dubačac	Aerial parts	Skin inflammations, strengthening of the immune system, anaemia	Infusion, decoction, tincture, elixirs, powder	Antioxidant, antiproliferative, antiinflammatory, antimicrobial and antiseptic	Large doses cause liver damage
Teucrium montanum L.	Mountain germander, trava iva, dubačac mali	Aerial parts	Respiratory inflammations, Strengthening of the immune system, stress, tiredness, stimulation of bile secretion, stimulation of digestion, loss of appetite, cleaning wounds	Infusion, powder, tincture	Antioxidant, antiviral, antimicrobial, anticancer, antispasmodic and antipyretic	Used in aromatising wines, liquers, and rakija Leaves are used to flavour meat
Teucrium polium L.	Felty germander, kalpooreh, pepeljuša	Aerial parts	Inflammations, diabetes, rheumatism, indigestion, abdominal pain, headache, kidney rocks	Infusion, powder	Antioxidant, antidiabetic, antiinflammatory, antispasmodic, antipyretic, antimicrobial, antifungal and antiviral	Large doses cause liver damage
Teucrium arduini L.	Croatian germander, planinska metvica	Aerial parts	Scarce data	Infusion	Antioxidant, antimicrobial and antiviral	No data
Teucrium botrys L.	Cut leaf germander, dubačac crveni	Aerial parts	Strengthening of the immune system, cold, enhancing digestion, kidney diseases	Infusion	Antioxidant, antimicrobial and anticancer	No data

Teucrium scordium subsp. *scordium*	Water germander, vodeni dubačac	Aerial parts	Infectious skin diseases, diseases of digestive tract, increased temperature and fever, respiratory disorders, poisoning, a cough, intestinal parasites	Infusion	Antioxidant, antiproliferative and antifungal	No data
Teucrium scordium subsp. *scordioides*	Water germander, lukovac	Aerial parts	Respiratory disorders	Infusion	Antioxidant, antimicrobial, anticancer and proapoptotic	No data
Teucrium fruticans L.	Tree germander, žbunasti dubačac	Aerial parts	Diuretic and depurative effects	Infusion	Antioxidant and antimicrobial	No data
Teucrium flavum L.	Yellow germander, žuti dubačac	Aerial parts	Diabetes, external treatment of skin traumas, cleaning of wounds	Infusion	Antiinflammatory, antidiabetic, anticancer, antioxidant and antimicrobial	No data

of the genus *Teucrium* are distributed in both southern parts of the continent and certain nearby islands (Fig. 11.1) (Meusel et al. 1978, Hollis and Brummitt 1992).

Out of the total number of 250 species of the genus *Teucrium,* approximately 200 species are found in the Mediterranean area. The widest distribution has been recorded on the Iberian peninsula whereas the smallest distribution has been observed on the Apennine peninsula. The areas of Northern Africa, Asia Minor and the Balkans are registered as having a significant number and wide variety of species. About ten species of the genus are found in Australia and America.

In terms of distribution, it is necessary to emphasise that there is a small number of widely distributed species, i.e., there are more moderately distributed or even endemic species. *Teucrium chamaedrys, Teucrium montanum* and *Teucrium scordium* are among the most frequently found species on the territory of Europe (Tutin and Wood 1972, Diklić 1974, Meusel et al. 1978).

The species of the genus *Teucrium* grow on all types of calcareous soil. However, seldom are these plants found on serpentine and silicate substrates. Saline soils are also regarded as unfavourable for the species in question. These species may be found in mountain, hill and lowland localities such as arid, termophilous habitats with high level of insolation. A small number of the species grows in humid lowland areas. Deciduous, mixed, and coniferous forests, communities of herbaceous vegetation, rocky habitats, meadows, pastures, continental steppes and sands have been confirmed as established habitats of the species. Ocassionally, the species may grow in hygrophilous communities along rivers as well as in ruderal habitats (Horvat et al. 1974, Diklić 1974, Ellenberg and Strutt 2009).

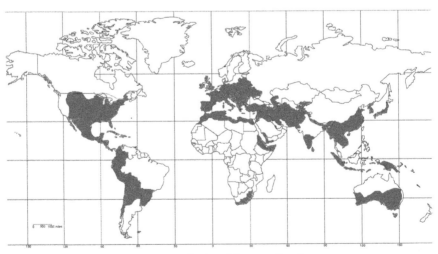

Fig. 11.1. Distribution of the genus *Teucrium.*

Secondary Metabolites of the Species of the Genus *Teucrium*

Analyses of quantitative and qualitative content of flavonoids in greater number of the species of the genus *Teucrium* have shown that flavonol glycosides are characteristic of this taxon whereas neo-clerodane diterpenes are regarded as taxonomic markers (Harborne et al. 1986, Bedir et al. 2004). When it comes to flavonoids, the highest concentrations of these compounds were those of luteolin and apigenin. Among several phenolic acids, the chlorogenic and rosemary acids are the ones most often observed. As for evaporable components, the species of this genus contain compounds from the group of monoterpenes and sesquiterpenes with germacrene, β-caryophyllene and β-pinene (Sundaresan et al. 2006, Monsef-Esfahani et al. 2010).

The General Characteristics, Ethnobotany and Phytotherapeutic Effects of the Selected *Teucrium* Species

Teucrium chamaedrys L.

Teucrium chamaedris (Fig. 11.2.1) is a semi-bushy perennial plant. It has well developed and partly woody root which serves as the basis for underground stolons. Plant habitus is upright, branchy and it can grow up to 30 cm. The branches are covered in grayish or purple hairs. The hairy, densely arranged leaves have rather

Fig. 11.2. Described *Teucrium* species: (1) *T. chamaedrys*; (2) *T. montanum*; (3) *T. polium*; (4) *T. arduini*; (5) *T. botrys*; (6) *T. scordium* subsp. *scordium*; (7) *T. scordium* subsp. *scordioides*; (8) *T. flavum*; (9) *T. fruticans*.

short petiole and dentate margins. Flowers are located in the basis of bracts. The calyx has mild dorsal protuberance and is covered in hairs. Mildly red corolla is twice or thrice as long as calyx. Stamens, located above flowers, are slightly bent. Fruits are ovoid nuts (Stanković et al. 2010, Stanković 2012a).

Teucrium chamaedrys is a Mediterranean and Middle European species distributed in whole Europe, save for the northern part of the continent. It can also be found in northern Africa and western Asia (Horvat et al. 1974, Diklić 1974, Stanković 2012a).

The species *Teucrium chamaedrys* mostly grows in arid rocky localities of calcareous origin. Rarely can it be found on serpentine and silicate substrate. The species may inhabit areas whose altitudes range from 200 to 1500 m (Horvat et al. 1974, Diklić 1974). It is widely distributed in deciduous, mixed, and coniferous forests as well as in communities of broadleaved evergreen and deciduous forests of bushy vegetation. Meadows, pastures, continental steppes and sands are also frequent habitats of *Teucrium chamaedrys* (Horvat et al. 1974, Diklić 1974, Ellenberg and Strutt 2009).

Teucrium chamaedrys is used for preparing teas, tinctures, wines and other therapeutic products which have bitter taste due to the presence of bitter active substances. These substances positively influence digestive system and may be used in the treatment of skin inflammation and anaemia as well as for the purpose of strengthening the immune system. The species is used as an ingredient of different elixirs, rakijas made with different medicinal herbs, and traditional drink vermouth. The tea is made out of dried herb and water. It is taken before meal 2 or 3 times a day. The tincture is made of dried herb and rakija. After 8 days, the mixture is filtered and 30 drops of the filtered content is used before every meal. Tincture powder is used to strengthen the organism in quantities of 5 to 10 g a day (Sarić 1988, Kojić et al. 1998, Keršek 2006, Stanković et al. 2012b).

Germander is widely applied as food supplement either in combination with other herbs or as capsules available in 200 mg dose (Barceloux 2012). Long term usage of the species *Teucrium chamaedrys* may cause liver damage as it contains secondary metabolites from the group of neo-clerodane diterpenes such as Teucrin A (Savvidou et al. 2007). On average, 100 ml of germander tea contains between 75 and 150 mg of Teucrin A (Barceloux 2012). Consequently, the usage of teas and other products made of this species must be moderate and carefully dosed (Khleifat 2002). US FDA has allowed metabolite Teucrin A as flavouring agent in alcoholic drinks such as aromatic wines, vermouth or bitter drinks (Barceloux 2012).

When it comes to phenolic content of the above ground parts of the species *Teucrium chamaedrys*, quantitative and qualitative analyses of flavonoids have shown high concentrations of luteoline and apigenin. Among other detected substances are phenylpropanoid glicosides, diterpenes and phenolic acids. The most important glicosides are teucroside, verbascoside, teucroside-3-O-methyl ether and teucroside-3,4-O-dimethyl ether. Among detected diterpenes are Teucrin A, Teucrin G, teuflin and dihydroteugin. With regard to phenolic acids, the presence of caffeine and rosemary acid has been confirmed (Gafner et al. 2003, Avula et al.

2003, Bosisio et al. 2004, Bruno et al. 2004, Kadifkova-Panovska et al. 2005, Lin et al. 2009). By the examination of quantity and content of evaporable components using GC/MC analysis, it has been established that the principal components of essential oil are β-caryophyllene and germacrene D (Bežić et al. 2011).

By following the influence of neo-clerodane diterpenes isolated from the species *Teucrium chamaedrys* on seed germination and growth and development of young plants of the species *Dactylis hispanica* Roth, *Petrorhagia velutina* (Guss.) Ball et Heyw., *Phleum subulatum* (Savi) Asch. et Gr., the allelopathic effect of secondary metabolites of the species has been observed (Fiorentino et al. 2009).

The extracts of the species *Teucrium chamaedrys* have antioxidant (Kadifkova-Panovska et al. 2005), proapoptotic (Stanković et al. 2011a), genotoxic, antimutagenic (Milošević-Djordević et al. 2013), antiinflammatory (Pourmotabbed et al. 2010), antimalarial (Tagarelli et al. 2010), antimicrobial and antiseptic (Vlase et al. 2014) activity in both *in vitro* and *in vivo* conditions. Apart from the previously stated effects, it has been established that *Teucrium chamaedrys* has a significant cytotoxic impact on HCT-116 cells after 24 hour exposure (Stanković et al. 2011a).

The previously conducted analysis showed that the species from serpentine habitats are important in terms of biological activity as it contains greater concentrations of phenols and flavonoids and shows high level of antioxidant activity. Therefore, *Teucrium chamaedrys* from these habitats is safe for use as it does not contain heavy metals (Zlatić et al. 2017).

Teucrium montanum L.

Teucrium montanum (Fig. 11.2.2) is a perennial plant with developed semi-woody root. The well spread branches either rise or cover the ground and may reach the length between 5 and 25 cm. The leaves are oppositely arranged, lanceolate, with entire margins, narrowed in the upper parts. Abaxial side is covered in white hair, whereas adaxial side may be hairless. The flowers form dichasium and their stalks are short. Corolla is of pale yellow colour. Stamens, reaching over corolla, are located between upper corolla lobes. The fruits are nuts of ovoid shape (Stanković et al. 2011b, Stanković 2012a).

Teucrium montanum grows in southern Europe as well as in southern parts of Western and Eastern Europe, Asia Minor and in northern Africa. It is widely distributed in Mediterranean areas as mountain species (Horvat et al. 1974, Diklić 1974, Stanković 2012a).

The habitats of *Teucrium montanum* are calcareous rocks, pine forests on calcareous soil, pastures and occasionally serpentine substrate. The species may inhabit areas whose altitudes range from 30 to 2000 m (Horvat et al. 1974, Diklić 1974, Stanković 2012a).

It grows in mixed, coniferous and deciduous forests, in serpentine rocky habitats, gorges and canyons as well as in the communities in sub-Mediterranean region, xerophytic and mountainous habitats (Horvat et al. 1974, Diklić 1974, Ellenberg and Strutt 2009).

For centuries, peoples of the Balkans have kept to traditional belief that this herb is medicinal and highly beneficial. *Teucrium montanum* is used in the treatment of inflammations of respiratory system and to strengthten the organism in stressful, tiring and physically demanding conditions. Teas, tinctures and wines stimulate bile secretion, suppress a cold, relieve digestive problems and enhance digestion. When it comes to gallbladder disorders, a tincture is used. Tincture is made of 20 g of dried herb and 100 ml of rakija. After 14 days, the compound is macerated and filtrated. The maximum amount to be taken is 20 drops per dose dissolved in water or tea of some other herb thrice a day. Tea is prepared in standard manner and daily intake is limited to less than 3 cups. The powder of plant may be used for similar purposes and the maximum amount of it is 1 to 3 g per dose thrice a day before meal (Sarić 1988, Kojić et al. 1998). *Teucrium montanum* is mixed with other herbs for purpose of making different elixirs and medicinal types of rakija (Keršek 2006). The species is widely used as a natural medicine for diseases related to disorders of bile secretion, liver and lack of appettite. As for external application, it is predominantly used for treating and cleaning wounds and other skin traumas.

HPLC analysis of phenols of the species *Teucrium montanum* confirmed that genistein acid is the predominant phenolic acid. Though quantitatively present to a lesser degree, it is important to bring up a variety of other detected acids such as chlorogenic, coumarin, syringic, gallic, vanillin, caffeine and ferulic acid (Tumbas et al. 2004, Čanadanović-Brunet et al. 2006). Quantitative analysis of etheric oil by means of GC/MC test showed that the main components are: β-cadinene, β-selinene, α-calacorene, β-caryophyllene, β-pinene and germacrene (Vuković et al. 2008, Bežić et al. 2011).

The extracts of the species *Teucrium montanum* have antioxidant (Čanadanović-Brunet et al. 2006), antiviral (Bežić et al. 2011), antimicrobial (Stanković et al. 2012b), genotoxic, antimutagenic (Milošević-Djorđević et al. 2013), antirheumatic, antispasmodic and antipyretic activity (Shah and Shah 2015). In terms of anticancer effect, *Teucrium montanum* extracts have shown a significant level of cytotoxic activity on HeLa and K562 cancer cell lines (Stanković et al. 2015).

The up to date research showed that *Teucrium montanum* from serpentine localities show high level of biological activity reflected in the total content of phenols and flavonoids as well as in intense antioxidant effect. Consequently, due to the undetected presence of heavy metals, *Teucrium montanum* from serpentine habitats is safe for wide application (Zlatić et al. 2017).

Teucrium polium L.

Teucrium polium (Fig. 11.2.3) is a perennial plant with semi-bushy above ground plant parts and well developed semi-woody root. The branches are upright, moderately developed and may grow up to 40 cm. Sessile hairy leaves are oppositely arranged, elongated, slightly oval and dentate at the top. The flowers are zygomorphic and form dichasium. The whole surface of calyx is covered in white hair. Corolla leaves are mostly white in colour. Upper labia make two oval oppositely arranged lobes. The length of anthers rarely surpasses the length of corolla. The fruits are ovoid nuts (Stanković 2012a).

The species *Teucrium polium* is distributed in southern Europe, northern Africa and Asia Minor. It inhabits localities at an altitude varying from 300 m to 1000 m. It grows mostly on arid and rocky terrains, in gorges and canyons as well as in sub-Mediterranean regions and xerophile communities (Tutin and Wood 1972, Horvat et al. 1974, Diklić 1974, Ellenberg and Strutt 2009, Stanković 2012a).

Teucrium polium is used in traditional medicine in the treatment of different physiological disorders such as inflammation, diabetes, rheumatism or diseases of gastrointestinal tract. The tea of the species is used to enhance digestion, to alleviate abdominal pain, to mitigate the symptoms of the most serious type of diabetes as well as in the treatment of cold, a headache, kidney stones and excessive sweating (Afifi et al. 2009, Vahidi et al. 2010, Jaradat 2015).

Quantitative and qualitative analyses of secondary metabolites in watery and alcoholic extract have shown that flavonoids apigenin and luteolin are the most frequently found secondary metabolites in the above ground plant parts of *Teucrium polium* (Safaei and Haghi 2004). Among other important and often detected active substances are rutin, cirsiliol, diosmetin, cirsimaritin and cirsilineol (Kadifkova Panovska et al. 2005, Sharififar et al. 2009, Stefkov et al. 2011). Examination of etheric oil from above ground plant parts demonstrated that the principal evaporable components of the species *Teucrium polium* are germacrene, linalool, ρ- cimen, carvon, β-caryophyllene, α-pinene, β-pinene, carvacrol and caryophyllene oxide (Afifi et al. 2009, Moghtader 2009, Menichini et al. 2009, Bežić et al. 2011).

The previous studies have shown that the extracts of *Teucrium polium* have antioxidant, antidiabetic, antiinflammatory, antispasmodic, antipyretic, antimicrobial, antifungal and antiviral activity (Jaradat 2015). When it comes to antioxidant effects of the species, the previously conducted research has shown that *Teucrium polium* extracts may be regarded either as equally or even more efficient than Ginkgo and Green tea extracts (Stanković et al. 2012c). As for the anticancer effect of *Teucrium polium*, metabolites from the extract of the species may be used in the treatment of prostate cancer as these inhibit signaling pathway which leads to metastasis of cancer (Kanduoza et al. 2010). In treating cell cultures of glioblastoma, alveolar adenocarcinoma, breast cancer and adrenal gland with extracts of the species *Teucrium polium,* it has been established that secondary metabolites of the plant show significant cytotoxic activity in *in vitro* conditions (Eskandary et al. 2007, Nematollahi-Machani et al. 2007). Analysis of the impact of water extract of *Teucrium polium* on NSCLS cell line of lung cancer demonstrated that the extract shows intense antiproliferative activity by inhibiting cell cycle and causing death of cell (Haidara et al. 2011). Ethanol extract of the species *Teucrium polium* enhances the effect of vincristine, vinblastine and doxorubicin which are widely used in cancer treatment (Rajabalijan 2008).

Teucrium arduini L.

Teucrium arduini (Fig. 11.2.4) is a perennial plant with moderately branched hairy and woody older shoots which may grow up to 50 cm in height. The leaves are oval, petiolated, with well developed leaf nervature, serrated margins and extended base. Flowers are densely arranged into clusters that look like raceme inflorescence. They open up from the basis up to the top of spike. Calyx is differentiated into upper and lower labia. Hairy corolla is mostly white in colour and its size surpasses the size of calyx. Stamens grow beyond corolla and have reddish anthers. The fruit is ovoid nut (Stanković 2012a).

Teucrium arduini is an endemic species native to the Balkan Peninsula, that is, it grows on the territory encompassing the belt from northwestern Albania to northwestern Croatia (Kremer et al. 2011).

Teucrium arduini grows on calcareous substrate. It inhabits arid, termophile rocky habitats located at an altitude varying between 50 and 1500 m. Calcareous rocks, cliffs, canyons and gorges of mountain regions are also well known types of habitat of the species (Tutin and Wood 1972, Horvat et al. 1974, Diklić 1974, Ellenberg and Strutt 2009, Stanković 2012a).

There is scarcity of data on quantitative and qualitative content of secondary metabolites of the species *Teucrium arduini*. Examination of the quantity and content of evaporable components in the above-ground plant parts of the species *Teucrium arduini* by means of GC/FID and GC/MS tests demonstrated that predominant components of etheric oil are β-caryophyllene, caryophyllene oxide and germacrene D. As for flavonoids, the presence of cirsimarin, characteristic of this species of the genus *Teucrium*, has been confirmed (Vuković et al. 2011, Dunkić et al. 2011).

The extracts of the species *Teucrium arduini* show high level of antimicrobial, antiviral and antioxidant activity (Šamec et al. 2010, Stanković et al. 2012b, Kremer et al. 2013). The most prominent proapoptotic effect of *Teucrium arduini* on HCT-116 cells was observed after a three-day long exposure (Stanković et al. 2011a).

Teucrium botrys L.

Teucrium botrys (Fig. 11.2.5) is an annual, rarely biannual herbaceous plant. The semi-woody root is well branched. The hairy shoots are upright, moderately branched, greenish or red in colour and may grow up to 40 cm in height. The oppositely arranged, hairy leaves are cut into 5 to 7 lobes with well developed stalk. Zygomorphic flowers are vertically set in comparison with the stem. They form dense clusters shorter than bracts. Tube-like, hairy calyx has dorsal protuberance of tubular part. Corolla is purple with dark streaks. Its external surface is hairy. Stamens and style of gynoecium are longer than corolla. Dark nut fruits are circular or ovoid in shape (Stanković 2012a).

Teucrium botrys is distributed in Middle Europe and in northwestern Africa. It grows on calcareous substrate and inhabits open, arid and rocky areas, formerly cultivated soil, rocky pastures, arid places beside roads and generally, localities at up to 1.500 m of altitude (Tutin and Wood 1972, Horvat et al. 1974, Diklić 1974, Ellenberg and Strutt 2009, Stanković 2012a).

This species, as all others which belong to the genus *Teucrium*, has medicinal properties. It is used to strengthen the immune system, to relieve symptoms of cold, in the treatment of kidney diseases and to enhance digestion (Stanković 2012a).

There is little data on the results of research on quantitative and qualitative characteristics of secondary metabolites of the species *Teucrium botrys*. Analyses of chemical composition of essential oil obtained from the above ground plant parts of the species have shown the presence of significant concentrations of neo-clerodane diterpenoids 19-deacetylteuscorodol and teubotrin as well as of diterpenes such as teucvidin, montanine D, teuchamedryn C and 6β-hydroxyteusoordin. 19-deacetylteuscorodol and teubotrin are significant for the species *Teucrium botrys* in chemotaxonomic terms (De La Torre et al. 1986).

Biological activity of extracts of the species *Teucrium botrys* is reflected in its antioxidant, antimicrobial and anticancer activity (Stanković et al. 2012b, Fajfarić 2013). With regard to antioxidant efficiency, *Teucrium botrys* extracts are proved to be either equally or even more efficient than Ginkgo and Green tea extracts (Stanković et al. 2014). When it comes to anticancer effects, *Teucrium botrys* extracts have shown a significant level of cytotoxic activity on HeLa and K562 cancer cell lines (Stanković et al. 2015).

Teucrium scordium L. subsp. *scordium*

Teucrium scordium subsp. *scordium* (Fig. 11.2.6) is a perennial herbaceous plant with stolons. The stem is upright, well branched or unbranched and it may reach 50 cm of height. It is purple and completely covered in white hair. The hairy, serrated leaves are densely and oppositely arranged on the stem. The flowers are arranged in clusters and grow on short petioles. The corolla is covered in rose hairs. Stamens grow beyond corolla and have reddish anthers. Fruits are dark nuts (Stanković 2012a).

The subspecies *Teucrium scordium* subsp. *scordium* can be found in most of Europe and in western Asia. *Teucrium scordium* subsp. *scordium* grows in forest communities, in mesophilous and flood meadows, on the edges of pools and swamps and on the coasts of rivers and channels (Tutin and Wood 1972, Horvat et al. 1974, Diklić 1974, Ellenberg and Strutt 2009, Lakušić et al. 2010, Stanković 2012a).

Teucrium scordium subsp. *scordium* is used in the treatment of infectious skin diseases, increased temperature, fever, breathing difficulties, poisoning, cough and intestinal parasites. The species is regarded for stimulating functioning of digestive tract organs. At the time of Hippocrates, the species was known as a highly beneficial medicine applied in the treatment of necrotic wounds and symptoms of plague (Petrović 1883, Redžić 2007).

Analysis of quantitative and qualitative content of etheric oil obtained from the above ground plant parts of *Teucrium scordium* has shown that principal evaporable components are β-caryophyllene, caryophyllene oxide, (E)-β-farnesene,

1,8-cineole and β-eudesmol (Morteza-Semani 2007). Quantitative and qualitative chromatographic examination of phenolic compounds has demonstrated that luteolin, apigenin, diosmetin, luteolin-7-O-glucoside, luteolin-7-O-rutinoside and diosmetin-7-O-glucoside are the principal secondary metabolites from group of flavonoids (Jakupović 1985, Kundaković et al. 2011).

The extracts of the subspecies *Teucrium scordium* subsp. *scordium* have antioxidant (Stanković et al. 2012b), antifungal, antiproliferative as well as cytotoxic effects against both ER+ and ER− cell lines, MDA-MB-361 and MDA-MB-453 (Kundaković et al. 2011).

Teucrium scordium L. subsp. *scordioides* (Schreber) Meire & Petimg.

Teucrium scordium subsp. *scordioides* (Fig. 11.2.7) is a perennial species with stolons. The branched root is well developed. The stem is upright and may reach the height of 50 cm. It is moderately branched, has side shoots and is entirely covered in thick hair. The leaves are ovoid, hairy, elongated, serrated and sessile. Flowers have stalk and are located at the bottom of leaves. Hairy calyx is wider in lower part. Corolla is twice as long as calyx and is of reddish colour with dark blue streaks. Stamens are longer than calyx and its anthers are of red colour. The fruits are ovoid nuts (Stanković 2012a).

Subspecies *Teucrium scordium* subsp. *scordioides* is distributed in southern and southeastern Europe, Asia Minor and northern parts of Africa. Wet soils, flood-meadows and swamps are found favourable for the growth and development of the species (Tutin and Wood 1972, Horvat et al. 1974, Diklić 1974, Ellenberg and Strutt 2009).

Using HPLC analysis of plant extracts of the species, the presence of both flavonoid aglycones such as luteolin, diosmetin, apigenin and glycosides of apigenin and luteolin (apigenin- and luteolin-7-O-glucoside, luteolin-7-Orutinoside and luteolin-7-sambubioside) was detected (Kundaković et al. 2011).

Extracts of the subspecies *Teucrium scordium* L. subsp. *scordioides* have various effects which point to its biological activity. Among the observed effects are antioxidant, anticancer and proapoptotic activity (Stanković et al. 2011a, 2012b). The detected anticancer activity is reflected in cytotoxic impact on HeLa and K562 cancer cell lines (Stanković et al. 2015). Antibacterial activity has been detected on *Pseudomonas aeruginosa, Klebsiella pneumoniae, Escherichia coli* and *Bacillus subtilis* (Kundaković et al. 2011).

Teucrium flavum L.

Teucrium flavum L. (Fig. 11.2.8) is a perennial, evergreen, bushy species with shoots reaching up to 60 cm. The stem is upright, hairy and branchy, with ovoid, mildly dentate leaves and small flowers of pale yellow colour. The species mostly inhabits xerophyll, calcareous and rocky areas located at up to 1000 m of altitude. It can be found in communities of maquis and garrigue in the Mediterranean (Menichini et al. 2009, Stanković and Zlatić 2017).

The majority of other species of the genus *Teucrium*, *Teucrium flavum* L. is used in traditional medicine as infusion made of above ground plant parts (Menichini et al. 2009). Infusion, obtained of flowers of this species, is traditionally used as antipyretic and antiseptic. Decoction of leaves is applied directly to the skin as a cicatrizant (Acquaviva et al. 2017). The species is widely used in the treatment of diabetes and for cleaning wounds (Menichini et al. 2009). As for the chemical composition of *Teucrium flavum*, dominant substances are diterpenoids such as fruticolone, isofruticolone, 8β-hydroxyfruticolone, 7β-hydroxyfruticolone, 11-hydroxyfruticolone, deacetylfruticolone, and 10-hydroxy-6-acetylteucjaponin B (Piozzi et al. 2005). Among detected evaporable compounds of this species are caryophyllene, 4-vinyl guaiacol, caryophyllene oxide and α-humulene (Menichini et al. 2009). Extracts of the species show antiinflammatory, anticancer (Menichini et al. 2009), antioxidant and antibacterial activity (Acquaviva et al. 2017).

Teucrium fruticans L.

Teucrium fruticans L. (Fig. 11.2.9) is a perennial, evergreen, bushy species with shoots reaching up to 2 m. The above ground plant part is branchy. The oval leaves of grey-green colour on adaxial side and grey-white on abaxial side have a thick indumentum. Blue flowers are quite small. The species grows in termophile calcareous habitats in western Mediterranean region. In other parts of the Mediterranean, the species is successfully cultivated (Frabetti et al. 2009, Stanković and Zlatić 2017).

Teucrium fruticans L. is appreciated to a great extent in central parts of Italy due to its pharmacological properties. The leaf of the species is used to prepare nfusion with diuretic and depurative effects. The extracts of this species contain neo-clerodane diterpenes germacrene D and β-caryophyllene. High amounts of β-pinene and β-myrcene as well as flavonoid compounds cirsilineol and cirsimaritin which show high level of biological activity were also detected (Kisiel et al. 2001, Frabetti et al. 2009, Acquaviva et al. 2017). The extracts of the species from Spain have antimicrobial and antioxidant effect and may be used in food, pharmaceutical and cosmetic industry (Stanković et al. 2017).

Conclusions

Last decades have brought new tendencies in the treatment of numerous disorders and diseases. The supremacy of pharmacological industry has been challenged by traditional medicine. Plant species have significant place in contemporary folk medicine. This chapter is a review of morphological, chemical and medicinal properties, of geographical distribution as well as of ethnobotanical and commercial application of 9 selected species of the genus *Teucrium*. The studies conducted so far have proved and confirmed the presence of metabolites and other beneficial substances of therapeutical importance. Various products made of the species such as teas, tinctures, rakija and so on have important position in contemporary medicine as well as in pharmaceutical and food industry. However, cautious and moderate usage of the species and its products is recommended as greater quantities of its active substances may have hepatotoxic effect.

References

Acquaviva, R., Genovese, C., Amodeo, A., Tomasello, B., Malfa, G., Sorrenti, V., Tempera, G., Addamo, A.P., Ragusa, S., Rosa, T., Menichini, F. and Giacomo, C.D. 2017. Biological activities of *Teucrium flavum* L., *Teucrium fruticans* L., and *Teucrium siculum* rafin crude extracts. Plant Biosyst. http://dx.doi.org/10.1080/11263504.2017.1330773.

Afifi, F.U., Abu-Irmaileh, B.E. and Al-Noubani, R.A. 2009. Comparative analysis of the esential oils of *Teucrium polium* L. grown in different arid & semi-arid habitats in Jordan. Jordan J. Pharm. Sci. 2: 42–52.

Avula, R., Manyam, R.B., Bedir, E. and Khan, I.A. 2003. Rapid separation and determination of four phenylpropanoid glycosides from *T. chamaedrys* by capillary electrophoresis method. Chromatographia 8: 751–755.

Barceloux, D.G. 2012. Medicinal toxicology of natural substances. Foods, fungi, Medicinal herbs, plants and venomous animal. John Wiley & Sons, Hoboken.

Bedir, E., Manyam, R. and Khan, I.A. 2003. Neo-clerodane diterpenoids and phenylethanoid glycosides from *Teucrium chamaedrys* L. Phytochemistry 63: 977–983.

Bezić, N., Vuko, E., Dunkić, V., Ruščić, M., Blažević, I. and Burčul, F. 2011. Antiphytoviral activity of sesquiterpene-rich essential oils from four Croatian *Teucrium* species. Molecules 16: 8119–8129.

Bosisio, E., Giavarini, F., Dell'Agli, M., Galli, G. and Galli, C.L. 2004. Analysis by high-performance liquid chromatography of teucrin A in beverages flavoured with an extract of *Teucrium chamaedrys* L. Food Addit. Contam. 21: 407–414.

Bruno, M., Piozzi, F., Rosselli, S., Maggio, A., Alania, M., Lamara, K., Al-Hillo, Y.R.M. and Servettaz, O. 2004. Phytochemical investigation of four *Teucrium* species. Rev. Soc. Quím. Mex. 48: 137–138.

Čanadanović-Brunet, J., Djilas, M.S., Ćetković, S.G., Tumbas, T.V., Mandić, I.A. and Čanadanović, M.V. 2006. Antioxidant activities of different *Teucrium montanum* L. extracts. Int. J. Food Sci. Tech. 41: 667–673.

De La Torre, C.M., Fernández-Gadea, F., Michavila, A., Rodríguez, B., Piozzia, F. and Savonaa, G. 1986. Neo-clerodane diterpenoids from *Teucrium botrys*. Phytochemistry 25: 2385–2387.

Diklić, N. 1974. *Teucrium* L. pp. 349–356. *In*: Josifović, M. (ed.). Flora of Federative Republic of Serbia, Belgrade.

Dunkić, V., Bezić, N. and Vuko, E. 2011. Antiphytoviral activity of essential oil from endemic species *Teucrium arduini*. Nat. Prod. Commun. 9: 1385–1388.

Ellenberg, H. and Strutt, K.G. 2009. Vegetation Ecology of Central Europe, 4th ed., Cambridge University Press.

Eskandary, H., Rajabalian, S., Yazdi, T., Eskandari, M., Fatehi, K. and Ganjooei, A.N. 2007. Evaluation of cytotoxic effect of *Teucrium polium* on a new glioblastoma multiforme cell line (REYF-1) using MTT and soft agar clonogenic assays. Int. J. Pharmacol. 3: 435–437.

Fajfarić, M. 2013. Antioxidant activity of the plant species *Teucrium botrys* L. Dissertation. Faculty of Pharmacy and Biochemisty. University of Zagreb. Croatia.

Fiorentino, A., D'Abrosca, B., Esposito, A., Izzo, A., Pascarella, M.T., D'Angelo, G. and Monaco, P. 2009. Potential allelopathic effect of neo-clerodane diterpenes from *Teucrium chamaedrys* (L.) on stenomediterranean and weed cosmopolitan speciec. Biochem. Syst. Ecol. 37: 349–353.

Frabetti, M., Gutiérrez-Pesce, P., Mendoza, D.E., Gyves, E. and Rudini, E. 2009. Micropropagation of *Teucrium fruticans* L., an ornamental and medicinal plant. *In vitro* Cell. Dev. Biol. 45: 129–134.

Gafner, S., Bergeron, C., Batcha, L.L. and Angerhofer, C.K. 2003. Analysis of *Scutellaria lateriflora* and its adulterants *Teucrium canadense* and *Teucrium chamaedrys* by LC-UV/MS, TLC, and digital photomicroscopy. J. AOAC Int. 86: 453–460.

Gostuški, R. 1979. Treatment with medicinal plants, Narodna knjiga, Belgrade.

Haïdara, K., Alachkar, A. and Moustafa, A.A. 2011. *Teucrium polium* plant extract provokes significant cell death in human lung cancer cells. Health 3: 366–369.

Harborne, B.J., Tomás-Barberán, A.F., Williams, A.C. and Gil, I.M. 1986. A chemotaxonomic study of flavonoids from European *Teucrium* species. Phytochemistry 25: 2811–2816.

Hollis, S. and Brummitt, R.K. 1992. World Geographical Scheme for Recording Plant Distributions. Hunt Institute for Botanical Documentation Carnegie Mellon University, Pittsburgh. Available at https://www.kew.org/tdwg-world-geographical-scheme-recording-plant-distributions. Accessed June 11, 2016.

Horvat, I., Glavač, V. and Ellenberg, H. 1974. Vegetation Südosteuropas, Fischer, VEB, p. 768.

Hosseinzadeh, S., Jafarikukhdan, A., Hosseini, A. and Armand, R. 2015. The application of medicinal plants in traditional and modern medicine: a review of *Thymus vulgaris*. Int. J. Clin. Med. 6: 635–642.

Jakupovic, J., Baruah, R.N., Bohlmann, F. and Quack, W. 1985. New clerodane derivatives from *Teucrium scordium*. Planta Med. 51: 341–342.

Jaradat, A.N. 2015. Review of the taxonomy, ethnobotany, phytochemistry, phytotherapy and phytotoxicity of germander plant (*Teucrium polium* L.). Asian J. Pharm. Clin. Res. 8: 13–19.

Kadifkova-Panovska, T., Kulevanova, S. and Stefova, M. 2005. *In vitro* antioxidant activity of some *Teucrium* species (Lamiaceae). Acta Pharmacol. 55: 207–214.

Kandouza, M., Alachkarb, A., Zhangc, L., Dekhila, H., Chehnab, F., Yasmeena, A. and Al Ala-Edin Moustafa. 2010. *Teucrium polium* plant extract inhibits cell invasion and motility of human prostate cancer cells via the restoration of the e-cadherin/catenin complex. J. Ethnopharmacol. 129: 410–415.

Keršek, E. 2006. Medicinal plants in wine and rakija. V.B.Z., d.o.o., Zagreb, Croatia.

Khleifat, K., Shakhanbeh, J. and Tarawneh, K. 2002. The chronic effects of *Teucrium polium* on some blood parameters and histopathology of liver and kidney in the rat. Turk. J. Biol. 26: 65–71.

Kisiel, W., Stojakowska, A., Piozzi, F. and Rosselli, S. 2001. Flavonoids from *Teucrium fruticans* L. Acta Soc. Bot. Pol. 70: 199–201.

Kojić, M., Stamenković, V. and Jovanović, D. 1998. Medicinal plants in southeastern Serbia. Zavod za udžbenike i nastavna sredstva. Belgrade.

Kremer, D., Randić, M., Koselec, I., Brkljačić, A., Lukač, G., Krušić, I., Ballian, D., Bogunić, F. and Karlović, K. 2011. New localities of the subendemic species *Berberis croatica*, *Teucrium arduini* and *Micromeria croatica* in the Dinaric Alps. Acta Bot. Croat. 70: 289–300.

Kremer, D., Kosir, I.J., Kosalec, I., Koncic, M.Z., Potocnik, T., Cerenak, A., Bezic, N., Stecec, S. and Dunkic, V. 2013. Investigation of chemical compounds, antioxidant and antimicrobial properties of *Teucrium arduini* L. (Lamiaceae). Curr. Drug Targets. 14: 1006–1014.

Kundaković, T., Milenković, M., Topić, A., Stanojković, T., Juranić, Z. and Lakušić, B. 2011. Cytotoxicity and antimicrobial activity of *Teucrium scordium* L. (Lamiaceae) extracts. Afr. J. Microbiol. Res. 5: 2692–2696.

Lakušić, B., Stevanović, B., Jančić, R. and Lakušić, D. 2010. Habitat-related adaptations in morphology and anatomy of *Teucrium* (Lamiaceae) species from the Balkan peninsula (Serbia and Montenegro). Flora. 205: 633–646.

Lin, L., Harnly, M.J. and Upton, R. 2009. Comparison of the phenolic component profiles of Skullcap (*Scutellaria lateriflora*) and Germander (*Teucrium canadense* and *T. chamaedrys*), a potentially hepatotoxic adulterant. Phytochem. Anal. 20: 298–306.

Maccioni, S., Baldini, R., Tebano, M., Cioni, P.L. and Flamini, G. 2007. Essential oil of *Teucrium scordonia* L. ssp. *scordonia* from Italy. Food Chem. 104: 1393–1395.

Menichini, F., Conforti, F., Rigano, D., Formisano, C., Piozzi, F. and Senatore, F. 2009. Phytochemical composition, anti-inflammatory and antitumour activities of four *Teucrium* essential oils from Greece. Food Chem. 115: 679–686.

Meusel, H., Jäger, E., Rauschert, S. and Weinert, E. 1978. Vergleichende Chorologie der Zentraleuropäischen Flora. VEB Gustav Fischer, Jena, Vol. 2.

Milošević-Đorđević, O., Stošić, I., Stanković, M. and Grujičić, D. 2013. Comparative study of genotoxicity and antimutagenicity of methanolic extracts from *Teucrium chamaedrys* and *Teucrium montanum* in human lymphocytes using micronucelus assay. Cytotechnology 65: 863–869.

Monsef-Esfahani, R.H., Miri, A., Amini, M., Amanzadeh, Y., Hadjiakhoondi, A., Hajiaghaee, R. and Ajani, Y. 2010. Seasonal variation in the chemical composition, antioxidant activity and total phenolic content of *Teucrium persicum* Boiss. essential oils. Res. J. Biol. Sci. 5: 492–498.

Moghtader, M. 2009. Chemical composition of the essential oil of *Teucrium polium* L. from Iran. Am. Eurasian J. Agric. Environ. Sci. 5: 843–846.

Morteza-Semnani, K., Saeedi, M. and Akbarzadeh, M. 2007. Essential oil composition of *Teucrium scordium* L. Acta Pharmacol. 57: 499–504.

Nematollahi-Mahani, S.N., Rezazadeh-Kermani, M., Mehrabani, M. and Nakhaee, N. 2007. Cytotoxic Effects of *Teucrium polium* on some established cell lines. Pharm. Biol. 45: 295–298.

Petrović, S. 1883. Medicinal plants of Serbia, Serbian archive for medicine, Belgrade.

Petrovska, B.B. 2012. Historical review of medicinal plants usage. Pharmacogn. Rev. 6: 1–5.

Piozzi, F., Bruno, M., Rosselli, S. and Maggio, A. 2005. Advances on the chemistry of furano-diterpenoids from *Teucrium* genus. Heterocycles 65: 1221–1234.

Pourmotabbed, A., Farshchi, A., Ghiasi, G. and Khatabi, M.P. 2010. Analgesic and anti-inflammatory activity of *Teucrium chamaedrys* leaves aqueous extract in male rats. Iran. J. Basic Med. Sci. 13: 119–125.

Rajabalian, S. 2008. Methanolic extract of *Teucrium polium* L. potentiates the cytotoxic and apoptotic effects of anticancer drugs of vincristine, vinblastine and doxorubicin against a panel of cancerous cell lines. Exp. Oncol. 30: 133–138.

Redžić, S.S. 2007. The ecological aspect of ethnobotany and ethnopharmacology of population in Bosnia and Herzegovina. Coll. Antropol. 31: 869–890.

Safaei, A. and Haghi, G. 2004. Identification and quantitative determination of flavonoids in the aerial parts of *Teucrium polium* by HPLC. Int. J. Pharm. Res. 3: 90.

Sarić, R.M. 1988. Medicinal plants of Federative Republic of Serbia, Serbian Academy of Sciences and Arts, Belgrade, pp. 640.

Savvidou, S., Goulis, J., Giavazis, I., Patsiaoura, K., Hytiroglou, P. and Arvanitakis, C. 2007. Herb-induced hepatitis by *Teucrium polium* L.: report of two cases and review of the literature. Eur. J. Gastroenterol. Hepatol. 19: 507–511.

Šamec, D., Gruzb, J., Strnadb, M., Kremerc, D., Kosalecc, I., Grubešić, J.R., Karlović, K., Lucica, A. and Piljac-Žegaraca, J. 2010. Antioxidant and antimicrobial properties of *Teucrium arduini* L. (Lamiaceae) flower and leaf infusions (*Teucrium arduini* L. antioxidant capacity). Food Chem. Toxicol. 48: 113–119.

Sharififar, F., Dehghn-Nudeh, G. and Mirtajaldini, M. 2009. Major flavonoids with antioxidant activity from *Teucrium polium* L. Food Chem. 112: 885–888.

Shah, M.M.S. and Shah, H.M.S. 2015. Phytochemical, antioxidant, antinociceptive and anti-inflammatory potential of the aqueous extracts of *Teucrium stockisianum* bioss. BCM Complement. Altern. Med. 15: 351.

Stanković, M.S., Topuzović, M., Solujić, S. and Mihailović, V. 2010. Antioxidant activity and concentration of phenols and flavonoids in the whole plant and plant parts of *Teucrium chamaedrys* L. var. *glanduliferum* Haussk. J. Med. Plants Res. 4: 2092–2098.

Stanković, M.S., Ćurčić, M.G., Žižić, J.B., Topuzović, M.D., Solujić, S.R. and Marković, S.D. 2011a. *Teucrium* plant species as natural sources of novel anticancer compounds: Antiproliferative, Proapoptotic and Antioxidant Properties. Int. J. Mol. Sci. 12: 4190–4205.

Stanković, M.S., Nićiforović, N., Topuzović, M. and Solujić, S. 2011b. Total phenolic content, flavonoid concentrations and antioxidant activity, of the whole plant and plant parts extracts from *Teucrium montanum* L. var. *montanum*, f. *supinum* (L.) Reichenb. Biotechnol. Biotechnol. Equip. 25: 2222–2227.

Stanković, M.S. 2012a. Biological Effects of Secondary Metabolites of the Species of the genus *Teucrium* L. of Serbian flora. Dissertation. Faculty of Science. University of Kragujevac. Serbia.

Stanković, M.S., Stefanović, O., Čomić, Lj., Topuzović, M., Radojević, I. and Solujić, S. 2012b. Antimicrobial activity, total phenolic content and flavonoid concentrations of *Teucrium* species. Cent. Eur. J. Biol. 7: 664–671.

Stanković, M.S., Nicifirović, N., Mihajlović, V., Topuzović, M. and Solujić, S. 2012c. Antioxidant activity, total phenolic content and flavonoid concentrations of different plant parts of *Teucrium polium* L. subsp. *polium*. Acta Soc. Bot. Pol. 81: 117–122.

Stanković, M.S., Jakovljević, D., Topuzović, M. and Zlatković, B. 2014. Antioxidant activity and content of phenolics and flavonoids in the whole plant and plant parts of *Teucrium botrys* L. Oxid. Commun. 37: 522–532.

Stanković, M.S., Mitrović, T.Lj., Matić, I.Z., Topuzović, M.D. and Stamenković, S.M. 2015. New values of *Teucrium* species: *in vitro* study of cytotoxic activities of secondary metabolites. Not. Bot. Horti. Agrobo. 43: 41–46.

Stanković, M.S., Stojanović-Radić, Z., Blanco-Salas, J., Vázquez-Pardo, F.M. and Ruiz-Téllez, T. 2017. Screening of selected species from Spanish flora as a source of bioactive substances. Ind. Crops Prod. 95: 493–501.

Stanković, M. and Zlatić, N. 2017. The total quantity of phenolic compounds and antioxidant activity of the selected specie of the genus *Teucrium*. XXII Counselling on Biotechnology, Faculty of Agronomy Čačak, University of Kragujevac, Serbia, pp. 515–520.

Stefkov, G., Kulevanova, S., Miova, B., Dinevska-Kjovkarovska, S., Møjgaard, P., Jäger, A.K. and Josefsen, K. 2011. Effects of *Teucrium polium* spp. *capitatum* flavonoids on the lipid and carbohydrate metabolism in rats. Pharm. Biol. 49: 885–892.

Sundaresan, P.R., Slavoff, S.A., Grundel, E., White, K.D., Mazzola, E., Koblenz, D. and Rader, L. 2006. Isolation and characterisation of selected germander diterpenoids from authenticated *Teucrium chamaedrys* and *T. canadense* by *HPLC, HPLC-MS* and *NMR*. Phytochem. Anal. 17: 243–250.

Tagarelli, G., Tagarelli, A. and Piro, A. 2010. Folk medicine used to heal malaria in Calabria (Southern Italy). J. Ethnobiol. Ethnomed. 6: 27.

Tumbas, T.V., Mandić, I.A., Ćetković, S.G., Djilas, M.S. and Čanadanović-Brunet, M.J. 2004. *HPLC* analysis of phenolic acids in Mountain Germander (*Teucrium montanum* L.) extracts. Acta Periodica Technologica 35: 265–273.

Tutin, T.G. and Wood, D. 1972. *Teucrium*. pp. 129–135. *In*: Tutin, T.G., Heywood, V.H., Burges, N.A., Moore, D., Valentine, D. and Walters, S. (eds.). Flora Europaea, 1st Ed. Cambridge University Press, Cambridge, England.

Vahidi, A.R., Dashti-Rahmatabadi, M.H. and Bagheri, S.M. 2010. The effects of *Tecrium polium* boiled extract in diabetic rats. Int. J. Diabetes Obes. 2: 27–31.

Vlase, L., Benedec, D., Hanganu, D., Damian, G., Csillaq, I., Sevastre, B., Mot, A.C., Silaghi-Dumitrescu, R. and Tilea, I. 2014. Evaluation of antioxidant and antimicrobial activities and phenolic profile for *Hyssopus officinalis*, *Ocimum basilicum* and *Teucrium chamaedrys*. Molecules 19: 5490–5507.

Vuković, N., Milošević, T., Sukdolak, S. and Solujić, S. 2008. The chemical composition of the essential oil and the antibacterial activities of the essential oil and methanol extract of *Teucrium montanum*. J. Serb. Chem. Soc. 73: 299–305.

Vukovic, N., Sukdolak, S., Solujic, S., Mihailovic, V., Mladenovic, M., Stojanovic, J. and Stankovic, M.S. 2011. Chemical composition and antimicrobial activity of *Teucrium arduini* essential oil and cirsimarin from Montenegro. J. Med. Plants Res. 5: 1244–1250.

Zlatić, N.M., Stanković, M.S. and Simić, Z.S. 2017. Secondary metabolites and metal content dynamics in *Teucrium montanum* L. and *Teucrium chamaedrys* L. from habitats with serpentine and calcareous substrate. Environ. Monit. Assess. 189: 110.

CHAPTER 12

Ethnobotanical Issues on Medicinal Plants from Paraguay

Rosa Luisa Degen de Arrúa,[1,]* *Yenny González*[1] and
Esteban A. Ferro B.[2]

Introduction

Paraguay is a land-locked country located in the sub-tropical region of South America. It has a area of 406,760 km², distributed in two big regions separated by Paraguay river. The country does not have any relevant heights, but it is gifted with a rich flora with about 5,500 to 6,000 vascular species. Among those, a considerable amount has medicinal uses in the country (Mereles 2014). This use of plants can be traced back to the oldest cultural traditions, and medicinal plants were established as the base of traditional medicine (Bertoni 1927), practices that are still present today. The recognition of these plants' value comprises their use for health preservation and treatment, becoming valuable sources of raw material for the extraction of bioactive molecules to be employed alone or following either chemical or microbial processing, and even more, providing lead structures for the development of new molecules directed to treat a variety of diseases (Ibarrola and Degen 2011).

The plants employed for medicinal purposes receive several names, such as diet supplements, traditional medicine or phytotherapeutic remedies, and the WHO supports the implementation of these therapeutic resources among the state members. The utilization of medicinal plants evokes different reactions among

---
[1] National University of Asunción, Faculty of Chemical Sciences, Research Department, Department of Botany, Paraguay.
[2] National University of Asunción, Faculty of Chemical Sciences, Research Department, Department of Phytochemistry, Paraguay.
* Corresponding author: degenrosa@gmail.com, rdegen@qui.una.py

different societies, going from the enthusiasm without any criticism to the extreme lack of credibility. Yet, their use as alternative and complement treatments is quickly increasing in developed countries; however, in developing countries, these resources remain widely applied, strongly enough to be the first therapeutic resource in several countries. This is a frequent practice in Latin American countries, including Paraguay. Some countries of the region have established legal frameworks and standardized criteria in order to fulfill safety, effectiveness and quality requirements for herbal remedies, in order to make reliable products available for the population, reducing these risks associated with the consumption of plant-derived medicines (Ibarrola and Degen 2011). It is considered as a priority for the country to have a national policy for medicinal plants and phytotherapeutic resources. Such policy would help to guarantee the safe and rational use of those products as valuable resources for primary health care by rural and urban populations (Degen et al. 2009).

The aim of this chapter is to review the background of the use of medicinal plants in Paraguay, their conservation and trading conditions, the advance in research efforts and their future perpectives.

Ethnobotanical Backgrounds and Studies on Medicinal Plants

The current application of plants in the medicinal practice-either preventive, therapeutic or palliative—comes from a huge mixture of knowledge, beliefs and plant-related practices of the native population with those introduced by people from different places, even continents (Vera 2011). In fact, the Guarani people—to name just one of the native ethnicities that inhabited this land when European conquerors arrived in Paraguay—had a deep knowledge about the flora and they made a smart use of it; they believed that each plant had a therapeutic property, and they also knew about the affinity and antagonism among them (González Torres 1992).

After the discovery of America, their natural resources—flora, wildlife and minerals—and their uses and applications were quickly recognized. Spaniards showed great interest to know and apply the native medical wisdom, and in 1579 they decided to send naturalists and physicians in order to study the plants for medical purposes, mainly to collect and deliver to Spain those herbs highly renowned as medicinal plants. One of the pioneers in this purpose was the Jesuit priest Father Ventura, or Buenaventura Suárez, who wrote in 1710 the book *Missionary Medical Matters*, a valuable summary of the medicinal plants from Brazil and the Missions area, which was lately attributed to Brother Pedro Montenegro, in 1888. In his work, he described plants as *yerba mate* (Fig. 12.1), *yvyrá ysy*, *caña fístula*, *contrayerba* and *santa lucía*, among many others. By the middle of 1700, Sigismund Asperger wrote an original work describing the plants and medicinal herbs from Paraguay; he is considered the author of several botanical and medical books containing the recipe for the "elixir or balm of the Missions" employing "aguará yvá", a species from *Schinus* genera; such preparation was traditionally employed for several physical ailments, mainly as a depurative and to treat wounds and boils. Father José Sánchez Labrador also wrote several books describing many plants and their

Fig. 12.1. *Ilex paraguariensis* var. *paraguariensis*, "yerba mate", medicinal plant.
(Photo: Germán González, Department of Botany).

uses. By 1877, Domingo Parodi prepared a list of plants using both Spanish and Guarani names, specifying botanical families and names, and their medical use. More recently, Moses Bertoni a naturalist from Switzerland, studied the Paraguayan flora. He settled in Paraguay by the end of 19th century and described the most used plants in Paraguay and the neighboring countries, in alphabetical order. Moreover, he recovered information from the aboriginals, and by the end of 1899 discovered the species recognized as 'sweet herb' or *ka'á he'ẽ* (Fig. 12.2), *Stevia rebaudiana* (González Torres 1992, MAG-DGPE 2008, Scavone 2010). Between centuries 19th and 20th, two botanists made great collections of the flora from Paraguay. They were the Swiss Dr. Emil Hassler, and the Paraguayan Teodoro Rojas, considered as the first botanist from Paraguay, since the previous botanists came from other

Fig. 12.2. *Stevia rabaudiana*, "ka'á he'ẽ", medicinal plant.
(Photo: Yenny González, Department of Botany).

countries. Teodoro Rojas worked hardly under the influence of great botanists, like the previously mentioned E. Hassler and M. Bertoni, among others. Rojas became an expert technician in systematic botany, collecting thousands of species, describing new ones and publishing several works (Basualdo 1993). His collections, performed in Paraguay, Bolivia and Argentina reached 12,837 records, during his 50 years of activity as a botanist collecting about 40,000 samples, counting both original samples and duplicates (Schinini 2005).

More recently, Prof. Dionisio González Torres published in 1970 the first edition of his book entitled "Catálogo de plantas medicinales y alimenticias y útiles usadas en el Paraguay" (Catalogue of medicinal, edible and useful plants employed in Paraguay). In this book, he presented 1,500 popular names of plants, alphabetically ordered, their scientific names, properties and attributes, and preparation procedures. Most of the named species were autochthonous, but exotic species were also included, following a vast review of the bibliography, with information collected by naturalists from centuries 16th to 19th, as a result of interviews performed in different areas of the country along several years. This book was considered as the only one with enough records on medicinal plants from Paraguay for a long time, and even at the present time it is considered as a valuable starting point to perform research and academic works.

Another outstanding researcher of Paraguayan natural environment is Dr. Carlos Gatti Battilana, who expanded his interests in social problems, focusing his attention on both Guarani civilization and the Paraguayan population of the 20th century. Such interest resulted in the publication of a Spanish–Guarani dictionary for medical use, where is recorded an important content of the medical Guarani knowledge, obtained from his patients (Benítez 1986). At this point, we must name pharmacist Rómulo Feliciángeli, who prepared in 1939 the project of the Paraguayan Pharmacopoeia. This work included an addendum, containing 33 plant species used at that time for medicinal purposes (Feliciángeli 1939).

In 1978, Prof. Isabel Basualdo and her all female research team from the Facultad de Química y Farmacia (Chemistry and Pharmacy School), today Facultad de Ciencias Químicas, FCQ (Chemical Sciences School), started their research works for the recovery of the popular knowledge on the use of medicinal plants in Paraguay, backed by the botanical identification of the species and the preservation of voucher specimens. This activity made possible to set up the Herbarium of the Chemical Sciences School of the Universidad Nacional de Asunción, which is internationally recognized as FCQ for its acronym in Spanish.

Between 1982 and 1985, the United Nations Development Program supported the training of human resources and provided basic equipment required to start chemical studies on medicinal plants from Paraguay. As a result of this project, chemical studies on species from Asteraceae and Celastraceae families were executed. Following these efforts, a joint project between the FCQ and the Medical and Pharmaceutical University of Toyama, Japan—with the support of the Japan International Cooperation Agency (JICA)—was executed with two main objectives: to record with documentary backing the medicinal flora from Paraguay, and to evaluate its biological activity in a bioassay panel. These works allowed to publish in 1987 a document summarizing both chemical and pharmacological findings from herbs collected in Paraguay. A few years later, following the same research line, a list of medicinal plants commercialized in popular markets in Asunción and its metropolitan area was published (Basualdo et al. 2004), and more recently the *Illustrated catalogue with 80 medicinal plants from Paraguay* was also published. This work was awarded with the National Science Award, granted by the National Congress of Paraguay to the researchers from FCQ in 2012 (Degen and González 2014b).

Several researchers have published lists of medicinal plants that include their popular and scientific names, and also their descriptions, applications, preparation mode and habitat, all backed by herbarium reference material. These names are usually related to some location, village or native community, as observed in the ethnobotanical works published by Pastor Arenas and Ricardo Moreno Azorero (1976, 1977). They started their research with a project on the genetic characteristics of indigenous populations, focusing initially on the Maká community characterized by its high level of isolation and the preservation of their language and manners. Later, they followed their studies in other native communities such as Nivakle, Ayoreo and others, getting relevant results. In Maká community, they found high inbreeding, high mortality and low fertility rates, which moved them to study how the natives controlled the fertility to find medicinal plants as natural resources for birth control. A total of 22 plant species were recorded as abortive, contraceptive, sterilizing and fecundating from ten different ethnic groups and native communities from Paraguay.

After that, Mereles and Degen (1994a, 1997) published a list containing 31 trees and bushes from 16 botanical families, growing in the Boreal Chaco region and employed by native populations. Basualdo et al. (1977) described useful plants growing at Cerro Corá National Park, at Amambay Department in the Northeastern region of Paraguay. Mereles (2001) mentioned the useful plants from Tebicuary-mí and Capiibary basin of Paraguay Eastern region, just to name other studies in the subject.

Different authors have published books on medicinal plants and related issues. We can mention Soria and Basualdo (2005), who recorded plants employed in a community from Caazapá, Pin et al. (2009) who presented a list of medicinal plants cultivated at the Asunción Botanical Garden, and other authors as Vera (2009, 2011), Ibarrola and Degen (2011), Scavone (2011), who described traditional application of medicinal plants in popular environment. Polini et al. (2013) published two books containing the description of traditional use of medicinal plants by the Enxet community, located in the western region of Paraguay. The later authors performed studies that allowed rescuing the traditional knowledge about medicinal plants, its distribution, conservation status, pharmacological properties and chemical composition, among other data. Finally, it should be mentioned that the two editions of the book about medicinal plants from Latin America, edited by Gupta (1995, 2008) include plants used as medicine in Paraguay.

The Ministry of Health and Social Welfare (MSPyBS) from Paraguay started in 2008 a process directed to include the traditional medical practices as a relevant component of the primary health services (APS) in order to make health services accessible to the communities. Within this concept, Soria and Ramos (2015) published a guide of medicinal plants aimed to ease the recognition of these species by the primary health practitioners in APS units, linking the traditional knowledge with the formal health services. Following this initiative, Maidana et al. (2015) have named the plant species employed by type 2 diabetes mellitus patients receiving attention within the framework of the Diabetes National Program from the Paraguayan Ministry for Health and Wellbeing.

The Common Names of the Plants

Plants, like the animals, are usually recognized by their common names, which result from different components like their appearance or their similarity with some object, organoleptic characteristics such as color or smell, or the place where the species grow, among others (Mereles and Degen 1994b). The common names are not universally recognized in the plant science, and they are usually inaccurate (Degen 1990); however, it is mandatory to recognize them because they are employed by the people who named the plants, especially those having popular uses. Names are also closely related to their applications, the medicinal use being the most relevant. It could be exemplified by *Tabebuia aurea*, popularly named as "paratodo" (plant for everything), or plant able to cure any disease (Mereles and Degen 1994a,b, 1997).

A relevant issue about popular names is their use in local trading, which mostly are expressed in Guaraní language.

Among the 269 plant species commercialized in the markets of Asunción, several of them have the same common name despite belonging to different families, examples of this being "kambará", ascribed to *Gochnatia polymorpha*, Asteraceae, and *Buddleja madagascariensis*, Buddlejaceae (Degen et al. 2005). This situation can also involve one common name to describe more than one species belonging to the same genus, as "jaguareté ka'á" (jaguar herb), which corresponds to several species of *Baccharis* genus (Fig. 12.3) (Soria 1993), or one plant species is recognized by more than one common name that could be exemplified by the names "palo azul" (blue stick) and "matorro negro" (black bush) for *Cyclolepis genistoides* (Fig. 12.4), or "palo borracho" (drunken tree) and "samu'u" for *Chorisia insignis* (Mereles and Degen 1994b), to name just a few. González Torres (1992) mentioned 1,500 common names in both Spanish and Guarani, with their corresponding scientific names, as a countrywide data collection, but these botanical determinations should be scientifically verified. For instance, different plants -even belonging to different plant families—with toxic properties are referred in Guarani as *ka'avó pochy* (angry herb). Besides that, those epiphytic or parasite plants are named *ka'avo tyre'y*, (orphan plant), being several species of different genii of the Viscaceae family. Another case to mention corresponds to those lianas and climbing plants, which are popularly referred as *ysypo* "liana" or

Fig. 12.3. *Baccharis trimera,* "jaguareté ka'a" (jaguar herb).
(Photo: German González, Department of Botany).

Fig. 12.4. *Cyclolepis genistoides,* "palo azul" (blue stick) and "matorro negro" (black bush).
(Photo: Giuseppe Polini, Intercultural Pharmacies Project).

"climbing liana", with several examples as *ysypo he'é* "sweet liana" (*Rynchosia senna*), *ysypo hú* "black liana" (*Adenocalymma marginatum*) and *ysypo kamby* "milky liana" (*Morrhenia odorata*). Trying to identify those species cultivated as *menta* (peppermint) and *menta'i* (peperina) in Paraguay, a project was performed where two species were recognized, *Mentha arvensis* and *M. spicata*, two hybrids

M. × *piperita* y *M.* × *rotundifolia* and the variety *M. piperita* var. *citrata* (Resquín et al. 2011); however, all of them are reported *Mentha* × *piperita*.

This background allows us to affirm that common names used for the plants creates a problem that must be solved though a standardization process based on taxonomical identification.

Medicinal Plants and Their Applications

Several plant species from Paraguayan flora are useful for humans, especially those from indigenous communities and rural settlements, where plants are consumed for medicinal, food, stimulant or building purposes. Industrial processes also took advantage of plant diversity, such as the extraction of essential oils from Palo Santo and *Bulnesia sarmientoir*, etc. and the exploitation of non-volatile oil extracted from coco or *mbokajá, Acrocomia aculeata*, just to give a few examples (Mereles 2014).

In Paraguay, medicinal plants are employed to treat several medical conditions or their symptoms. Generally speaking, medicinal species are applied to prevent or to treat mostly chronic ailments, diabetes and arterial hypertension (Maidana et al. 2015) or to regulate fertility (Arenas and Moreno Azorero 1976, 1977). The most reported species for medical purposes are those employed to treat digestive problems like stomach ache, overeating, abdominal cramps or to improve digestion, and some others used for respiratory ailments, both for upper and lower respiratory tract: colds, cough, stuffy nose, bronchitis or asthma. Another popular use of medicinal plant species is like components of refreshing beverages; those plants, known as "refrescante", are employed as cold water macerates to prepare *tereré* (Fig. 12.5), the national beverage made from *Ilex paraguariensis*, for refreshing and stimulating purposes, mainly in the hot season.

Several lists of medicinal plants from Paraguay have been prepared as the result of surveys and periodical visits performed at selling stalls located at public

Fig. 12.5. Tereré, traditional drink of Paraguay.
(Photo: Germán González, Department of Botany).

markets in different cities (Fig. 12.6). The people trading in such stands, known as "yuyeros" o "yuyeras" (herb sellers), are also recognized as key informants about the use of plants. Following this methodology, Basualdo and Soria (1996a) reported 20 species of aromatic plants employed for medicinal use, including whole plants and plant organs like roots, leaves, rhizomes, aerial parts and flowering tops. In the same year, these authors published a report on 24 species employed to treat respiratory tract diseases (Basualdo and Soria 1996b). As a continuous effort in the same way, Basualdo et al. (2004) published a list containing 266 plant species regularly commercialized in popular markets from Asuncion metropolitan area. This list included plants from 82 botanical families, with 85 medical indications for different plant organs. The list also included plants employed for preparing refreshing beverages, as *Gomphrena decumbens* and *Euphorbia serpens*, among others. Pin et al. (2009) recorded 510 medicinal species in the markets of Asunción, San Lorenzo and Luque, belonging to 94 botanical families; 309 out of those species were included in the book *Plantas Medicinales del Jardín Botánico de Asunción* (Medicinal Plants from Asuncion Botanical Garden), which contains the basic description in order to identify them. Degen and González (2014a), retrieving information through regular visits to the main markets from Asunción, San Lorenzo, Luque and Caacupé, reported 37 different species and one variety among those was employed as anti-inflammatory; they belonged to 27 botanical families, and included species like *Verbena litoralis*, *Sida cordifolia* and *Schinus weinmannifolius*.

Discrepancies in the use of medicinal plants were reported when different communities or country regions were surveyed, as mentioned by Mereles (2001), who reported 51 species with their applications, among those growing in the basins of rivers Tebicuary-mí and Capiibary, comprising the departments of Guairá, Caazapá, Paraguarí and a little area from Caaguazú department. These locations have experienced heavy deforestation resulting in the disappearance of some

Fig. 12.6. Medicinal plant from Paraguay commercialized in popular markets from Asuncion. (Photo: Yenny González, Department of Botany).

species, and the concomitant loss of the knowledge about their use. It is relevant to mention that among such species, two ferns were scarcely mentioned previously as medicinal like *Microgramma vaccinifolia* (Polypodiaceae) and *Lycopodiella cernua,* (Lycopodiaceae).

Another relevant contribution was provided by Soria and Basualdo (2005), who mentioned 55 species employed in a rural community from Caazapa departament. They recorded plant species collected in the surroundings, including modified soils, such as *Chenopodium ambrosioides*, *Solanum sisymbrifolium*, and *Senna occidentalis*, while other species were obtained from the wild forests, like *Allophylus edulis* (Fig. 12.7), *Macfadyena unguis-catis, Sorocea bonplandii,* and *Eugenia uniflora.* Besides those, other species were grown in gardens and orchards, like avocado (*Persea Americana*), wormwood (*Artemisia absinthium*), anise (*Pimpinela anisum*), rye (*Hordeum vulgare*), lemongrass (*Cymbopogon citratus*) and *Plectranthus barbatus.*

Vera (2009) reported 60 species, comprising both autochthonous and exotic plants among those employed in the buffering zone of the three protected wild areas, belonging to the private natural reserve "Tapytá", located within San Rafael and Caazapá National Parks.

In a recent survey performed in the communities Itá Azul (Fig. 12.8) and San Gervasio, located in a protected area of Ybyturuzu, Guaira department, the inhabitants mentioned 68 medicinal plant species, distributed in 35 botanical families. Most of the named plants belong to the following families: Asteraceae, Rutaceae, Fabaceae, and Lamiaceae. The named medical conditions related to these species were grouped into 11 categories, the digestive and respiratory ailments being the most frequently mentioned. In this study, a quite unusual report was observed: 2 species of Orquidaceae family, *Aspidogyne kuczynskii* and *Miltonia flavescens*,

Fig. 12.7. *Allophylus edulis*, "koku", medicinal plant.
(Photo: Germán González, Department of Botany).

Fig. 12.8. Itá Azul Community located in a protected area of Ybyturuzú, Guairá Department. (Photo: Germán González, Department of Botany).

were mentioned in this area as medicinal, probably collected in the wild (Degen and González 2014b).

González et al. (2013) reported a list comprising 15 medicinal species belonging to 15 different botanical families from a survey performed at Pikysyry, Cordillera department (Fig. 12.9). The ailments for which these species are employed are diverse: four species were reported to treat respiratory conditions, three for digestive problems, two for lowering cholesterol, two as hypotensive, two for liver diseases, and one species for diabetes, diarrhea, cancer, heart ailments, wound cleaning,

Fig. 12.9. Pikysyry, Cordillera Department. (Photo: Germán González, Department of Botany).

herpes wounds and appendicitis, respectively. Leaves and aerial parts were the parts most frequently employed.

Mereles (2006) mentioned the use of medicinal plants thriving in wetlands areas as a current practice in Paraguay. One of the species growing in slow-flowing surface waters is *Victoria cruziana*, of which leaves are used to prepare infusions to treat respiratory tract diseases. Soria et al. (2006) had recorded 25 medicinal species that grow in wetlands, damped and exposed to flood soils, swamps and flooded channels. The conservation of these species is at risk due to dramatic modification of soils, mainly related to the road building and rice plantations.

Concerning those plants growing in the Paraguayan Western region or *Chaco*, Mereles and Degen (1994a, 1997) quoted 29 plant species from 16 botanical families popularly employed for medicinal purposes, as referred by native communities, rural inhabitants, and even some immigrants living in the surrounding of Mennonite settlements in Central Chaco. Recently, a research study performed on the Enxet community from that region identified 165 medicinal species referred by shamans, chieftains, native leaders, elderly people of both sexes, men and women. Thirty one out of these species are exclusively employed by shamans to treat ailments, and 19 plants are used for the reproductive function in women (Polini et al. 2013).

Vera (2011) quoted 30 plant species employed to prepare phytomedicines, presented as syrups, ointments or capsules. Among many others, it is common to find *Aloe vera*, *Cecropia pachystachya*, *Cymbopogon citratus*, *Eucalyptus* sp., *Piper regnellii* and *Rosmarinus oficinalis*.

To describe a view beyond the informal trading of medicinal plants, Degen et al. (2004) named 108 different species that are currently commercialized by local enterprises comprising activities like distribution, fractioning and packaging of medicinal and aromatic plants, either as powders or coarsely crushed dry plant material; among such species could be mentioned "koku", *Allophyllus edulis, Scoparia dulcis, Stevia rebaudiana, Menta x piperita, Maytenus ilicifolia* (Fig. 12.10), and *Cecropia pachystachya*.

It is remarkable to name one of the foremost published works about the consumption of medicinal plants by patients. In the present case, Maidana et al. (2015) reported that 78.1% out of 61 type-2 diabetes mellitus patients used medicinal plants; 33 species from 25 botanical families were identified, mostly consumed along with prescription drugs. The most frequently species named in that work are *Moringa oleifera, Artemisia absinthium, Tithonia diversifolia, Baccharis trimera* and *Stevia rebaudiana*. Besides this, Soria and Ramos (2015) named 23 plant species that, based on scientific evidence, could be useful in primary health care; mostly employed for this purpose are *Menta x piperita, Artemisia absinthium, Eugenia uniflora, Lippia alba* and *Allophylus edulis*. The authors also cited some species considered harmful to humans.

A landmark concerning the use of medicinal plants by humans in Paraguay was the controlled clinical trial reported by Achucarro et al. (2011), describing the effect of tablets prepared with *Stevia rebaudiana* leaf powder on intestinal parasites.

Fig. 12.10. *Maytenus ilicifolia*, "cangorosa", medicinal plant.
(Photo: Germán González, Department of Botany).

Consumption of Medicinal Plants

Medicinal plants are consumed in Paraguay in different ways, either fresh or dried, the last being as powder or coarsely crushed. Different parts are employed, even the whole plant too, especially leaves, but parts such as roots, rhizomes, flower tops and the aerial parts are also widely employed. Fresh plants are preferred, and even until now they are macerated with water in order to prepare "tereré", the traditional refreshing beverage of Paraguay (MAG-DGPE 2008, Vera 2009, Degen and González 2014b). To prepare *tereré*, a glass of water is required, or better, a cone shaped container or "guampa" elaborated with a cow horn or a piece of wood, with about 150 mL of capacity, where the leaf powder of processed yerba mate, *Ilex paraguariensis* (Aquifoliaceae), is placed. The liquid is poured into the container and drunk through a metal straw, inserted in the leaf powder, and acting as a filter. It's brewed with cold water, or more frequently, with a cold macerate prepared by crushing fresh medicinal plants in a mortar, usually made with wood of *palo santo* tree, *Bulnesia sarmientoi* (Degen and González 2014b). Considering the popular consumption of "tereré", it provides more than the phytochemicals of the macerated medicinal plants, by extracting chemicals from the *yerba mate,* mainly saponins, polyphenols and methylxanthines, and also measurable quantities of vitamins and minerals (Vera García et al. 2005, Heck and De Mejia 2007). Another way to consume medicinal plants in our country is along with "mate" (Degen et al. 2004, Vera 2009, Degen and González 2014a), the traditional beverage from Río de la Plata region, which is very similar to the *tereré* described above, but prepared with hot water. Other procedures quite widely employed to prepare medicinal plants for human consumption are infusions and decoctions (Basualdo et al. 2003, 2004, Degen et al. 2004), which could also be consumed, either cold or hot as substitutes of drinking water (Degen and González 2014b). Some species are also employed in different ways, to perform steam inhalations from boiled eucalypt leaves, *Eucalyptus* sp., as macerates to be applied or rubbed on the skin like rue, *Ruta chalepensis;* for wound washing like *Acanthospermum australe*; as poultices with chamomile, *Matricaria chamomilla*, or to prepare hot baths with bitter orange leaves, *Citrus aurantium*, respectively (Basualdo et al. 2003, Degen and González 2014a). More recently, the consumption of processed composite *yerba mate, Ilex paraguariensis*—a mixture of *yerba mate* with other plants—and the *"flavored composite yerba mate"* which contains one or more essential oils, became popular. Peppermint, boldo, and lemon are the herbs and flavors most frequently added (Degen et al. 2011).

Even at the present time, it is quite usual to find in Asuncion downtown and other Paraguay main cities medicinal plant sellers in street stalls, who also recommend different medicinal plants to the clients, according to their ailments or requirements. These plants are offered fresh, usually comprising mixtures of two or more species, crushed in a mortar and packed in a plastic bag, ready to be mixed with water for "tereré". It is remarkable that the clients of such stalls are either young or elder people, from different social status, which affirms that "tereré" consumption is a deeply rooted tradition in the Paraguayan population. Along with the promotion of processed *yerba mate* and other plant-related products through mass media, younger generations are acquiring the practice of "tereré" consumption which is also a relevant way for medicinal plants use. In such a way, two popular practices like the consumption of both the traditional beverage "tereré" and the medicinal plants are being consolidated as a part of Paraguayan traditions in the whole society, instead of being lost over the years (Degen and González 2014a).

Origin, Trading and Conservation Status of Medicinal Plants

Most of the medicinal plants commercialized in Paraguay—both in formal or informal trading-come from their harvest in the wilderness, without possible replacements. This fact restricts the effectiveness of their use due to the lack of control about the harvest time and the post-harvest procedures, and at the same time threatens their conservation status and limits their mainstream use (Degen and González 2014a, MAG-DGPE 2008). There is also a demand of plant material for international trade, which is satisfied with some cultivated species and others collected in their environment (Vera 2011). The pivotal role of women in the access to the plant resources is generally recognized, either for collecting firewood, or the harvest and use of medicinal plants. Moreover, native women from indigenous populations are the ones who own the best knowledge about the plant's properties, which is observed to a less extent in mestizo women (Masulli et al. 1996).

Pin et al. (2009) quoted that medicinal plants in Paraguay come from three sources: introduced plants, plants existing naturally in the region, and those autochthonous from Paraguay. As an example, among those plants employed as anti-inflammatory, Degen and González (2014a) have mentioned 25 native species, 9 exotic and 2 acclimated plants.

At the local level in popular markets, the trade of medicinal plants—either native or introduced—is performed both as fresh or dried plants. Dealers and packing enterprises also trade medicinal plants as crushed material or dried powders. Besides this, processing industries for both food products and phytomedicines prepare syrups, ointments and capsules based on medicinal plants.

From the 108 medicinal species named by Degen et al. (2004), 66% were native, 9% imported and 25% cultivated; out of the 41 cultivated species, 13 were native and 27 acclimatized. The species with the highest demand were *Allophyllus edulis*, *Scoparia dulcis*, *Stevia rebaudiana*, *Menta* x *piperita*, *Maytenus ilicifolia*, *Cecropia pachystachya*, *Cassia angustifolia* and *Peumus boldus*. Some of the species showed high demand, with seasonal variations; chamomile (*Matricaria chamomilla*), and flaxseed (*Linum usitatissimun*), were sold throughout the whole year; however, others like borage (*Borago officinalis*) were just demanded in the cold weather season. The use of some species has also followed fashion tendencies like *Macfadiena unguis-cati*, or *Uncaria* sp. and *Moringa oleifera*, just to mention a few (Degen et al. 2004, Degen and González 2014a).

The medicinal plants business usually gets raw material from collectors who buy the productions from family groups, for whom it represents an important source of income. Agroindustrial producers usually claim select species where just the leaves or another aerial part are employed, but it's quite common to notice the trading of underground parts like roots or rhizomes and roots barks, as observed in *Boerhavia diffusa*, *Cayaponia espelina*, *Herreria montevidensis*, *Jatropha isabelli*, among many others, with a high risk of overexploitation. The use of other parts like the aerial parts or flower tops allows the reproduction of such species (Degen et al. 2004).

The work performed on Itá Azul and San Gervasio communities revealed that the medicinal plants employed by the population have several sources, i.e., 50% and more were autochthonous species mostly consumed as fresh herbs and collected from the wild; following this group are introduced species which are cultivated around the houses only for family consumption, and finally are those imported species locally available as dried herbs in the stores. They also mentioned a few native species being grown because of their shortage in the wild, like *Piper regnellii*, which was included in the list of threatened species elaborated by the National Environmental Secretary (SEAM) (2006), or *Stevia rebaudiana*, being currently absent in its natural habitat, but widely cultivated for industrial purposes. Another species severely compromised as a consequence of overexploitation is *Maytenus ilicifolia,* since its root bark is the most requested part. Collectors harvest very small plants, which results in very low development, placing the species in a critical conservation status. *Achyroclyne alata* can also be mentioned, since this native species is present only in certain regions of the country, which moves the inhabitants of rural communities to buy it for personal use (Degen and González 2014b, Degen et al. 2004).

The continuous pressure on medicinal plants in their habitat motivated research efforts in order to initiate acclimation studies to contribute to their preservation and to encourage sustainable production. As a result of such efforts, the successful development in controlled conditions of three medicinal species considered as endangered species, *Cyclolepis genistoides*, *Equisetum giganteum*, and *Herreria montevidensis*, was accomplished (Céspedes et al. 2007, Céspedes et al. 2014). Going in the same direction, Resquín (2011) published a proposal for the sustainable production of peppermint, *Mentha arvensis* and *M.* x *piperita*, within the framework of familiar agricultural practices, showing productive technologies for these species, facing their high demand for the elaboration of mixes with processed *Ilex paraguariensis*. The objective of the study was to organize and disseminate the knowledge and the methodology that could also be applied to preserve other endangered species. González and Degen (2015) have summarized different techniques aimed to preserve the soil and to improve its productivity in order to reduce the negative effects of deforestation. Furthermore, the National School of Agronomic Science of Asuncion National University (Facultad de Ciencias Agrarias—Universidad Nacional de Asunción) had developed several research projects, including one on the production of three medicinal species: *Aloysia polystachya*, peppermint and oregano (Vargas 2013), followed by graduate dissertations on the acclimating processes pursuing the improvement of the productivity of medicinal species (Armadans 2010).

Morphoanatomy of Plant Drugs Commercialized in Paraguay as an Identification Tool

Morphoanatomy studies on herbal drugs make a relevant contribution to the right identification of plant material for trading purposes, regularly available in popular markets of Paraguay. The propper identification of the commercial samples is a critical issue, either for those intended for direct consumption, or in quality assessment protocols aplayed to material directed to the preparation of phytomedicines. These studies are focused on the recognition of the drugs based on their micro and macroscopic characteristics, relevant for the identification of the parts of each plant and for quality evaluation. In Paraguay, pioneer studies about the anatomy of medicinal plants were performed in the Botany Department of the FCQ, at the Universidad nacional de Asunción, in 1990. The first work employing this methodology was published by Ortiz et al. (1993), and it was directed to analyze the leaf glandular trichomes of *Heliotropium procumbens*. In 1997, Ortiz and Delmás have published the anatomical study of three medicinal plants frequently employed in Paraguay: *Cecropia pachystachya, Gochnatia polymorpha* and *Piper regnellii*. Ten years later, these studies were taken up with the descriptions of the species *Killinga brevifolia, K. odorata, K. vaginata* and *Scleria distans* (Cyperaceae), *Lippia alba* (Verbenaceae), *Jatropha isabelliae* (Euphorbiaceae), *Ilex paraguariensis* var. *paraguariensis* (Aquifoliaceae), *Genipa americana* (Rubiaceae), *Sorocea bonplandii* (Moraceae), *Acanthospermum australe, A. hispidum* (Asteraceae), *Aloysia polystachya* (Verbenaceae), *Begonia cucullata* (Begoniaceae), and *Phyllanthus orbiculatus*, published by González and Degen (2008), González et al. (2008), González et al. (2009a), Riveros et al. (2009),

González et al. (2009b), González et al. (2011), Degen et al. (2011), Degen et al. (2012a), Degen et al. (2012b), González et al. (2014) and González et al. (2016).

This kind of research has also been performed in the Biology Department of the Faculty of Exact and Natural Sciences (Facultad de Ciencias Exactas y Naturales, UNA), resulting in the anatomical description of *Stevia rebaudiana, Tagetes minuta, Pterocaulon polystachyum* (Asteraceae), and *Laurus nobilis* (Lauraceae), among others (Benítez et al. 2010, Pereira et al. 2011, González and Pereira 2012, González et al. 2015).

Based on the number of plant species currently employed for medicinal purposes in Paraguay, it is evident that much work still needs to be done concerning morphology and anatomy studies of our plants.

An Approach to Medicinal Plants' Evaluation

In our country, the initial projects on medicinal plants were focused on the recovery of the knowledge about their use, but they were not accompanied by research on their chemical composition nor their pharmacological effects, nor toxicological issues in order to support their safe and effective use. The agronomic issues required to secure the provision of raw material with uniform quality were not deeply evaluated yet. Such information is only available for the production of *yerba mate, Ilex paraguariensis*, which is under commercial exploitation since colonial times (17th century). Since 1980, the research staff of the Faculty of Chemical Sciences (FCQ) has continued strengthening its research staff, steadily contributing to the best knowledge of our cultural heritage related to medicinal plants in its different aspects: botanical, phytochemical and pharmacological (Ibarrola and Degen 2011). These efforts were also aimed to study some toxicological issues with sanitary impact for both human beings and livestock (Schmeda Hirschmann et al. 1987, Basualdo et al. 1992).

It is relevant to mention the results from phytochemical studies achieved in the Phytochemistry Department of FCQ, about species of Celastraceae and Hippocrataceae families (Alvarenga et al. 1999, 2000) in aromatic plants as the composition of the essential oil from *Aristolochia gibertii* (Canela et al. 2004), the composition and antifungal activity of *Piper regnellii* (*P. fulvescens*) essential oil (Freixa et al. 2001), and from *Lippia alba* essential oil (Alvarenga et al. 2008). There were former published results on the *in vitro* biological activity on enzymes, such as xanthine oxidase (Theoduloz et al. 1988) and aldose reductase (Ferro and Degen 2011), and the insecticide and feeding deterrent activity of secondary metabolites from *Maytenus* (Avilla et al. 2000), just to mention a few of them. The establishment of the Pharmacology Department at FCQ allowed to expand the knowledge on the activity of medicinal plants, through studies of cardiovascular activity as the hypotensive effect of *Solanum sisymbriifolium, Syagrus romanzoffiana, Gomphrena globosa* and *Bromelia balansae*. Soon afterwards, the same research group started to study the effects of medicinal plants on the central nervous system and the animal behavior in the search of molecules acting as sedative, anxiolytic, antidepressant, antipsychotic and anticonvulsant. Such studies started with the analysis of biological properties of *Kyllinga brevifolia* and *Aloysia polystachya*, and more recently, they kicked off the evaluation of activity on the digestive tract, either gastroprotective or hepatoprotective in *A. polystachya* and *Sida rhombifolia*. Some efforts were also directed to perform the experimental validation of species traditionally employed to treat diabetes, like *Prosopis ruscifolia* and *Cyclolepis genistoides* (Ibarrola and Degen 2011).

A very good example of a multidisciplinary evaluation process on medicinal plants was achieved by the researchers from the FCQ, which have obtained relevant results from preclinical essays with *Solanum sisymbriifolium*, demonstrating low toxicity, little effects on animal behavior and an important decrease of arterial pressure (Ibarrola et al. 1996, 2006); furthermore, bioassay-guided isolation was conducted for identification of two closely related spyrostannic monodesmosidic saponins: isonuatigenoside (Fig. 12.11)—a new chemical structure (Ferro et al. 2005), and nuatigenoside (Fig. 12.12). Recently, these are recognized as the most active isolated metabolites, allowing to establish a close correlation between the activity of the metabolites and the popular use of the plant (Ibarrola and Degen 2011).

Facing the required social and productive projection of the knowledge acquired from medicinal plants, new routes to catch the interest of local pharmaceutical and nutraceutical enterprises were explored through innovation projects. A good example was the development of a standardized protocol for the sustainable production of a phytomedicine based on the experimental evaluation of *Aloysia polystachya* as an antidepressant, which included botanical, chemical and pharmacological evaluations.

At the present time, bioprospective research projects on medicinal plant species are being conducted, along with their chemical, pharmacological and toxicological evaluations, comprising species from Euphorbiaceae, Solanaceae, Verbenaceae, Moraceae and Lorantaceae families, among others.

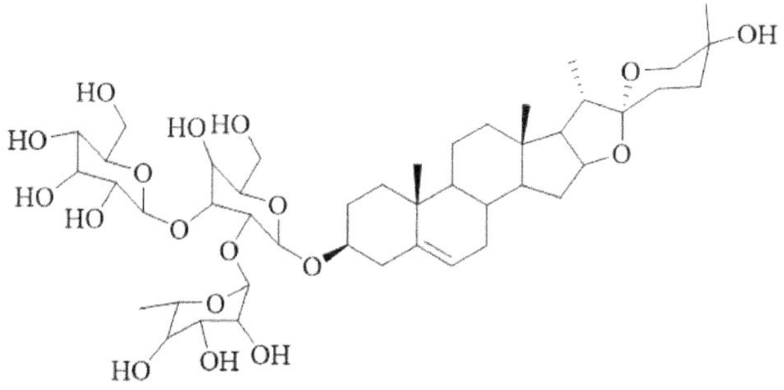

Structure 1

Fig. 12.11. Isonuatigenoside (Ferro et al. 2005, Structure elaborated by Amner Muñoz Acevedo).

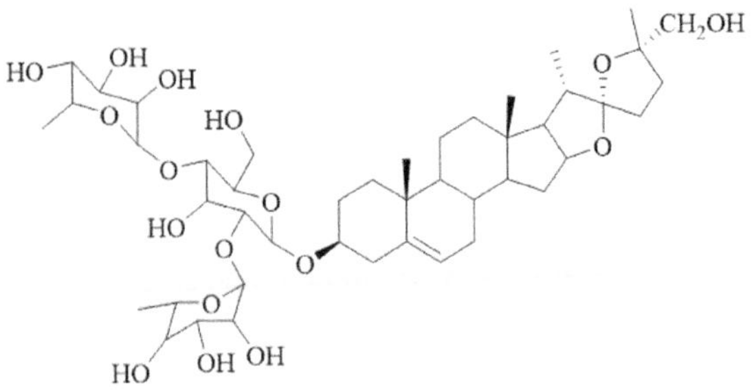

Structure 2

Fig. 12.12. Nuatigenoside (Mimaki et al. 1995, Structure elaborated by Amner Muñoz Acevedo).

Conclusions

Paraguay is a country that has very rich flora, with a variety of native traditions from both indigenous and European origin: some are autochthonous and others are shared with other Latin America countries. Among these traditions, those related to medicinal plants are remarkable and closely related to the native heritage. Medicinal plants are employed either as fresh or dried plant material, consumed together with "mate" or "tereré" as refreshment beverages and to treat different ailments, especially those from digestive and respiratory tracts, but also to treat chronic health conditions like diabetes or hypertension. This mode of employment reaches the whole population, across different social classes, and it is getting stronger rather than being lost. Medicinal plants are a relevant option to develop innovative drugs, which could result in safer and more effective treatments compared to synthetic drugs. However, the challenges concerning medicinal plants go further than validation of traditional knowledge, and should face their transformation into safe drugs. These challenges include critical issues, like those related to the sustainable exploitation of the natural resources preserving the availability of highly demanded plant material without the risk of extinction. Other issues to be addressed are those related to the recognition of the intellectual rights of native populations, in order to return a part of the economic benefits and royalties resulting from the trading of drug developed from traditional medicinal plants, the tracing of raw material in the productive processes, and the strengthening of multidisciplinary research teams, with both the human and material resources required for the production of new knowledge. Such efforts are required to succeed in the race against the menace of losing traditional heritage and biodiversity, just to mention some critical issues.

Acknowledgments

The authors thank the institutions that have supported the research on medicinal plants at the Facultad de Ciencias Químicas, UNA, as the Research Direction of the Universidad Nacional de Asunción, JICA, CONACYT, Tropical Forest Conservation Fund-Paraguay, CYTED, FIDA and FUNDAQUIM.

They also thank José Luis Martinez for the invitation to participate in this project, building up an important platform to disseminate the knowledge related to medicinal plants, and to Lic. Gabriela Ferro for her kind revision of the English translation.

References

Achucarro, C., Ferro, E.A., Richer, Y., Salazar, M.E., Ciciolli, S., Ortiz, I., Campos, S., Sckell, C., Samudio, M., Alborno, R.M., Meza, B., Varela, I., Losanto, J. and Pedrozo, J.R. 2011. Evaluación clínica preliminar del efecto antiparasitario de *Stevia rebaudiana* Bertoni (ka`a he ê) en adultos y niños. An. Fac. Cienc. Méd. 44(2): 35–46.

Alvarenga, N.L., Velázquez, C.A., Gómez, R., Canela, N.J., Bazzocchi, I.L. and Ferro, E.A. 1999. A new antibiotic nortriterpene quinone methide from *Maytenus catingarum*. J. Nat. Prod. 62(5): 750–751.

Alvarenga, N.L., Ferro, E.A., Ravelo, A.G., Kennedy, M.L., Maestro, M.A. and González, A.G. 2000. X-ray analysis of Volubilide, a new decacyclic Diels-Alder C20–C30 adduct from *Hyppocratea volubilis* L. Tetrahedron 56: 3771–3774.

Alvarenga, N., Canela, N.J. and Torio, H. 2008. Composición química y actividad antifúngica del aceite esencial de hojas de *Lippia alba*, Verbenaceae. Rojasiana 8(1): 39–42.

Armadans, A. 2010. Estudiantes de la FCA presentaron en público trabajos de investigación y tesis. Congreso Nacional de Ciencias Agrarias. Available at: http://www.agr.una.py/Difusion/imagen/boletines/boletinfca10.pdf. Consultado el 10 de abril de 2016.

Arenas, P. and Moreno Azorero, R. 1976. Plantas usadas en la medicina folklórica paraguaya para regular la fecundidad. Rev. Soc. Cient. Paraguay 16: 21–43.

Arenas, P. and Moreno Azorero, R. 1977. plants used as jeans of abortion, conception, sterilization and fecundation by paraguayan indigenous people. Econ. Bot. 31: 306–306.

Avilla, J., Teixidó, A., Velázquez, C., Alvarenga, N., Ferro, E.A. and Canela, R. 2000. Insecticidal activity of *Maytenus* species (Celastraceae) nortriterpene quinone methides against codling moth, *Cydia pomonella* (L.) (Lepidoptera: Tortricidae). J. Agricultural & Food Chemistry 48: 88–92.

Basualdo, I., Soria, N., Ortiz, M., Acosta, L., Degen, R. and Eliceche, A. 1992. Plantas tóxicas para el ganado en los Departamentos de Concepción y Amambay, Paraguay. Editorial EDUNA. Asunción-Paraguay, pp. 129.

Basualdo, I. 1993. Teodoro Rojas, un ilustre botánico paraguayao. Rojasiana 1(1): 1–2.

Basualdo, I. and Soria, N. 1996a. Plantas aromáticas de la medicina folclórica paraguaya. Anales de SAIPA. Sociedad argentina para la investigación de productos aromáticos 14: 57–62.

Basualdo, I. and Soria, N. 1996b. Farmacopea herbolaria paraguaya: especies de la medicina folklórica utilizadas para combatir enfermedades del tracto respiratorio (Parte I). Rojasiana 3(2): 197–238.

Basualdo, I., Soria, N., Keel, S. and Rivarola, N. 1997. Recursos fitogenéticos, Parque Naiconal Cerro Corá – Amambay. Plantas útiles. Dirección de Parques Nacionales y Vida Silvestre, Facultad de Ciencias Químicas – Universidad Nacional de Asunción, The Nature Conservancy (USA). Asunción. 75 pp.

Basualdo, I., Soria, N., Ortiz, M. and Degen, R. 2003. Uso medicinal de plantas comercializadas en los mercados de Asunción y Gran Asunción, Paraguay. *In*: Revista de la Sociedad Científica del Paraguay 14: 5–22.

Basualdo, I., Soria, N., Ortiz, M. and Degen, R. 2004. Plantas medicinales comercializados en los mercados de Asunción y Gran Asunción. Rojasiana 6(1): 95–114.

Benítez, L.G. 1986. Breve historia de grandes hombres. Industrial Gráfica Comuneros, Asunción, Paraguay. 390 pp. Disponible en: http://www.portalguarani.com/1484_luis_g_benitez/14151_breve_historia_de_grandes_hombres_obra_de_luis_g_benitez_.html. Consultado el 1 de abril de 2016.

Benítez, B., Pereira, C., González, F., Molinas, C. and Bertoni, S. 2010. Morfología y micrografía del ka'a he'e, *Stevia rebaudiana* (Bertoni) Bertoni, provenientes de cultivares de Concepción, Paraguay. Steviana 2: 55–67.

Bertoni, M. 1927. La medicina Guaraní. Conocimientos científicos. Imp. y Edit. Ex-Sylvis, Alto Paraná, Paraguay. 299 pp. Disponible en: http://www.mag.gov.py/bina/dato/La%20Medicina%20Guarani. pdf. Consultado el 5 de mayo de 2016.

Canela, N., Alvarenga, N.L., Ferro, E.A., Vila, R. and Cañigueral, S. 2004. Chemical composition of the essential oil of *Aristolochia giberti* Hooker from Paraguay. J. Essential Oil Research 16: 566–567.

Céspedes de Zárate, C., González, G., Delmás de Rojas, G., Vogt, C. and Quiñonez, P. 2007. Aclimatación de tres especies de uso en medicina popular con rango de amenaza. Rev. Investigaciones y estudios en la UNA 3: 63–74.

Céspedes de Zárate, C., González, G. and Delmás de Rojas, G. 2014. Producción de raíces de *Herreria bonplandii*, lecomte y *H. montevidensis* Klotzch ex Griseb., "zarzaparrilla", especies utilizadas en medicina popular, Paraguay. Rojasiana 13(2): 25–33.

Degen, R. 1990. Los nombres vulgares y científicos en la taxonomía botánica. La Revista Crítica 2(4): 60–65.

Degen, R., Basualdo, I. and Soria, N. 2004. Comercialización y conservación de especies vegetales medicinales en Paraguay. Revista Fitoterapia 4(2): 129–138.

Degen, R., Soria, N., Ortiz, M. and Basualdo, I. 2005. Problemática de nombres comunes de plantas medicinales comercializadas en Paraguay. Dominguezia 21(1): 11–16.

Degen de Arrúa, R.L., González, Y.P. and Amarilla, A. 2009. Legislación sobre Plantas Medicinales y Fitoterápicos en Paraguay: una tarea pendiente. Bol. Latinoam. Caribe Plant. Med. Aromat. 18(1): 12–16.

Degen, R., González, Y. and González de García, M. 2011. Análisis de la yerba mate elaborada compuesta, comercializada en Asunción y Gran Asunción, Paraguay. Rojasiana 10(2): 81–91.

Degen de Arrúa, R., González, Y., González de García, M. and Delmás de Rojas, G. 2012a. Morfoanatomía comparativa de dos especies de Acanthospermum (Asteraceae). Rojasiana 11(1-2):67–78.

Degen, R., Mercado, M.I., Coll Aráoz, M.V., Ruiz, A. and Ponessa, G.I. 2012b. Morfología y anatomía de dos variedades de *Begonia cucullata* (Begoniaceae) comercializadas como «agrial» en Paraguay. Lilloa 49(2): 87–97.

Degen, R. and González, Y. 2014a. Plantas medicinales utilizadas en la medicina popular paraguaya como antiinflamatorias. Bol. Latinoam. Caribe Plant. Med. Aromat. 13(3): 213–231.

Degen de Arrúa, R. and González, Y. 2014b. Plantas medicinales utilizadas en las comunidades de Itá Azul y San Gervasio, (Paraguay). Revista de Fitoterapia 14(2): 33–47.

Feliciangeli, R. 1939. Proyecto de la Farmacopea Paraguaya. La Colmena S.A.: Asunción. Tomo III: 168–182.

Ferro, E.A., Alvarenga, N.L., Ibarrola, D.A., Hellion-Ibarrola, M.C. and Ravelo, A.G. 2005. A new steroidal saponin from *Solanum sisymbriifolium* roots. Fitoterapia 76: 577–579.

Ferro, E. and Degen de Arrúa, R.L. 2011. Actividad Inhibitoria de Extractos de Plantas Medicinales de Paraguay sobre Aldosa Reductasa de Cristalino de Rata. Rojasiana 10(2): 31–42.

Freixa, B., Vila, R., Ferro, E.A., Adzet, T. and Cañigueral, S. 2001. Antifungal principles from *Piper fulvescens*. Planta Med. 67(9): 873–5.

Gonzalez Torres, D. 1992. Catálogo de plantas medicinales (Utiles y Alimenticias) usadas en Paraguay. Asunción. 456p

González, F. and Pereira, C. 2012. Morfoanatomia foliar y caulinar de *Tagetes minuta* L. (suico) comercializada como medicinal. Libro de resúmenes de las I Jornadas Paraguayaas de Botánica. Pág.: 18.

González, F., Benítez, B. and Soria, N. 2015. Morfoanatomía cualitativa foliar y caulinar de *Pterocaulon polystachyum* DC. (Asteraceae), de uso medicinal en Paraguay. Steviana 7(supl.): 25.

González, G. and Degen de Arrúa, R. 2015. Capítulo V: Fotalecimiento y uso sostenible (Vivero). Capitulo V. Conservación, Fortalecimiento y uso sostenible de la Flora de Itá Azul, Colonia Independencia, Paraguay. Fondo de Conservacion de Bosques Tropicales, Facultad de Ciencias Químicas, FUNDAQUIM. 85–105 p.

González, M., González, Y. and Degen, R. 2009a. "Yerba mate", *Ilex paraguariensis* A. St-Hil. var. *paraguariensis* (Aquifoliaceae) caracteres exo-endomorfológicos y farmacognósticos. Rojasiana 8(2): 39–51.

González, M., Brítez, L., González, Y. and Degen, R. 2011. Morfoanatomía comparativa de *Genipa americana* L. (Rubiaceae) y *Sorocea bonplandii* (Baill.) W.C. Burger, Lanj. & Wess. Boer (Moraceae). Rojasiana 10(1): 93–101.

González, Y. and Degen, R. 2008. Morfoanatomía comparativa de las especies comercializadas como "kapi'i kati" en los mercados de Asunción y San Lorenzo, Paraguay. Rojasiana 8(1): 43–47.

González, Y., Degen, R. and Delmás, G. 2008. Estudio morfoanatómico de "salvia" *Lippia alba* (Miller) N.E. Brown (Verbenaceae). Rojasiana 8(1): 93–95.

González, Y., Mercado, M.I., Degen, R. and Ponessa, G. 2009b. Morfoanatomía y etnobotánica de rizoma, tallo y escapo de "kapi'i kati", *Kyllinga odorata* (Cyperaceae) y sus sustituyentes, de Asunción y alrededores, Paraguay. Lilloa 46(1-2): 58–67.

González, Y., Degen, R., González, G. and Delmás, G. 2013. Especies medicinales, su estado de conservación y usos, de la Compañía Pikysyry, Departamento de Cordillera, Paraguay. Rojasiana 12(1-2): 105–115.

González, Y., Degen de Arrúa, R., Delmás de Rojas, G. and González de García, M. 2014. Etnofarmacobotanica foliar de burrito, *Aloysia polystachya* (Griseb.) Moldenke (Verbenaceae) cultivado en Paraguay. Rojasiana 13(1): 31–41.

González, Y., González de García, M., Delmás de Rojas, G. and Degen de Arrúa, R. 2016. Morfología y anatomía de *Phyllanthus orbiculatus* (Phyllanthaceae) comercializada como «para para'í» en Paraguay y sus posibles sustituyentes. Lilloa 53(2): 217–228.

Gupta, M. 1995. 270 plantas medicinales iberoamericanas. Convenio Andrés Bello-CYTED. Ed. Presencia Ltda. Santa Fe de Bogotá, Colombia, pp. 617.

Gupta, M. 2008 (ed.). 500 Plantas Medicinales iberoamericanas. CYTED. Convenio Andrés Bello. 1003 pp.

Heck, C.I. and De Mejia, E.G. 2007. Yerba Mate Tea (*Ilex paraguariensis*): A comprehensive review on chemistry, health implications, and technological considerations. J. Food Sciences 72(9): R138–R151.

Ibarrola, D.A., Hellión, M.C. and Ferro, E.A. 1996. The hypotensive effect of the crude root extract of *Solanum sisymbriifolium* Lam. In normo and hypertensive rats. Journal of Pharmaceutical Biology 44(5): 378–381.

Ibarrola, D.A., Hellión, M.C., Ferro, E.A., Hatakeyama, N., Shibuya, N., Yamazaki, M., Momose, Y., Yamamura, S. and Tsuchida, K. 2006. Cardiovascular action of nuatigenosido from Solanum sisymbriifolium. Journal of Pharmaceutical Biology 44(5): 378–381.

Ibarrola, D.A. and Degen de Arrúa, R.L. (eds.). 2011. Catálogo ilustrado de 80 plantas medicinales del Paraguay. Facultad de Ciencias Químicas-UNA y Agencia de Cooperación Internacional del Japón-JICA, Asunción, Paraguay, pp. 1–178.

MAG-DGPE. 2008. Informe Final. Elaboración del Estudio de la Situación de la Recolección, Producción y Comercialización de Plantas Medicinales y Aromáticas de Paraguay. Programa Regional de Apoyo a la Red de Desarrollo de Fitoterápicos en el MERCOSUR. 1–189.

Maidana, M., Gonzalez, Y. and Degen de Arrúa, R. 2015. Plantas medicinales empleadas por pacientes diabéticos en Paraguay. Infarma 27(4): 218–224.

Masulli, B., Mereles, F., Aquino, A., Gamarra, I., Medina, F., Rossato, V., Sottoli, S. and Vera, V. 1996. El rol de la mujer en la utilización de los recursos naturales en el Paraguay: Un enfoque multidisciplinario. 1–247.

Mereles, F. and Degen, R. 1994a. Leñosas de uso popular en chaco boreal. Revista forestal del Paraguay, Universidad Nacional de Asunción 10(1): 14–19.

Mereles, F. and Degen, R. 1994b. Los nombres vulgares de los árboles y arbustos del Chaco Boreal, Paraguay. Rojasiana 2(2): 67–101.

Mereles, F. and Degen, R. 1997. Contribución al conocimiento de los árboles y arbustos indígenas utilizados como medicinales en el Chaco Boreal (Paraguay). Parodiana 10 (1-2): 75–89.

Mereles, F. 2001. Recursos Fitogenéticos: Plantas útiles de las cuencas del Tebicuary mi y Capiíbary, Paraguay Oriental. Proyecto Sistema Ambiental de la Región Oriental, (SARO). Rojasiana, Vol. Esp. 144 pp.

Mereles, F. 2006. La diversidad, los usos y la conservación de las especies vegetales en los humedales del Paraguay. Rojasiana 7(2): 171–185.

Mereles, F. 2014. Una aproximación al estado ambiental del Paraguay. pp. 127–140. *In*: Segundo Simposio internacional. Hacia nuevas políticas culturales. Centro Cultural de la República. Paraguay.

Mimaki, Y., Nakamura, O., Sashida, Y., Nakaido, T. and Ohimoto, T. 1995. Phytochem. 385: 1279–1286.

Ortiz, M., Degen, R. and Benítez, M.C. 1993. Tricomas glandulares en hojas de *Heliotropium procumbens* Miller. Rojasiana 1(1): 16–20.

Ortiz, M. and Delmás, G. 1997. Anatomía foliar de tres especies utilizadas en la medicina folklórica paraguaya. Rojasiana 4(1): 1–10.

Pin, A., González, G., Marin, G., Céspedes, G., Cretton, S., Christen, P. and Rouget, D. 2009. Plantas medicinales del Jardín Botánico de Asunción. Municipalidad de Asunción, AEPY y Universite de Geneve. 1–441.

Pereira, C., González, F. and Benítez, B. 2011. Micrografía foliar de *Laurus nobilis* L. (Lauraceae) como herramienta para el control de calidad de muestras comerciales. Boletín de la Sociedad Argentina de Botánica 46(Supl.): 114.

Polini, G., López Ramírez, A., Degen de Arrua, R., Quarti, A., Delmás de Rojas, G., González, Y. and Aquino, O. 2013. Comer Del Monte, Plantas Medicinales Del Chaco Central. ISBN 978-99967-611-2-6. 352 pp.

Resquin Romero, G.A. 2011. Producción sostenible de Menta *Mentha arvensis* y *Mentha* x *piperita* en sistemas de agricultura familiar campesina de la Región Oriental, Paraguay. FCA-UNA/CONACYT/CETEC. 235 p.

Resquín, G., Degen de Arrúa, R., Delmás de Rojas, G. and Macchi Leite, G. 2011. Las especies de *Mentha* L. cultivadas en Paraguay. Rojasiana 10(1): 77–91.

Riveros, R., González, Y., González, M. and Degen, R. 2009. Etnofarmacobotánica de "Jagua rová", *Jatropha isabelli* Mull. Arg. Rojasiana 8(2): 25–30.

Scavone, C. 2011. Plantas medicinales: un poco de historia. Artículo en ABC Edic. Impresa. 27 de abril de 2010. Disponible en: http://www.abc.com.py/articulos/plantas-medicinales-un-poco-de-historia-95752.html. Consultado el 15 de abril 2016.

Schinini, A. 2005. Teodoro Rojas, biografía, viajes y especies dedicadas. Rojasiana 7(2): 101–149.

Schmeda, G. and Silva, M. 1982. The flavonoids of Eupatorium laeve DC. Compositae. Publicaciones FCQ N° 5.

Schmeda Hirschmann, G., Ferro, E.A., Franco, L., Recalde, L. and Theoduloz, C. 1987. Pyrrolizidine Alkaloids from *Senecio brasiliensis* Populations. J. Nat. Prod. 50(4): 770–772.

SEAM (Secretaría del Ambiente). 2006. Resolución 524/06 por la cual se aprueba el listado de especies de flora y fauna amenazadas del Paraguay.

Soria, N. 1993. Las especies aladas de Baccharis utilizadas como medicinales en Paraguay. Rojasiana 1(1): 3–12.

Soria, N. and Basualdo, I. 2005. Medicina herbolaria de la Comunidad Kavajú Kangué, Departamento de Caazapá. 1–138.

Soria, N., Basualdo, I. and Ortiz, M. 2006. Las especies medicinales de los humedales en Paraguay. Rev. Soc. Cient. Paraguay 20: 94–112.

Soria, N. and Ramos, P. 2015. Guía para el uso de plantas medicinales en atención primaria de salud, Paraguay. Editorial Académica Española. 1–127.

Theoduloz, C., Franco, L., Ferro, E.A. and Schmeda-Hirschmann, G. 1988. Xanthine oxidase inhibitory activity of paraguayan Myrtaceae. J. Ethnopharmacology 24: 179–183.

Vargas Lehner, F. (Coord.). 2013. Producción de plantas medicinales: burrito, menta, orégano. Proyecto Financiado por la comunidad europea. Facultad de Ciencias Agrarias. San Lorenzo.

Vera García, R., Peralta, I. and Caballero, S. 2005. Fraction of minerals extracted from Paraguayan yerba mate *Ilex paraguariensis*, by cold tea, maceration and hot tea infusion as consumed in Paraguay. Rojasiana 7(1): 21–25.

Vera Jiménez, M. 2009. Plantas medicinales de tres áreas silvestres protegidas y su zona de influencia en el sureste de Paraguay. Fundación Moisés Bertoni. Asunción, Paraguay. 1–160.

Vera Jiménez, M. 2011. 30 plantas medicinales utilizadas en fitoterapia. Fundación Moisés Bertoni, Mancomunidad Mbaracayú y Asociación Etnobotánica Paraguaya. 1–40.

CHAPTER 13

An Overview of *Vetiveria zizanioides* (Linn.) Nash (Poaceae)

Traditional Uses and Products

Shubhangi N. Ingole

Introduction

The world is endowed with a rich wealth of medicinal plants and India is sitting on a gold mine of well recorded and traditionally well-practiced knowledge of herbal medicine. More than 1500 herbal preparations are sold as dietary supplements or ethnic traditional medicines (WHO 2000). India is perhaps the largest producer of medicinal herbs and is rightly called the botanical garden of the world (Parrotta 2001). India is well known as an emporium of medicinal plants (Rao et al. 2016). The herbal drugs provide an alternative and effective treatment for chronic disorders (Astin 1998, Cupp 1999).

Man cannot survive on this earth for a long time without the plant kingdom because the plant products and their active constituents play an important role. The herbal medicines as the major remedy in traditional medical systems have been used in practice for thousands of years. Herbs have always been the principal form of medicine in India and presently they are becoming popular throughout the world. These have made a great contribution in maintaining human health. Plants as a source of medicine have been playing an important role in health services around the globe and constitute the backbone of the herbal medicine (Shankar and Liao 2004).

But India despite its rich traditional knowledge, heritage of herbal medicines and large biodiversity has a dismal share in the world market due to export of crude extracts and drugs (Kamboj 2000, Desmet 2002, Dubey et al. 2004). Hence, today

Department of Botany, Bar. R.D.I.K. and N.K.D. College Badnera-Amravati, Maharashtra, India.
Email: shubhaingole@gmail.com

the physicians, research team and pharmaceutical industries of many countries have once again turned their attention to the sources of natural raw materials and to medicinal plants as sources for the isolation and production of irreplaceable drugs. Drugs are becoming important and highly profitable for industry and agriculture. Cure without side effects is the key feature of herbal medicine and treatments. For these reasons, medicinal plants and their active ingredients or phytochemicals are again claiming the attention of whole world.

Herbal medicines are available as single or poly herbal preparations. Because of consumption of these herbal preparations by large masses of developed as well as developing countries. As use of plant materials as raw materials for the pharmaceutical industry and as it represents a substantial proportional of the world drug market, there is need to control and assure the quality of such preparations through systematic scientific studies. The quality of herbal drugs is sum of all factors which contribute directly or indirectly to the safety, effectiveness and acceptability of product.

India is inhabited by diverse tribal populations, dwelling in forest surroundings and depending on its resources. Out of several plants used by tribals, khas grass— *Vetiveria zizanioides* (L.) Nash (Poaceae), occupies the leading position, particularly in North Indian plains.

Habitat and Distribution

Khas grass is found throughout the plains and lower hills of India, particularly on the riverbanks and in rich marshy soil. It grows wild in almost all plains of all states in India up to an elevation of 1200 m. Only in some pockets of South India the grass is systematically cultivated but the yield from the cultivated crops meets only a small percentage of requirements. The grass grows luxuriantly in areas with an annual rainfall of 1000–2000 mm and temperature ranging from 22 to 43°C. Marshy riverbeds with sandy loam are best suited for this grass. It possesses a unique rooting pattern. Unlike most grasses which form horizontally spreading mat-like root systems, Vetiver's roots grow downward, 2–4 meters in depth. This typical pattern has made it widely used in erosion control systems. Vetiver is closely related to other fragrant grasses such as Lemon Grass, Citronella, and Palmarosa. Because of its aromatic properties, Vetiver is widely cultivated in the tropical regions of the world (Dashori and Gosavi 2013). It is found in aquatic and sub-aquatic plains of India.

The plant has unique characteristics of being xerophytic but it survives under long seasonal flooding, it tolerates extreme temperature and grows over a wide range of soil pH.

Taxonomy and Ecology

Systematic Position (According to Bentham and Hooker):
Division: Angiosperms

Sub-division: Monocotyledons
Series: Glumaceae
Family: Poaceae
Sub-family: Andropogoneae
Genus: *Vetiveria*
Spp.: *zizanioides*

Synonyms

Andropogon muricatus Retz., *A. squarrosus* Hook. f., non Linn. F., *Anatherum zizanioides* (Linn.) Hitchcock and Chase.

Sanskrit Synonyms

- Nalada, Amrnala, Veerana, Virani, Veeratara, Valaka, Balaka Truna Valukam, Veniga Mulakam Abhaya, Ranapriya, Virataru, Haripriya Jalvasa, Jalaamoda, grows in moist places.
- Amrunala, Mrunalaka – Appears similar to Lotus stalk.
- Sugandhika, Samagandika, Sugandhamula – Has a pleasant odor.
- Bahumoola – Bushy roots.
- Shishiram, Sheetmoola – root is a coolant.
- Sevya – suitable for consumption.

There are many ecotypes of vetiver. The vetiver called VS-01 (or sunshine) is a south Indian variety, has robust root system, and has 1% aromatic oils in its roots. Other varieties like VS-03 have less aromatic oil content (around 0.3%). Research has shown that the aromatic oil is produced in the roots by certain bacteria from precursors released by the vetiver. It is just to say that all vetiver roots are not the same.

Vetiver roots treat the mother earth too. It is perhaps mentioned in Krishi Shastra. Vetiver roots increase the ground water recharge capability of soils. It increases the biomass content in the soil up to 3 meters depth. It will be very difficult to add leaves and manure at that depth. Vetiver roots add the biomass as roots, and they decay after 18 months. Vetiver root oil is termite-repelling. The vetiver root system increases the aeration in the soil, it also play ecological role by helping to stabilize soil, protects it against erosion and prevents nematode (small worms that attack the roots and other plants) infection. Because of these effects, vetiver is grown as rows along vegetable farms, and around trees in orchard which improves the health and yield of the cultivated crops. Thus, growing neem, tulsi and vetiver in farms and around houses is good for the health of the plants and animals in the earth.

Two species of *Vetiveria* are found in India, of which *Vetiveria zizanioides* is the common source of the well known oil of Vetiver, which is used in medicine and in perfumery. Khas grass grows wild in many states, namely Haryana, Uttar Pradesh,

Table 13.1. Some Vernacular names for Khas Grass in India.

Sr. No.	Language	Vernacular Name
1	Sanskrit	Ushira, Amrnala, Veerana, Virani, Veeratara, Valaka, Balaka
2	Hindi, Bengali	Khas, Khas-Khas, Khus-Khus, Khus
3	Gujarati	Valo
4	Marathi	Vala
5	Telugu	Kuruverru, Vettiveellu, Vettiveerum
6	Tamil	Vattiver
7	Kannad	Vattiveru, Laamancha, Kaddu, Karidappasajj Hullu
8	Malyalam	Ramaccham vettiveru
9	Ayurvedic name	Ushira
10	Fijian name	Mulimuli
11	French name	Chiendent odorant, Petiver
12	Portuguese (Brazil) name	Patchuli-falso
13	Spanish name	Zacate violeta, Pacholi
14	Tongan name	Ahisiaina

Rajasthan, Gujarat, Bihar, Orissa and Madhya Pradesh and throughout South India. It is systematically cultivated in the North Indian states of Rajasthan, Uttar Pradesh and Punjab and in the South Indian states of Kerala, Tamil Nadu, Karnataka and Andhra Pradesh. The yield from the cultivated crops, however, meets only a very small percentage of the requirements of the country. The bulk of the roots used for cooling purposes and for the extraction of the oil are obtained from the wild.

It is also cultivated for the fragrant essential oil distilled from its roots and used in high end perfumes. Worldwide production is estimated at about 250 tons per annum. The oil is amber brown and rather thick.

History

It has been known to India since the ancient times. It has been considered as a high class perfume. Copper plate inscriptions listing the perfume as one of the articles used by royalty have been discovered. In Ayurvedic literature, it is called 'Suganti-mulaka' (meaning sweet smelling) and 'sita-mulaka' (having cool roots). All over India, the roots are made into scented mats, fans, ornamental baskets and many other articles and also burnt as a fumigatory. The aromatic roots are made into bundle and kept in water vase to impart aroma to water and getting cooling effect. It is also put into cold beverages like unripe mango juice or Indian cold drinks to impart its taste and coolness.

Morphological Description

It is densely tufted grass. The culms are arising from an aromatic rhizome. The grass is stout, up to and over 2 m tall, in dense tufts, with stout spongy aromatic roots. The leaves are narrow, erect, keeled, glabrous and its margins are scabrid. The inflorescence is panicle of numerous slender racemes in whorls on a central axis. The spikelets are grey-green or purplish in color and in pairs. One is sessile and other is predicelled. Both are alike in form and size, different in sex and 2-flowered. The lower floret is reduced to lemma, and upper is bisexual in the sessile. Male is in the pedicelled spikelet, glumes armed with short, tuber debased spines, and lemmas are without awns, palea minute.

Fig. 13.1. *V. zizanioides* whole plant.

Fig. 13.2. *V. zizanioides* roots.

Chemical Constituents

Principal constituents of oil are d and p vetivone, zizanal and epizizizanal, vetiselimenol, Khisumol, Allokhusiol, Benzoic acid, Cyclocapacamphene, Epikhusinol, 2-epizizanone, B-eduesmol, Eugenol, Iskhusimol, Isokhusinoloxide, Isovalencenol, Isovalencic, Khusimyl acetate, Khusinodiol, Khusinol, Khusitoneol, Laevojujenol, Levojunenol, Vanillin, Vertiselinenol, B-& J. Vetivene, Vetivenic acid, vetiverol, zizaene, Zizanol, etc. which were isolated from the oil. Due to the presence of these chemicals, it has its pharmacological properties (Mishra et al. 2013, Pareek and Kumar 2013).

Pharmacology

Zizanal and epizizanal exhibited insect repellent activity.

Parts Used

Roots, vetiver oil.

Doses

Powder 3–6 gm, decoction 15–100 ml, infusion 25–50 ml.

Medicinal Properties

Rasa (taste) – Tikta (bitter), Madhura (Sweet), Guna (qualities), Rooksha (dryness), Laghu (Lightness), Vipaka – Katu – undergoes pungent taste conversion after digestion, Veerya – sheeta – cold potency effect on tridosha – pacifies vata and pitta.

Uses

Khas grass has variety of uses from household to therapeutic.

Ethnomedicinal Uses

About 70% of rural folk depend on medicinal plants for their healthcare. Since ancient times, ethnobotanical use of plants has been known and use of traditional medicine and medicinal plants in most developing countries as therapeutic agents for the maintenance of good health has been widely observed and accepted (Rao et al. 2016).

Various tribes use the different parts of the grass for many of their ailments such as mouth ulcer, fever, boil, epilepsy, burn, snakebite, scorpion sting, rheumatism, fever, headache, etc. The Santhal tribes of Bihar and West Bengal use the paste of fresh roots for burn, snakebite and scorpion sting, and a decoction of the roots as a tonic for weakness; in West Bengal, the tribal people use the root paste for headache, rheumatism and sprain, and a stem decoction for urinary tract infection; in Madhya Pradesh, the leaf juice is commonly used as anthelmintic; the tribes of the Varanasi district inhale the root vapor for malarial fever. The root ash is given to patients for acidity by the Oraon tribe. Likewise, there are many different applications of the plant for different ailments among different ethnic tribes (Singh and Maheshwari 1983, Jain 1991).

Decoction of the rhizome is taken internally to treat blood pressure and stomach ache problems in Kumargiri Hills, Tamil Nadu (Kam and Alagesaboopathi 2009). A decoction of leaves is recommended as a diaphoretic. When locally applied in rheumatism, lumbago and sprain, it is good ambrocation and affords relief. In M.P., the plant is used as an anthelmintic for children. The oil is reported to be used as a carminative in flatulence, colic and obstinate vomiting. It is regarded as stimulant, refrigerant and antibacterial and when applied externally, it removes excess heat from the body and gives a cooling effect. A decoction of leaves in recommended as a diaphoretic used in therapy, as blood purifier, for calming effect on nervous system, in ringworm treatment, for treating many skin disorders, in indigestion and loss of appetite.

Vetiver Essential Oil Uses

It is widely used in perfumes and for scenting soaps. It blends well with the oils of sandalwood and rose. Vetiver oil is used for external application in rheumatism, sprains and arthritis. The health benefits of Vetiver essential oil can be attributed to its properties like anti-inflammatory, antiseptic, aphrodisiac, cicatrisant, nervine, sedative, tonic and vulnerary. This Essential Oil is very popular in aroma therapy and has many medicinal properties, which are described in brief below.

Anti Inflammatory

The very soothing and cooling effect of this essential oil calms and pacifies all sorts of inflammations. But it is particularly good in giving relief from inflammations in circulatory system and nervous system. It is found to be an appropriate treatment for inflammations caused by sun stroke, dehydration and loo (name given to very hot and dry winds prevalent during summers in the dry regions of India and few neighboring countries).

Antiseptic

In tropical countries like India and its neighbors, microbes and bacteria grow very fast due to their favorable hot and humid climate found in this region. Then it becomes obvious that your wounds are most likely to get septic in these places since there are plenty of bacteria here. But Mother Nature is very kind and she has provided the remedies too, right in those places. One such remedy is this Vetiver and the essential oil extracted from it. This oil efficiently stops the growth of *Staphylococcus aureus*, the bacteria responsible for causing septic, and eliminates them, thereby helping cure septic and giving protection against it. Being totally safe, this oil can be applied externally on wounds or taken orally, to protect wounds as well as internal organs from septic.

Aphrodisiac

Mixed in sorbets and beverages as a flavoring agent, this oil has an aphrodisiac effect. It enhances libido and gives arousals. Since sex has more to do with the psychology (brain) than the physiology, remedy for most of the sexual disorders like frigidity, lack of libido, impotence, etc. lays in the brain. Certain components of this oil stimulate those portions of brain and the problems are over.

Cicatrisant

Cicatrisant is a property by virtue of which a substance speeds up the eradication or disappearance of the scars and other marks from the skin. It promotes growth of new tissues in the affected places which replace the dead and discolored tissues and helps achieve a uniform look. This is also useful for the post delivery stretch marks, fat cracks, after spots left by pox, burns, etc.

Nervine

A tonic for the nerves is called a nervine, like our Essential Oil of Vetiver is. It takes care of the nerves and maintains them in good health. It also heals the damages done to the nerves by shock, fear, stress, etc. Further, it helps get rid of nervous disorders, afflictions, epileptic and hysteric attacks, nervous and neurotic disorders such as Parkinson's Disease, lack of control over limbs, etc.

Sedative

The Essential Oil of Vetiver is a well known sedative. It sedates nervous irritations, afflictions, convulsions and emotional outbursts such as anger, anxiety, epileptic and hysteric attacks, restlessness, nervousness, etc. and even benefits patients of insomnia.

Tonic

The effect of a tonic on the body is quite similar to that of overhauling and servicing on a vehicle. A tonic tones up every system functioning in the body, namely the digestive system, respiratory system, circulatory system, excretory system, immune system, endocrinal system, nervous system and the neurotic system. Thus, in a nutshell, it keeps the metabolic system in order, rejuvenates the body, gives strength and boosts immunity.

Vulnerary

This property of Vetiver Essential Oil helps heal wounds by promoting growth of new tissues at the wounded place and also by keeping it safe from infections by inhibiting growth of microbes and promoting crowding of leucocytes and platelets at the place.

Healing

Vetiver essential oil helps in the formation of new tissue and is also used to accelerate the healing and recovery of skin wounds as well to remove stains, marks on the skin and the scars themselves. Also, it is used to repair the cracks and grooves in the skin caused by different circumstances such as pregnancy, diets, allergies, and burns.

Calming

In addition to various beverages for culinary purposes and aphrodisiacs, vetiver essential oil is used to make soothing infusion which is used to relax and recover from severe strain. It helps to overcome situations of shock, fear, high levels of stress, panic, etc.

Other Benefits

Other benefits that we tend to award to the use of vetiver essential oil are, for example, the strengthening of bones, the treatment of rheumatism, gout, arthritis, muscle aches, dryness, cramps and dry skin (Balasankar et al. 2013).

Ayurvedic Uses

- Pachana – Digestive, relieves Ama dosha,
- Stambhana – blocks, dries up channels,
- Dahahara – relieves burning sensation as in gastritis, neuropathy, burning sensation in eyes, etc.,
- Madahara – relieves intoxication,
- Jwarahara – useful in fever,
- Trushnahara – relieves thirst,
- Asrajit – useful in blood disorders such as abscess,
- Vshahara – antitoxic,
- Daurgandnyahara – relieves bad odor,
- Mutrakrichrahara – relieves dysuria, urinary retention, acts as diuretic,
- Kushtanut – useful in skin disorders,
- Baminut – relieves vomiting,
- Vranahara – heals wounds quickly,
- Klantihara – relieves tiredness, fatigue,
- Vataghna – Useful in treating disorders of vata dosha imbalance such as neuralgia, paralysis, constipation, bloating, etc.,
- Mchanut – useful in urinary tract disorders and diabetes,
- Vetiver water – Vetiver is used to make potable water. A few grams of vetiver root powder is added to water and this water is filtered and used after 2–3 hours. Once the vetiver is put into water, the water should be replaced with fresh ones. Such prepared water will have almost all the qualities of Khas-usheer, as explained above, but in slightly lesser intensities. (www.easyayurveda.com)

External Application

To apply externally as paste, to relieve burning sensation, skin disorders and to relieve excessive sweating, there is nothing better than khas (Charaka Sutrasthana

25). The root powder is made into paste and applied to relieve burning sensation, burn wounds, to relieve excessive sweetening, in skin disorders and to improve skin complexion.

Mode of Administration

- Its water can be used as potable water.
- Its powder or water decoction is also used in medicine. It is administered along with cow ghee, honey, etc., based on disease.
- Its powder is applied externally.

Commercial Products

- Used, Renalka, anti-wrinkle cream, baby powder, gentle baby shampoo, oil clear mud face pack, purifying neem foaming face wash.
- Vetiver essential oil is used as ingredient in many Ayurvedic medicines such as Borototal cream.
- Ayurvedic medicines with Khas (Usheera) as ingredient.
- Usheerasava—a fermented medicine used in skin diseases, intestinal worms, bleeding disorders, etc.
- Shadanga Paniya—A herbal drink used to treat fever.
- Gopanganadi Kashayam—Used in pitta type of fever.
- Nisosiradi oil—An effective Ayurvedic herbal oil used to treat diabetes, car buncles and abscesses.
- The annual market allotment of vetiver oil.

Country	Percentage
USA	40
France	20
Switzerland	12
England	10
Japan	04
Germany	2.4
The Netherlands	02
Countries of vetiver oil origin	12–16

Bioactive Compounds

Roots and leaves show presence of alkaloid, anthracene glycosides, aucubins, coumarin, emodin, fatty acid, flavonoid, polyuronoid, starch, tannin, reducing compounds, saponin, steroids, triterpenoids, volatile oil, and protein indicating the presence of their respective chemical compounds (Kaikade and Ingole 2014, Savanur 2017).

The GC-MS analysis of roots of *Vetiveria zizanioides*, which are medicinal part, revealed the presence of various compounds (phytochemical constituents) at different retention time that could contribute to the medicinal quality of the plant. The identification of the phytochemical compounds was confirmed based on peak area, retention time and molecular formula. Its root's GC-MS analysis shows presence of dibutyl phthalate ($C_6H_{22}O_4$), benzene-dicarboxylic acid ($C_6H_{22}O_4$), diisooctyl phthalate ($C_{24}H_{38}O_4$)—all possessing antimicrobial and antifouling properties; pthalic acid ($C_{16}H_{26}O_4$); bis (2-methyl propyl) ester; and butyl 2-methyl-propyl ester. Roots also show presence of Cedren-13-ol ($C_{15}H_{24}O$) possessing antioxidant property; antimicrobial, immune-modulatory, anti tumor property; spathylenol ($C_{15}H_{24}O$), 12-epoxide ($C_{15}H_{22}O_2$) as building blocks for the synthesis of a number of pharmaceutical compounds; and H-2-Indenol ($C_{13}H_{22}O_2$).

The phytochemicals identified through GC-MS analysis possess many biological properties (Kaikade and Ingole 2014).

Table 13.2. List of expected compounds at RT – 13.98.

S. N.	Component	Retention Time (RT)	Expected Compound Name	Molecular Formula
1	1	13.98	Naphthalene	$C_{12}H_{22}$
		13.98	Decahydro-1,1-dimethyl-cyclohexane	$C_{13}H_{24}$

Table 13.3. List of expected compounds at RT – 16.47.

S. N.	Component	Retention Time (RT)	Expected Compound Name	Molecular Formula
1	2	16.47	2-(3,4-Methylenedioxyphenyl) cyclohe-xanone	$C_{13}H_{14}O_3$
		16.47	4-Methoxymethyl-4-phenyl-1-butene	$C_{12}H_{16}O$
		16.47	n-Propl cinnamate	$C_{12}H_{14}O_2$

Various Known Pharmacological Activities

- **Antioxidant Activity**

  Free radicals induce numerous diseases by lipid peroxidation and DNA damage. It has been reported that some of the extracts from plants possess antioxidant properties capable of scavenging free radicals *in vivo*. *Vetiveria zizanioides* is a densely tufted grass which is widely used as a traditional plant for aromatherapy, to relieve stress, anxiety, nervous tension and insomnia (Devi et al. 2010, Hewawasam and Jayatilaka 2017).

- **Antifungal Activity**

  The antifungal activity was reported against *Aspergillus nigar*, *Aspergillus clavatus* and *Candida albicanus* using ethanolic and aqueous extract of Vetiver (Dev Prakash et al. 2011).

- **Antimicrobial Activity**

  It showed antimicrobial activity, especially vetiverin, and more likely a result of additive and synergistic effect of several compounds (Nantachit et al. 2010).

- Research showed that the Vetiver oil possessed a strong free radical scavenging activity when compared to standard antioxidants such as butylated hydroxytoluene (BHT) and α-tocopherol (Kim et al. 2005).

- **Antiproliferative Activity**

  Research showed that Vetiver oil exhibited strong antiproliferative activity in a pre-inflammed human dermal fibroblast model cells and significantly inhibited the production of collagen, an important molecule for skin and tissue remodeling processes (Han and Parker 2017).

- **Antibacterial Activity**

  The antibacterial activity is measured by zone of inhibition (mm). Totally four bacterial strains (two gram positive *S. aureus, B. subtilis* and two gram negative bacteria *P. aeurogenosa, E. coli*). Ethanolic extract of Vetiveria Zizanioides is known to posses flavonoids, alkaloids, terpenoids, saponins, tannins and phenols which, either individually or through combination, exert antimicrobial activity. The study showed that EEVZ inhibited gram negative bacteria than grampositive bacteria. Flavonoids are found to be effective antimicrobial substance against a wide range of microorganisms, probably due to their ability to complex with extra cellular and soluble proteins and to complex with bacterial cell wall; more lipophilic flavonoids may disrupt microbial membrane. Antibacterial activity of tannins may be related to their ability to inactivate microbial adhesion enzymes and cell envelope transport proteins; they also complex with polysaccharides. The presence of tannins present in the roots of *Vetiveria zizanioides* implied that tannin may be the active compound which may be responsible for *in vitro* antibacterial activity in this study. Tannin in the plant extract was found to possess antibacterial activity (Devi et al. 2010).

- **Hepatoprotective Activity**

  Methanolic extract of *Vetiveria zizanioides* Linn is hepatoprotective at the dose 300–500 mg/kg. The damage induced by ethanol 20% at the dose of 3.75 gm/kg (Chaudhary 2010).

- **Antitubercular Activity**

  The ethanolic extract of intact as well as spent root showed potent antituberculosis activity at a minimum concentration of 500 µg/mL (Saikia et al. 2012).

- **Mosquito Repellent Activity**

  In the laboratory oviposition deterrent test, the root extract of *Vetiveria zizanioides* at each concentration greatly reduced the number of eggs deposited by the gravid *Anapheles stephens* (Arthi and Murgan 2011).

- **Antihyperglycemic Activity**

  The effect of root extract of *Vetiveria zizanioides* in normal fasted rats after multiple doses showed significant antidiabetic activity at 2nd and 4th hour after administration compared to diabetic control (Karam et al. 2012).

- **Antidepressant Activity**

  The ethanolic extract of *Vetiveria zizanioides* possesses antidepressant activity and the combination of Fluxetine and ethanolic extract of *Vetiveria zizanioides* is effective in tail suspension test, and force swim test induced depressive behavior (Josephine 2012).

- **Other Uses**

  Apart from medicinal uses, there are multiple uses of khas grass in India.
  - The culms along with the panicles form a good broom.
  - Traditional medicine.
  - Roots as water flavoring agents.
  - Root mats for door, window screen during summer.
  - Socio-economic life of the rural population in India.
  - Dried roots for scenting clothes.
  - Dried roots for brooms and for thatching.
  - Pulp of the plant for paper and straw board.
  - Culms and leaves are also extensively used by tribals and villagers for thatching their huts and mud walls.
  - Some Kerala tribes use the mats of the roots and leaves as a bed for a cooling effect.

Domestic Use

Mats made by weaving Vetiver roots and binding them with ropes/cords are used in India to cool rooms in a house during summer. The mats are typically hung in the doorway and kept moist by spraying with water periodically. It acts like an air-cooler when wind from a fan or outside hits it. It also adds a pleasant aroma in the house which is commonly described as "cool" and "refreshing".

Commercial Uses

The commercial applications of the grass mainly pertain to the extraction of Vetiver oil through distillation of the roots. Vetiver oil is one of the most valuable and important raw materials in perfumery and finds extensive applications in the soap and cosmetic industries, for pharmaceutical companies and as antimicrobial and antifungal agent (Singh et al. 1978, Dikshit and Husain 1984). Over 150 compounds have been isolated and characterized from Vetiver oil so far. A major portion of oil consists of sesquiterpene alcohol (Thakur et al. 1989). A major application of the roots of Vetiver, particularly in North Indian plains, pertains to the preparation and sale of mats/screens for windows, doors and desert coolers during summer months when the temperature goes up to as high as 45°C.

As there is no systematic cultivation of the grass in North India, several villagers and rural folk collect the roots of the grass in large quantities from the wild growth and flock to the cities where they sell the root mats and loose roots for flavoring water to city dwellers. 3–4 months of livelihood of several rural families is sustained by the sale of roots of Vetiver grass playing important role in socio-economic sector of rural life.

It also plays an ecological role by helping to stabilize soil, protects it against erosion and also fields against pests and weeds.

Multiple Benefits of Vetiver System

Disaster risk management

- Natural defense against hazards.
- Vulnerability reduction.
- Eco-friendly and cost effective mitigation.

Ecosystem management

- Land stabilization.
- Carbon sequestration.
- Pollution control.

Social development

- Employment generation.
- Strengthen coping capacity.
- Community participation (Joseph et al. 2017).

Uses in Variety of Areas

Vetiver's stunning mass of deep, strong, fibrous roots and thick thatch of stiff leaves have led to its extensive use in a variety of areas:

- Findings show potential of Vetiveria grass in phytoremediation for heavy metals' removal in water, thus providing significant implication for treatment of metal contaminated water (Ashton et al. 2017).
- As a nurse crop—Vetiver stabilizes and replenishes nutrients in highly degraded areas. Rehabilitated sites welcome the return of native plants.
- As a privacy barrier—Vetiver forms a dense barrier that defeats pryingeyes and creates a serene green paradise. It creates a beautiful, economical perimeter on small, urban lots.

- Private companies and municipalities use Vetiver systems to protect and heal degraded environments. Vetiver roots absorb pollutants and clarify water.
- As a grass wall and boundary marker, Vetiver hedges are so stable that surveyors rely on them to establish property lines.
- Between slender rows of Vetiver, farmers can grow crops that benefit from the accumulation of silt and plant nutrients. Vetiver's vertical roots nurture adjacent crops.
- As an excellent batch material—Mature leaves produce long-lasting absorbent mulch that reduces evaporation and helps mycorrhizae to accumulate.
- As a bios wale—A Vetiver grass channel is an attractive alternative to traditional concrete drainage ditches, and effectively filters and attenuates storm water runoff.
- As a constructed wetland- Installed as a leach field, Vetiver absorbs nutrients generated by cesspools, piggeries, dairy and poultry farms. Vetiver clarifies effluent and eliminates odors.
- As livestock feed—Vetiver's nutritional value is similar to Napier grass (*Pennisetum pupureum*).
- As a carbon sink—Given the concern regarding global warming and CO_2 emissions, 44,500 acres of land protected by Vetiver hedges will provide a CO_2 sink for the carbon produced by 100,000 cars traveling 12,500 miles a year.
- As biofuels—Leaves can be used as a substrate for ethanol production through alkali pretreatment followed by enzyme hydrolysis and yeast fermentation, which generates an ethanol yield of 13% after one-cycle column distillation.
- As a food additive—Vetiver is used domestically in cooking; it's infused in tea and also used in baking.
- Handicrafters use Vetiver leaves and roots to create an extensive range of beautiful woven handicrafts. Like its sister, bamboo, which creates luxurious textiles, Vetiver would seem suited to produce soft, durable fabric.
- Vetiver is a beautiful ornamental plant for gardens, patios, decks, etc. It forms a dense, uniform and attractive hedge under tropical and subtropical climates. It also forms an aesthetically beautiful barrier to unsightly view.
- Agriculture-Related Activities: Used as a Mulch, compost, animal feed, substrate for mushroom cultivation, botanical pesticides, fungicides, agricides and weed control (Balasankar et al. 2013).

Contraindications, Interactions and Side Effects

Vetiver has abortifacient, emmenagogue, and uterotonic (induce contraction or greater tonicity of the uterus; uterotonics are used both to induce labor, and to reduce postpartum haemorrhage) properties. So its use is contraindicated in pregnancy. It is not suitable for babies and children under twelve years (http://www.bimbima.com).

Conclusion and Future Perspectives

Vetiver grass is a versatile grass possessing unique characteristic features and is a source of several bioactive compounds which are responsible for its wide array of medicinal and pharmacological activities, proving the wisdom of ethnic tribes to use it for several of their ailments and keeping it at leading position among several medicinal plants they use. There lies a great scope for new drugs' formulations. Owing to its multiple uses, establishment of vetiver industry can solve various problems like health related, poverty reduction, employment, ecological management particularly water and soil related. It has a great scope in pharmacological industries and in agroforestry.

In view of its tremendous use and its declining populations, large scale systematic cultivation of this grass in the plains of Indian states counting both the grass and the root system is recommended.

Because of various pharmacological activities, medicinal properties and presence of bioactive molecules, *Vetiveria zizanioides* has vast applications in medicinal sector to find new drugs and there is an increase in interest in the health and wellness benefits of herbs and botanicals. This is with good reason as they might offer a natural safeguard against the development of certain conditions and be a putative treatment for some diseases. Because of many properties of Vetiveria grass, it has found vast applications making it a green treasure (Mishra et al. 2013).

Considering its environmental implications, socio-economic aspects and industrial potential, Vetiver grass should be planted on large scale and utilization of Vetiver as a medicinal plant to produce pharmaceutical products on a commercial scale has great potential for development. A new concept, that of growing vetiver as an income generating plant, can be launched and it will prove to provide a very good income to the farmers if grown specifically for its roots.

Vetiveria zizanioides has wide ecological amplitude and this trait of the species must be exploited for eco-developments of the region devoid of biodiversity. The grass with its tuft-forming habit and thick root system greatly helps in checking soil erosion. It can be recommended for fallow areas and waste places including sodic soils. This will not only boost the economic conditions of local farmers but also improve the soil ecology.

Small scale village-level industries based on Khas grass could be established for extraction of vetiver oil, and for manufacture of straw board and handmade paper from pulp of the aerial parts of the grass. This can, to some extent, reduce the stress on bamboo resources, which are also declining. As this grass is invariably used by most of the tribes, the tribal and other village womenfolk should be encouraged (with suitable subsidies) to cultivate the species near their huts. Suitable arrangement can be made to collect the excess harvest left from these tribal pockets for trade or for local oil industries so that tribal families can also supplement their income. While this would bring the economy to the poor villagers, it would also help in the conservation and eco-development of the region.

Vetiver and agroforestry for poverty reduction in third world and natural resources protection large vetiver nurseries can be developed by farmers for the sale of planting material to the other farmers and construction agencies for Vetiver handicraft component industry. Use of vetiver for sustainable environment particularly in relation to land and water can be recommended.

Vetiver can produce up to 100 tonnes/hectare of biomass (dry weight) within 8 months. The grass is only harvested as a biomass fuel and the perennial vetiver grass is left in the ground and allowed to grow back year after year lasting forever.

Ecosystem based regional development programs are emerging worldwide and hence green technologies and systems for the prevention and mitigation of natural hazards need to be developed. There is a need to highlight the importance of unique grass like vetiver that has many unique characteristics.

References

Antiochia, R., Campanella, L., Ghezzi, P. and Movassaghi, K. 2007. The use of vetiver for remediation of heavy metal soil contamination. Anal. Biochem. 388: 947–956. [Google Scholar]

Arctander, S. 1960. Vetiver. Columns. pp. 649–653. *In*: Perfume and Flavour Materials of Natural Origin. Elizabeth, New Jersey. Published by the author; printed by Det Hoffensbergske Etablissement, Denmark.

Arthi, N. and Murgan, K. 2011. Effect of *Vetiveria zizanioides* L. Asian Pacific Journal of Tropical Disease 154–158.

Ash, R. and Truong, P. 2003. The use of Vetiver grasswetland for sewerage treatment in Australia. Proc. Third International Vetiver Conf. China, October 2003.

Ashton Lim, Suelee Sharifah, Nur Munirah, Syed Hasan, Fara diella Mohd. Kusin, Ferdaus Mohamat Yusuff and Zelina Zaiton Ibrahim. 2017. Phytoremediation potential of Vetiver grass (*Vetiveria zizanioides*) for treatment of metal contaminated water. Springer International Publishing, Switzerland 228: 158.

Astin, J. 1998. Why patients use alternative medicines. Results of National Study. J. Ann. Med. Assoc. 279: 1548–1553.

Bajpai, P.N., Singh, I., Tiwari, L.P., Chaturvedi, O.P. and Singh, J.P. n.d. Varietal performance of khus (*Vetiveria zizanioides* Stapf.). The Punjab Horticultural Journal n.v.: 208–211.

Balasankar, D., Vanilarasu, K., Selva Preetha, P., Rajeshwari, S., Umadevi, M. and Bhowmik Debjit. 2013. Traditional and medicinal uses of vetiver. Journal of Medicinal Plants Studies 2(3): 191–200.

Bertea, C.M. and Camusso, W. 2002. Anatomy, biochemistry and physiology. pp. 19–43. *In*: Maffei, M. (ed.). Vetiveria. The Genus Vetiveria. London and New York: Taylor and Francis Publishers. [Google Scholar]

Bharad, G.M. and Bathkal, B.C. 1990. Role of vetiver grass in soil and moisture conservation. *In*: Proceedings of The Colloquium on the Use of Vetiver for Sediment Control. April 25, 1990. Watershed Management Directorate, Dehra Dun, India.

Bhatwadekar, S.V., Pednekar, P.R. and Chakravarti, K.K. 1982. A survey of sesquiterpenoids of vetiver oil. pp. 412–426. *In*: Atal, C.K. and Kapur, B.M. (eds.). Cultivation and Utilization of Aromatic Plants. Regional Research Laboratory, Council of Scientific & Industrial Research, Jammu-Tawi, India.

Bisagacarya. Girindranath. Mukhopadhyaya. 1994. History of indian medicine, New Delhi: Munishiram Manoharlal publishers pvt LTD, Vol III, 496–499 pp.

Bowen, H.J.M. 1979. Plants and the Chemical Elements. London: Academic Press. [Google Scholar]

Chadha, K.L. 2011. Hand Book of Horticulture, Vetiver, ICAR, New Delhi, 631–632.

Chen, Y., Shena, Z. and Lib, X.D. 2004. The use of vetiver grass (*Vetiveria zizanioides*) in the phytoremediation of soils contaminated with heavy metals. Appl. Geochem. 19: 1553–1565.

Chomchalow, N. and Hicks, P.A. 2001. Health Potential of Thai Traditional Beverages. Paper presented at the 34th AISFT Annual Convention 2001, Adelaide, Australia, 1–4 July 2001; also published in AU J.T. 5: 20–30.

Chomchalowm, N. 2002. Review and Update of the Vetiver System R&D in Thailand. Proc.

Chopra, R.N., Nayar, S., Chopra, I.C. 1956. Glossary of Indian medicinal plants, NISCAIR, 1st edition 1956: 254.

Chou, S.T., Lai, C.P. Lince and Shin, Y. 2012. Study of the Chemical composition, antioxidant activity and anti-inflammatory activity of essential oils from *Vetiveria zizanoides* (L.) Nash. Food Chemistry 134: 262–268.

Clark, L.G. and Fisher, J.B. 1987. Vegetative morphology of grasses: shoots and roots. pp. 37–45. *In*: Soderstrom, T.R. (ed.). Grass Systematics and Evolution. Smithsonian Institution Press, Washington.

Council of Scientific and Industrial Research (CSIR). 1976. Vetiveria. pp. 451–457. *In*: The Wealth of India, Vol. X. Publications & Information Directorate, CSIR, New Delhi.

Cupp, M. 1999. Herbal remedies: Adverse effects and drug interaction. Ann. Fam. Physician 59: 1239–1244.

Dalziel, J.M. 1937. Useful Plants of West Tropical Africa. Crown Agents for the Colonies, London.

Dashora, K. and Gosavi, K.V. 2013. Grasses: An underestimated medicinal repository. Journal of Medicinal Plant Studies 1(3): 151–157.

DeSmet. 2002. Herbal remedies. New Engl. J. Med. 347: 2046–2056.

Dikshit, A. and Husain, A. 1984. Antifungal action of some essential oil against animal pathogens. Fitoterapia 55: 171–176.

Dubey, N.K. and Tripathi, P. 2004. Global promotion of herbal medicine: India's opportunity. Curr. Sci. 86: 37–41.

Elatler, E. and McCann, C. 1928. Revision of the flora of the Bombay Presidency. Gramineae. Journal of the Bombay Natural History Society 32: 408–410.

Gogte, V.M. 2000. Ayurvedic pharmacology and therapeutic uses of medicinal plants, 1st ed., Mumbai: Bharatiya vidya Bhavana 2000: 319–320 pp.

Gould, F.W. and Clark, C.A. 1983. Grass Systematics. 2nd edition. Texas A&M University Press, College Station, Texas.

Greenfield, J.C. 1987. Vetiver Grass (*Vetiveria zizanioides*). A Method of Soil and Moisture Conservation. 1st ed. The World Bank, New Delhi.

Greenfield, J.C. 1988. Vetiver Grass (*Vetiveria zizanioides*). A Method of Soil and Moisture Conservation. 2nd ed. The World Bank, New Delhi.

Greenfield, J.C. 1989. Novel grass provides hedge against erosion. *VITA News*: 2014–15 Greenfield, J.C. 1989. *Vetiver Grass* (*Vetiveria* spp.): The Ideal Plant for Vegetative Soil and Moisture Conservation. Asia Technical Department, Agriculture Division, The World Bank, Washington.

Greenfield, J.C. 2002. Vetiver Grass: An essential grass for conservation of planet earth. Haverford, PA: Infinity Publishing Co. [Google Scholar]

Han Xuesheng and Tory, L. Parker. 2017. Biological Activity of Vetiver (Vetiveria zizanioides) essential oil in human dermal fibroblasts Cogent medicine, 4 (http://dxdoi.org101086/233120 4x2017.1298176).

Hart, B., Cody, R. and Truong, P. Efficacy of vetiver grass in the hydroponic treatment of post septic tank effluent. October 6–9, Guangzhou, China. Proc. Third Int. Vetiver Conf. (ICV-3).

Hoang, T.T.T., Tu, T.C.L. and Dao, P.Q. 2007. Progress and results of trials using vetiver for phytoremediation of contaminated canal sludge around Ho Chi Minh City. Hanoi, Vietnam. Proc. Vetiver Workshop, Vietnamese.

Hooker, J.D. 1975. Flora of British India. Bishen Singh Mahendra Pal Singh, Dehra Dun, India.

http://interscience.org.uk/index.php/archive/47-volume-7-issue-3-may-june-2017/221-phytochemical-study-of-usheera-vetiveria-zizanioides-linn-nash.

http://www.bimbima.com.

http://www.easyayurveda.com.

http://www.tandfonline.com/doi/ref/10.1080/15226510902787302?scroll=top.

http://www.vetiver.org/TVN_refs.htm.

Hyun-Jin Kim, Feng Chen, Xi Wang, Hau Vin Chung and Zhengyu Jin. 2005. Evaluation of antioxidant activity of vetiver (*Vetiveria zizanioides* L.) Oil and identification of its antioxidant constituents. J. Agric. Food Chem. 53(2): 7691–7695.

Jain, S.K. 1991. Dictionary of Indian Folk Medicine and Ethnobotany. Deep Publ., New Delhi, India.

Joseph, J.K., Haridasan, A., Akhildev, K. and Pradeep Kumar, A.P. 2017. Applications of vetiver grass (*Chrysopagon zizanioides*) in Ecosystem based disaster risk reduction studies from Kerala state of India. J. Geogr. Nat. Disast. 7: 192, doi: 10.4172/2167-0587.1000192.

Juliard, C. 2001. Manuscript of Vetiverim 16. In Letters to the Editor, Vetiverim 17: 15.

Kaikade, R.S. and Ingole, S.N. 2014. Study of Anatomical and Phytochemical Biomarkers of Some Medicinal Plants mentioned in Atharvaveda, Ph.D. Thesis submiteed to S.G.B. Amravati University, Amravati.

Kam, A.J. and Alagesaboopathi, C. 2009. Ethnomedicinal plants and their utilization by villagers in Kumaragiri Hills of Salem District of Tamil Nadu, India. African Journal of Traditional, Complementary and Alternative Medicines 6(3): 222–227.

Kamboj, V.P. 2000. Herbal Medicine. Curr. Sci. 78(1): 35–39.

Khesorn Nanachit, Manasnant Bunchoo, Banyong Khantava and Chantana Khamvan. 2010. Antimicrobial activity of roots of *Vetiveria zizanoides* (L.) Nash. Thai Pharmaceutical and Health Science Journal 5(2): 99–1026.

Kirtany, J.K. and Paknikan, S.K. 1971. North Indian vetiver oils: comments on chemical composition and botanical origin. Science and Culture 37: 395–396.

Kirtikar and Basu. 2000. Indian medicinal plants, 3rd ed., Delhi: Sri Satguru publications, 3699 pp.

Kokate, C.K., Purohit, A.P., Gokhale, S.B. and Furia, D.K. 1998. Pharmacognosy. 10th Edition, Nirali Publication, Pune, pp. 319–320.

Liu, P., Zheng, C., Lin, Y., Luo, F., Lu, X. and Yu, D. 2003. Dynamic State of Nutrient Contents of Vetiver Grass. Proc. Third International Vetiver Conf. China, October 2003.

Mabberley, D.J. 1989. The Plant-book: A portable dictionary of the higher plants. Reprinted with corrections, Cambridge University Press, Cambridge.

Maistrello, L. and Henderson, G. 2001. Vetiver grass: Useful tools against Formosansubterranean termites. Vetiverim 16: 8.

Manzoor-i-Khuda, M., Faruq, M.O., Rahman, M., Yusuf, M., Wahab, M.A. and Chowdhury, J. 1984. Studies on the essential oil bearing plants of Bangladesh. Part 1. A preliminary survey of some indigenous varieties. Bangladesh Journal of Scientific and Industrial Research 19(1-4): 151–169.

Martinez, J., Rosa, P.T.V., Menut, C., Leydet, A., Brat, P., Pallet, D. and Meireles, M.A.A. 2004. J. Agr. Food Chem. 52: 6578–6584.

Mishra Snigdha, Sharma Satish, Kumar Mohapatra Sharmistha and Chauhan Deepa. 2013. An overview on *Vetiveria zizanioides*. Research Journal of Pharmaceutical, Biological and Chemical Sciences 4(3): 777–783.

Murti, K.S. and Moosad, C.R. 1949. South Indian vetiver root study. American Perfume and Essential Oil Record 54: 113–115.

Nair, E.V.G., Channamma, N.P. and Kumari, R.P. 1982. Review of the work done on vetiver (*Vetiveria zizanioides* Linn.) at the Lemongrass Research Station, Odakkali. pp. 427–430. *In*: Atal, C.K. and Kapur, B.M. (eds.). Cultivation and Utilization of Aromatic Plants. Regional Research Laboratory, Council of Scientific and Industrial Research, Jammu-Tawi, India.

Nanachit Khesorn, Manasnant Bunchou, Banyong Khantava and Chantava Khamuan. 2010. Antimicrobial activity of alkaloids from roots of *Vetiveria zizanoides* (L.) Nash. Thai Pharmaceutical and Health Science Journal 5(2): 99–1026.

National Research Council. 1993. Vetiver Grass: A Thin Green Line Against Erosion. National Academy Press, Washington.

National Resource Council. 1995. Vetiver Grass: A Thin Green Line Against Erosion. Washington, DC: National Academy Press.

Pareek Archana and Ashwani Kumar. 2013. Ethnobotanical and Pharmaceutical Uses of *Vetiveria zizanoides* (Linn) Nash: A medicinal plant of Rajasthan. International Journal of Life Sciences and Pharma Research 3(4): 1–12.

Pichai, N.M.R., Samjiamjiaras, R. and Thammanoon, H. 2001. The wonders of a grass, Vetiver and its multifold applications. Asian Infrastruct. Res. Rev. 3: 1–4.

Rao Srinivasa, G.M., Narasimha Rao and Prayaga Murthy. 2016. Diversity and indigenous uses of some ethnomedicinal plants in papikondalu wildlife sanctuary, Eastern Ghats of Andhra Pradesh, India. American Journal of Ethnomedicine 3(2): 6–10.

Sastry, K.N.R. 1998. Socio-economic dimensions of vetiver in rainfed areas of Karnataka, India. Proc. ICV-1, Chiang Rai, Thailand, 243–248.

Sellar, W. 1992. The Directory of Essential Oils. C.W. Daniel Co Ltd., Great Britain.

Sethi, K.L., Chandra, V. and Singh, A. 1976. Adaptability of vetiver hybrid clones to saline-alkali soils. pp. 166–169. *In*: Proceedings of Second Workshop on Medicinal and Aromatic Plants. Gujarat Agricultural University, Anand, India.

Singh, G., Singh, B.S. and Kumar, B.R. 1978. Antimicrobial activity of essential oils against keratinophilic fungi. Indian Drugs 16(2): 43–45.

Singh, K.K. and Maheshwari, J.K. 1983. Traditional phytotherapy amongst the tribals of Varanasi district, U.P.J. Econ.Tax. Bot. 4: 829–838.

Singh, B. and Sankhala, K.S. 1957. From khas roots to rooh khas. Indian Forester 83(5): 302–306.

Smyle, J.W. and Magrath, W.B. 1990. Vetiver grass—a hedge against erosion. Paper presented at the annual meeting of the American Society of Agronomy, October 2, 1990, San Antonio, Texas.

Thakur, R.S., Puri, H.S. and Akhtar, H. 1989. Major medicinal plants of India. pp. 521–527, CIMAP, Lucknow, India.

Truong, P. 2002.Vetiver grass technology. pp. 114–132. *In*: Maffei, M. (ed.). Vetiveria. The Genus Vetiveria, London and New York: Taylor and Francis Publishers.

Truong, P. and Smeal, C. 2003. Research, Development and Implementation of Vetiver System for Wastewater Treatment: GELITA Australia. Pacific Rim Vetiver Network Tech. Bull. 3.

Virmani, O.P. and Data, S.C. 1975. *Vetiveria zizanioides* (Linn.) Nash. Indian Perfumer 19: 35–73.

Watson, L. and Dallwitz, M.J. 1988. Grass Genera of the World (with microfiche and data disks). Australian National University Printing Service, Canberra.

WHO. 2000. General guidelines for methodology on research and evaluation of traditional medicines, World Health Organization, Geneva.

World Bank. 1993. Vetiver Grass: The Hedge Against Erosion. 4th ed. The World Bank, Washington.

CHAPTER 14

Native Medicinal Plants Used for the Treatment of Nervous System Ailments in Chile and the Current State of Its Scientific Studies

Arline Martínez,[1] *Luisauris Jaimes,*[2] *Raul Vinet,*[3,4] *Tiare Segura,*[5]
Claudio Laurido[2,†] and *José L. Martínez*[6,*]

Introduction

Some plants produce beneficial pharmacologic effects on the human body and have been used for treating illnesses since the dawn of time. The therapeutic properties are due to the secondary metabolites that naturally synthesize and accumulate in these plants like alkaloids, sterols, terpenes, flavonoids, saponins, glycosides, cyanogenic, tannins, resins, lactones, quiniles, volatile oils, etc. (Motaleb 2011). These compounds are identified by the rigor of scientific, pharmacological, toxicological and clinical tests in search of active ingredients that explain the rational therapeutic use of these plants and promote the validity of its use (Muñoz et al. 1999).

[1] Facultad de Agronomía, Pontificia Universidad Católica de Chile.
[2] Facultad de Química y Biología, Universidad de Santiago de Chile.
[3] Centro de MicroBioinnovación, Facultad de Farmacia, Universidad de Valparaíso.
[4] Centro Regional de Estudios en Alimentos Saludables (CREAS).
[5] Facultad de Ciencias, Universidad de Santiago de Chile.
[6] Vicerrectoría de Investigación, Desarrollo e Innovación, Universidad de Santiago de Chile.
† In memoriam
* Corresponding author: joseluis.martinez@usach.cl

Chile has been characterized by its impressive geographical contrasts. It extends for 4,337 km along the south-western margin of South America from the Altiplano highs at 17°35S to Tierra del Fuego, the Islas Diego Ramírez and Cape Horn at 56°S (Moreira-Muñoz 2011). The Andes Mountains, the Pacific Ocean and the Atacama Desert are the natural limits and responsible for the high degree of endemism (approximately 50%) of Chilean endemic plants. About 300 native species make up the traditional pharmacopeia of Chile, but not all species have been studied chemically. Many of the studies carried out have a limited scope and have not been guided by pharmacological bioassays (Niemeyer 1995).

Due to the variety of vegetation zones, Chile has developed a diverse flora that includes 184 families (18 Pteridophyta, 4 Gymnospermae, 132 Dicotyledoneae and 30 Monocotyledoneae), with a total of 5082 species, of which 5012 are native species, and 2561 endemic (Marticorena 1992 in Niemeyer 1995). According to Mellado et al. (1996), its catalog of Chilean medicinal plants consists of 464 species. The record shows that 123 species are used in the northern, 164 in the central and 296 in the southern regions. To treat problems of nervous origin, the registry shows 8 species in the northern, 8 in the central and 18 in the southern regions. Among these species, 12 are native species, 4 reported in the north, 1 in the central and 7 in the southern region. On the other hand, Muñoz et al. (1999) reported the medicinal use of 38 native plants, of which 9 are used in the treatment of conditions associated with nervous system diseases.

The aim of this chapter is to report the native medicinal plants used for the treatment of problems of nervous origin in Chile, as well as the current state of scientific studies that have been developed for these plants. This investigation consisted in the revision of data bases including thesis research book articles and journals. The data presented below were obtained using SciELO, PubMed, Science Direct, Scopus, and others.

Plants Used in Nervous System Ailments

A total of 30 native plants were determined with reported use for any condition related to the nervous system: *Acaena magellanica, Aristotelia chilensis, Buddleja globosa, Centaurium cachanlahuen, Chenopodium ambrosioides, Fabiana imbricata, Parastrephia lepidophylla, Proustia pyrifolia, Quillaja saponaria, Salix chilensis* Molina (it is synonymy of *Salix humboldtiana* Willd) and *Ugni molinae* with analgesic or antinociceptive activities; *Artemisia copa, Azara microphylla, Cryptantha hispida, Fabiana imbricata, Lampayo officinalis, Latua pubiflora, Oenothera acaulis, Peumus boldus, Salix chilensis, Sophora macnabiana* and *Wahlenbergia linarioides* for nerve disorders including sedative, anxiolytic, antidepressant and analgesic activities; *Araucaria araucana, Laurelia sempervirens,*

Peumus boldus, Salix chilensis, Solanum ligustrinum and *Maytenus boaria* to treat internal, dental, rheumatic or menstrual pain, as well as neuralgia or headache; *Lysimachia serrulana, Salix chilensis* and *Sphacele salviae* to treat paralysis; *Artemisia copa* and *Baccharis linearis* (Ruiz & Pav.) Pers. (it is synonym of *Baccharis rosmarinifolia* Hook. & Arn.) to treat convulsions; *Chenopodium ambrosioides* and *Salix chilensis* for intestinal cramps; *Latua pubiflora* with narcotic activity; *Geum quellyon* and *Senecio eriophyton* to treat impotence or contribute to the erection; *Senecio eriophyton* and *Chenopodium ambrosioides* as stimulant or for fatigue treatment; *Senecio eriophyton* as aphrodisiac and to chills (Fig. 14.1, Table 14.1). Some adverse effects (toxic, carcinogenic and respiratory) were reported for *Chenopodium ambrosioides, Laurelia sempervirens* and *Solanum ligustrinum* (Muñoz et al. 1999).

On the other hand, 12 of the 30 species were studied for their pharmacological action (*Acaena magellanica, Aristotelia chilensis, Artemisia copa, Buddleja globosa, Chenopodium ambrosioides, Fabiana imbricata, Haplopappus baylahuen, Latua pubiflora, Laurelia sempervirens, Peumus boldus, Proustia pyrifolia* and *Ugni molinae*), and another twelve were studied chemically (*Araucaria araucana, Azara microphylla, Centaurium cachanlahuen, Geum quellyon, Lampayo officinalis, Maytenus boaria, Parastrephia lepidophylla, Quillaja saponaria, Salix chilensis, Senecio eriophyton, Solanum ligustrinum* and *Sphacele alviae*). Finally, no scientific studies were found for 6 of these species (*Baccharis rosmarinifolia, Cryptantha hispida, Lysimachia sertulata, Oenothera acaulis, Sophora macnabiana* and *Wahlenbergia linarioides*).

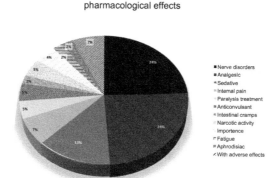

Native plants with different pharmacological effects

- Nerve disorders
- Analgesic
- Sedative
- Internal pain
- Paralysis treatment
- Anticonvulsant
- Intestinal cramps
- Narcotic activity
- Importence
- Fatigue
- Aphrodisiac
- With adverse effects

Fig. 14.1. Pharmacological uses and effects associated with Chilean medicinal plants. The pie chart highlights the use in nervous disorders and the analgesic and sedative properties.

Table 14.1. Scientific name, traditional medicinal applications, used parts, mode of administration, experimental scientific studies and reference used and scientific studies of the medicinal plants with report of use for treatment of diseases of nervous origin.

	Scientific Name	Traditional Medicinal Applications	Used Parts	Mode of Administration	Experimental Scientific Studies	Reference Used And Scientific Studies
1	*Acaena magellanica* (LAM.) VAHL	Analgesic	Whole plant	Infusion	A 20% w/v infusion administered orally at 16 mL/kg presented analgesic effect in the acetic acid-induced abdominal constriction test in mice.	Feresin et al. (2002)
2	*Araucaria araucana* (Molina) K. Koch, *	Headache	Resin	Patches	Five diterpenes were isolated from this plant, but its biological function is not known enough.	Muñoz et al. (1999)
3	*Aristotelia chilensis* (Molina) Stuntz*	Analgesic activity	Leaves	Infusion	Leaves have alkaloids and tannins which act as analgesics. In several topical analgesia models, dichloromethane, methanol, aqueous extract and a crude mixture of alkaloids were active in the formalin assay. In tail flick test, a crude mixture of alkaloids and methanol extract were the most active (58.2% and 55.2%, respectively). In the tail formalin assay, the methanol extract (74.1%) was the most active.	Muñoz et al. (1999), (2011), Misle et al. (2011)
4	*Artemisia copa* Phil*	Sedative, anxiolytic and anticonvulsant	Aerial parts	Infusion	At doses up to 1.5 g/kg, a dose-dependent sleep induction and potentiation of sub-hypnotic and hypnotic doses of pentobarbital were produced. Also, it produced increase and decrease in the spontaneous motor activity (0.5–1.5 g/kg) and a significant increase in the latency time and a decrease in the duration of seizures and mortality induced in mice (1.5 g/kg).	Miño et al. (2010)

5	*Azara microphylla* Hook.f.*	Antidepressant and analgesic	Leaves and stems	Infusion	The leaves have yielded a new flavonoid glycoside which has been called azamicroside.	Sagareishvili et al. (1983), Mellado et al. (1996), Tesauro Regional Patrimonial (2016a)
6	*Baccharis rosmarinifolia* Hook. & Arn.	Anticonvulsant	Leaves	Infusion	No studies found.	Houghton and Manby (1985), Mellado et al. (1996)
7	*Buddleja globosa* Hope*	Analgesic	Leaves	Infusion	The hexane extract showed 41.2% of analgesic effect at 600 mg/kg, inhibited by 47.7 and 79.0% of the arachidonic acid and 12-deoxyphorbol-13-decanoate induced inflammation at 3 mg/20 L/ear, respectively, in mice. The fraction obtained from dichloromethane defatted extract at 300 mg/kg also showed analgesic activity (38.7%). The metanol extract at 600 mg/kg *per os* showed analgesic activity (38.5%). Extracts showed a dose-dependent analgesic activity in all assays. Seasonal influence was observed since autumn extract resulted to be less active.	Backhouse et al. (2008a), Backhouse et al. (2008b)
8	*Centaurium cachanlahuen* (Mol) Rob.*	Analgesic	Whole plant	Macerated or decoction	Several xanthones have been isolated from this plant. It is likely that when biological studies will be carried out, it is concluded that the properties of plant come from xanthones.	Muñoz et al. (1999), Estomba et al. (2006)

Table 14.1 contd. ...

...*Table 14.1 contd.*

	Scientific Name	Traditional Medicinal Applications	Used Parts	Mode of Administration	Experimental Scientific Studies	Reference Used And Scientific Studies
9	*Chenopodium ambrosioides* L.	Stimulating, analgesic and antispasmodic	Root	Infusion	It contains thiamine, riboflavin, niacin, vitamin C, ascaridol, geranidol, 1-limonene, myrcene, p-cymene, d-alcamphor, butactic acid, spinasterol, terpinene, long chain alkyls, methyl salicylate, alkaloids, saponins and various glycosides. Analgesic effect was observed with the hotplate device maintained at 55°C as well as on the early and late phases of formalin-induced paw licking in rats. The 10% infusion has mild antispasmodic activity in rats. Infusions may be toxic ($DL_{50} > 1.0$ g/kg in mice).	Muñoz et al. (1999), Ibironke and Ajiboye (2007)
10	*Cryptantha hispida* (Phil.) Reiche*	Falling Asleep	-	-	No studies found.	Mellado et al. (1996)
11	*Fabiana imbricata* Ruiz and Pavon*	Sedative and analgesic	Bark and logs	Infusion	Derivatives of anthraquinones, terpene alkaloids, sugars and flavonoids have been isolated of this plant. The hydroalcoholic extract showed analgesic effects.	Muñoz et al. (1999)
12	*Geum quellyon* Sweet*	Impotence	Root	Decoction	It was detected that it contains hydrolyzable tannins, flavonoids, gallic acid and eugenol, but there is no biological-chemical information that supports its medicinal use.	Muñoz et al. (1999)
13	*Haplopappus baylahuen* J.Remy*	Aphrodisiac and chills treatment	Leaves and stems	Infusion or decoction	Anthraquinone glycosides, preniletine and flavonoids have been isolated from the aerial parts and stems. Some animal tests showed that the infusion of the plant has relaxing effects on the uterus.	Muñoz et al. (1999)

14	*Lampayo officinalis* F. Phil. ex Murillo*	it produces sleep	Leaves	Infusion	Extract showed higher concentrations of phenols and flavonoids.	Mellado et al. (1996), Garrido et al. (2013), Morales and Paredes (2014)
15	*Latua pubiflora* (Griseb.) Baill.*	Narcotic, anxiolytic and sedative activities	Leaves and flowers	Infusion	It contains alkaloids. The anxiolytic effect of an alkaloid fraction was corroborated. The alkaloid extract has sedative and anxiolytic properties involving the $GABA_A$ receptor. The non-alkaloid extract presents a sedative property not associated with $GABA_A$.	Plowman et al. (1971), Rojas (2002), Muñoz and Casale (2003), Ulloa (2004)
16	*Laurelia sempervirens* (Ruiz and Pav.) Tul.*	Headache	Leaves	Infusion	It contains complex alkaloids of aporphine type and bisbenzylisoquinolyl, and safrole. I.V. administration of a hydroalcoholic extract to rats elicited a hypotensive response of (−27.0%) in blood pressure of normotensive animals at a dose of 5 mg crude extract. Carcinogenic effects have been detected in experimental animals due to the use of the essential oil of the leaves (high safrole 91%).	Muñoz et al. (1999), Schmeda-Hirschmann et al. (1994)
17	*Lysimachia sertulata* Baud*	Paralysis	-	-	No studies found.	Mellado et al. (1996)
18	*Maytenus boaria* Molina	Dental and internal analgesic	Leaves and seeds	Infusion or decoction	A variety of sesquiterpenes, triterpenes, polyphenols and flavonoids have been isolated from the aerial. Sesquiterpenes are antiinflammatory and polyphenols are effective antioxidant in association with proteins and amino acids normalizing the cardiovascular function.	Muñoz et al. (1999)
19	*Oenothera acaulis* Cav*	Antidepressant and analgesic	-	-	No studies found.	Mellado et al. (1996)

Table 14.1 contd. ...

...*Table 14.1 contd.*

	Scientific Name	Traditional Medicinal Applications	Used Parts	Mode of Administration	Experimental Scientific Studies	Reference Used And Scientific Studies
20	*Parastrephia lepidophylla* (Wedd.) Cabrera	Analgesic	Leaves and branches	Infusion	Two benzofurans (tremetone 1 and methoxytremetone 6) were isolated. Tremetone 1 exhibited a morphine-like analgesic property.	Benites et al. (2012), Hierba medicinal (2016)
21	*Peumus boldus* Molina*	Sedative, antineuralgic and antirheumatic	Leaves	Infusion and cataplasma	The essential oil contains terpene hydrocarbons, common flavonol glycosides, aporophinoid alkaloids and non-aporphinoid alkaloids. The wood contains laurolitsina and several unidentified bases. The bark has boldine, benzylisoquinolinic alkaloids and coclaurine. Boldine exerts *in vitro* relaxing effect on rat musculature, directly interfering with the cholinergic mechanism involved. Antiinflammatory activity of ethanolic extracts has also been demonstrated.	Muñoz et al. (1999)
22	*Proustia pyrifolia* DC	Analgesic	Leaves and roots	Infusion and showers	The evaluation of the topic antiinflammatory activities induced by arachidonic acid, and phorbol 12-myristate 13-acetate of the different extracts showed that this species possesses active constituents that could diminish cyclooxygenase and lipoxygenases activities, the enzymes that allow the synthesis of proinflammatory endogenous substances as prostaglandin E2 and leukotrienes, respectively.	Delporte et al. (2005), Tesauro Regional Patrimonial (2016b)

23	*Quillaja saponaria* Molina*	Analgesic	Bark	Infusion or macerated	Bark contains a high percentage of triterpene saponins. The topical and systemic analgesic effects of a commercial partially purified saponin extract showed activity in both analgesic tests in a dose-dependent manner. The saponins, quillaic acid, its methyl ester, and one of the oxidized derivatives of the latter, elicit dose-dependent antinociceptive effects in two murine thermal models. Also, it has anti inflammatory activity.	Muñoz et al. (1999), Arrau et al. (2011), Tesauro Regional Patrimonial (2016c)
24	*Salix chilensis* Molina	Rheumatic and menstrual pain, headache, antispasmodic, sedative and Paralysis treatment	Bark shavings and leaves	Infusion	It contains salicin, of which there is a wide range of studies that confirm its analgesic use.	Mellado et al. (1996), MINSAL (2009), Waizel-Bucay (2011)
25	*Senecio eriophyton* Remy*	Fatigue and contribute to the erection	Aerial parts	Infusion	Extracts (dichloromethane, methanol, aqueous) of this plant showed *in vitro* relaxation effect on the guinea pig corpus cavernosum. It has phosphodiesterase inhibitors activity.	Mellado et al. (1996), Rahimi et al. (2009), Singh et al. (2013), Tesauro Regional Patrimonial (2016d)
26	*Solanum ligustrinum* Lodd*	Headache	Leaves and stems devoid of bark	Infusion	From the leaves and branches have been extracted scopoletin, β-sitosterol glycosides solasodine, solasonine and alkaloids. These have antiinflammatory properties, smooth muscle relaxant and hypotensive, among others.	Muñoz et al. (1999)
27	*Sophora macnabiana* Graham	Nerves	-	-	No studies found.	Mellado et al. (1996)

Table 14.1 contd. ...

...*Table 14.1 contd.*

	Scientific Name	Traditional Medicinal Applications	Used Parts	Mode of Administration	Experimental Scientific Studies	Reference Used And Scientific Studies
28	*Sphacele salvia* (Lindl). Briq.*	Paralysis treatment	Leaves	Infusion and tonic	In this plant were found abetes and ursolic acid as main secondary metabolites. Also, camosol and rosmadial were found. Antitumorigenic activity in oncology screens and antioxidant activity have been documented.	Mellado et al. (1996), Flagg (2000), Escuder et al. (2002)
29	*Ugni molinae* Turcz.*	Analgesic	Leaves	Infusion	Dichloromethane (DCM), ethyl acetate (EA) and methanol (ME) leaf extracts were assessed in mice showing a dose-dependent antinociceptive activity in all the assays under different administration routes.	Delporte et al. (2007)
30	*Wahlenbergia linarioides* Schrad. ex Roth	Sedative	Leaves and wood	Infusion	No studies found.	Mellado et al. (1996), Montes et al. (1974)

*endemic

Conclusions and Future Perspectives

For a long time, there have been no ethnobotanical studies on the use of medicinal plants in Chile. On this basis, our research group has been collecting bibliographic information about the medicinal use of Chilean plants against various diseases. Part of this research is presented here.

We study the medicinal plants that are recommended for the nervous system, finding two ailments that occupy almost 50% (nervous disorders and analgesic). In general, 19 endemic species of the 30 found in the literature are used in problems related to the central nervous system. Of the studies that have been conducted with medicinal species, it is found that there are 40% that focus on chemical studies and 40% on biological studies. This is due to the lack of a state policy that allows a systematic study of the medicinal plants recommended for use by the population. In addition, there is a lack of studies that demonstrate adequate use through pharmacological studies, demonstrating that a particular plant has a beneficial effect for the disease that is recommended. In this way, the appropriate governmental policies should contribute to increasing the therapeutic arsenal of new drugs, allowing studies and research based on chemical compounds isolated from medicinal plants.

Although the Chilean health authority periodically publishes documents recommending the proper use of medicinal plants, it should also provide economic resources to research on medicinal plants. These studies should focus on pharmacological studies (pharmacokinetics and pharmacodynamics), which allow knowing the availability of compounds isolated from medicinal plants in different tissues of the human body, as well as their mechanisms of action. An adequate policy would help the population to consume medicinal plants with the due security and scientific basis of their benefits and possible adverse effects. Also, an appropriate policy would increase the therapeutic arsenal of new drugs that would help not only the Chilean population but also the world population.

References

Arrau, S., Delporte, C., Cartagena, C., Rodríguez-díaz, M., González, P., Silva, X., Cassels, B.K. and Miranda H.F. 2011. Antinociceptive activity of Quillaja saponaria Mol. saponin extract, quillaic acid and derivatives in mice. J. Ethnopharmacol 133(1): 164–167. http://doi.org/10.1016/j.jep.2010.09.016.

Backhouse, N., Delporte, C., Apablaza, C., Farías, M., Goïty, L., Arrau, S., Negrete, R., Castro, C. and Miranda, H. (2008a). Antinociceptive activity of Buddleja globosa (matico) in several models of pain. J. Ethnopharmacol. 119(1): 160–165. http://doi.org: 10.1016/j.jep.2008.06.022.

Backhouse, N., Rosales, L., Apablaza, C., Goïty, L., Erazo, S., Negrete, R., Theodoluz, C., Rodríguez, J. and Delporte, C. 2008b. Analgesic, anti-inflammatory and antioxidant properties of Buddleja globosa, Buddlejaceae. J. Ethnopharmacol. 116(2): 263–269. http://doi.org: 10.1016/j.jep.2007.11.025.

Benites, J., Gutierrez, E., López, J., Rojas, M., Rojo, L., Costa, M.C., Vinardell, M.P. and Buc Calderon, P. 2012. Evaluation of analgesic activities of tremetone derivatives isolated from the Chilean altiplano medicine Parastrephia lepidophylla. Nat. Prod. Commun. 7(5): 611–4.

Delporte, C., Backhouse, N., Erazo, S., Negrete, R., Vidal, P. and Silva, X. 2005. Analgesic-antiinflammatory properties of Proustia pyrifolia. J. Ethnopharmacol. 99: 119–124. http://doi.org/10.1016/j.jep.2005.02.012.

Delporte, C., Backhouse, N., Inostroza, V. and Aguirre, M.C. 2007. Analgesic activity of Ugni molinae (murtilla) in mice models of acute pain. J. Ethnopharmacol. 112: 162–165. http://doi.org/10.1016/j.jep.2007.02.018.

Escuder, B., Torres, R., Lissi, E., Labbé, C. and Faini, F. 2002. Antioxidant capacity of abietanes from Sphacele salviae. Nat. Prod. Lett. 16(4): 277–281. http://.org/10.1080/10575630290020631.

Estomba, D., Ladio, A. and Lozada, M. 2006. Medicinal wild plant knowledge and gathering patterns in a Mapuche community from North-western Patagonia. J. Ethnopharmacol. 103(1): 109–119. doi:10.1016/j.jep.2005.07.015.

Feresin, G.E., Schmeda-hirschmann, G., Feresin, G.E., Tapia, A., R., A.G., Delporte, C., ... Schmeda-hirschmann, G. 2002. Free radical scavengers, anti-inflammatory and analgesic activity of Acaena magellanica analgesic activity of Acaena magellanica. J. Phar. Pharmacol. 54: 835–844. http://doi.org/10.1211/0022357021779014.

Flagg, M.L. 2000. Bioprospecting, chemical investigations and drug discovery from Chilean plants. (Unpublished Ph.D degree thesis). University of Arizona, Tucson, E.E.U.U. Retrieved from http://arizona.openrepository.com/arizona/bitstream/10150/284167/1/azu_td_9972115_sip1_m.pdf.

Garrido, G., Ortiz, M. and Pozo, P. 2013. Fenoles y flavonoides totales y actividad antioxidante de extractos de hojas de Lampaya medicinalis F. Phil. J. Pharm. Pharmacogn Res. 1(1): 30–38.

Hierba Medicinal. 2016. Thola para la tos y expectorar en resfriados y gripes. Retrieved from http://www.hierbamedicinal.es/thola-para-la-tos-y-expectorar-en-resfriados-y-gripes.

Houghton, P.J. and Manby, J. 1985. Medicinal plants of the Mapuche. J. Ethnopharmacol. 13(1): 89–103. http://doi:10.1177/1533210110391077.

Ibironke, G.F. and Ajiboye, K.I. 2007. Studies on the anti-inflammatory and analgesic properties of Chenopodium ambrosioides leaf extract in rats. Int. J. Pharmacol. 3(1): 111–115. http://doi.org/10.3923/ijp.2007.111.115.

Mellado, V., Medina, E. and San Martín, C. 1996. Herbolaria médica de Chile, diagnóstico de su estado actual y perspectivas futuras para la medicina oficial chilena. Santiago, Chile: Ministerio de Salud.

Miño, J.H., Moscatelli, V., Acevedo, C. and Ferraro, G. 2010. Psychopharmacological effects of Artemisia copa aqueous extract in mice. Pharm. Biol. 48: 1392–1396. http://doi.org/10.3109/13880209.2010.486407.

MINSAL (Ministerio de Salud). 2009. Medicamentos herbarios tradicionales. Retrieved from http://www.minsal.cl/portal/url/item/7d9a8480e0821613e04001011e01021b.pdf.

Misle, E., Garrido, E., Contardo, H. and González, W. 2011. Maqui [Aristotelia chilensis (Mol.) Stuntz]—the Amazing Chilean Tree: a review.

Montes, M., Wilkomirsky, T. and Ubilla, H. 1974. Quelques aspects de la medication vegetale populaire dans la region de Concepcion (Chili). Planta Medica 25(02): 183–192.

Morales, G. and Paredes, A. 2014. Antioxidant activities of Lampaya medicinalis extracts and their main chemical constituents. BMC Complem. Altern. M. 14(1): 1. http://doi.org/: 10.1186/1472-6882-14-259.

Moreira-Muñoz, A. 2011. Plant Geography of Chile. (Plant and Vegetation, Volume 5). London, New York: Springer.

Motaleb, M.A. 2011. Selected Medicinal plants of chittagong hill tracts. Retrieved from http://cmsdata.iucn.org/downloads/medicinal_plant_11_book.pdf.

Muñoz, O., Montes, M. and Wilkomirsky, T. 1999. Plantas medicinales de uso en Chile: química y farmacología. (2da ed.) Santiago, Chile: Editorial Universitaria S.A.

Muñoz, O. and Casale, J.F. 2003. Tropane alkaloids from Latua pubiflora. Zeitschrift für Naturforschung C, 58(9-10): 626–628. http://doi.orh/doi: 10.1515/znc-2003-9-1003.

Muñoz, O., Christend, P., Crettond, S., Backhouse, N., Torres, V., Correa, O., Costa, E., Miranda, H. and Delporte, C. 2011. Chemical study and anti-inflammatory, analgesic and antioxidant activities of the leaves of Aristotelia chilensis (Mol.) Stuntz, Elaeocarpaceae. J. Phar. Pharmacol. 63: 849–859. http://doi.org/10.1111/j.2042-7158.2011.01280.x.

Niemeyer, H.M. 1995. Biologically active compounds from Chilean medicinal plants. pp. 137–153. *In*: Arnason, J.T., Mata, M. and Romeo, J.T. (eds.). Phytochemistry of Medicinal Plants (Volume 29). Nueva York: Plenum Press.

Plowman, T., Gyllenhaal, L.O. and Lindgren, J.E. 1971. Latua pubiflora magic plant from Southern Chile. Bot. Mus. Leafl. 23(2): 61–92. Retrieved from http://www.jstor.org/stable/41762272.

Rahimi, R., Ghiasi, S., Azimi, H., Fakhari, S. and Abdollahi, M. 2010. A review of the herbal phosphodiesterase inhibitors; future perspective of new drugs. Cytokine 49(2): 123–129. http:// doi.org doi: 10.1016/j.cyto.2009.11.005.

Rojas, M.A. 2002. Caracterización química de extractos de "palo brujo" (*Latua pubiflora* (Griseb.) Phil.), con propiedades depresoras del sistema nervioso central, y su caracterización farmacológica en ratones. (Unpublished Thesis for the title of pharmaceutical chemist). Universidad Austral de Chile, Valdivia, Chile. Retrieved from http://cybertesis.uach.cl/tesis/uach/2002/fcr741c/doc/fcr741c.pdf.

Sagareishvili, T.G., Alaniya, M.D. and Kemertelidze, É.P. 1983. A new flavonol glycoside from Azara microphylla. Chem. Nat. Compd. 19(3): 275–278. http://doi.org:10.1007/BF00579757.

Schmeda-Hirschmann, G., Loyola, J., I. and Rodriguez, J. 1994. Hypotensive effect of Laurelia semperviren (Monimiaceae) on normotensive rats. Phytother. Res. 8: 49–51. http://doi.org/10.1002/ ptr.2650080112.

Singh, S., Ali, A., Singh, R. and Kaur, R. 2016. Sexual Abnormalities in Males and Their Herbal Therapeutic Aspects. Pharmacologia, 1–11. http://doi.org/10.5567/pharmacologia.2013.265.275.

Tesauro Regional Patrimonial. 2016a. Azara microphylla, corcolén. Retrieved from http://www. tesauroregional.cl/trp/tesauro/default.asp?a=338&Element_ID=2212.

Tesauro Regional Patrimonial. 2016b. Proustia pyrifolia, tola blanca Retrieved from http://www. tesauroregional.cl/trp/tesauro/default.asp?a=338&Element_ID=2079.

Tesauro Regional Patrimonial. 2016c. Quillaja saponaria, quillay. Retrieved from http://www. tesauroregional.cl/trp/tesauro/default.asp?a=338&Element_ID=2288.

Tesauro Regional Patrimonial. 2016d. Senecio eriophyton, chachacoma. Retrieved from http://www. tesauroregional.cl/trp/tesauro/default.asp?a=338&Element_ID=2057.

Ulloa, G.E. 2004. Estudio químico de un extracto alcaloideo de *Latua pubiflora* (Griseb) (Latúa) y evaluación farmacológica de los compuestos alcaloideos (Unpublished Thesis for the title of pharmaceutical chemist). Universidad Austral de Chile, Valdivia, Chile. Retrieved from http:// cybertesis.uach.cl/tesis/uach/2004/fcu.42e/pdf/fcu.42e-TH.4.pdf.

Waizel-Bucay, J. 2011. Plantas y compuestos importantes para la medicina: los sauces, los salicilatos y la aspirina. Rev. Fitoter 61–75.

CHAPTER 15

Medicinal Plants Used by the Tharu Communities in Nepal

Shandesh Bhattarai

Introduction

Traditional knowledge on the use of medicinal plants to treat human diseases still exists in many parts of the globe (Katewa et al. 2004, Mahishi et al. 2005). Safe, effective and inexpensive indigenous remedies are gaining popularity and communities in Nepal are still practicing the use of medicinal plants to treat a variety of ailments (Chaudhary 1998, Bhattarai 2009, Manandhar 2002, Baral and Kurmi 2006). Tharus are a culturally and linguistically diverse ethnic group of Nepal Tarai having a long history of knowledge on the use of medicinal plants.

In the rural areas of Tharu communities, even today healthcare among the Tharus is checked by the local *Gurewa* (traditional healer), who performs primeval rites of protection, blessings and healing by using local medicinal herbs (Manandhar 2002). Tharu people mainly settled in Surkhet, Dang, Kailali, Kanchanpur, Banke, Kapilvastu, Rupandehi, Nawalparasi, Chitwan, Parsa, Bara, Rautahat, Sindhuli, Udyapur, Morang and Jhapa districts in Nepal (Fig. 15.1).

The fertile land of Tarai is called rice basket of Nepal (Sapkota 2014, Bhattachan 2016). Tharu people live in villages in houses plastered with mud and cow dung (Pyakuryal 1982). The houses are large and communal (Meyer and Deuel 1999). The Tharus are recognized as an official nationality by the Government of Nepal (Lewis et al. 2014), whereas the Government of India recognizes the Tharu people as a scheduled tribe (Verma 2010).

Nepal Academy of Science and Technology, Khumaltar, Lalitpur.
Email: shandeshbhattarai@gmail.com

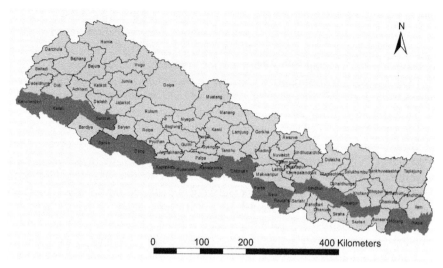

Fig. 15.1. Districts showing the main settlements by the Tharu communities in Nepal.

The ethnobotanical knowledge of Tharu is scattered in different literatures (Manandhar 1985, 2002, Dangol and Gurung 1991, Thapa 2001, Chapagain et al. 2004, Ghimire and Bastakoti 2009, Bhattarai et al. 2009, Acharya and Acharya 2009, Muller-Boker 1991, 1993, 1999a,b). Hence, the first attempt to review the ethnomedicinal usage of Tharu communities of Nepal was undertaken. It is hoped that this paper will be very much useful for the future ethnobotanical research in the Tharu communities of Nepal.

History, Tradition and Culture

There are two major groups in Nepal based on origin: Tibeto-Nepali and Indo-Nepali. The Tarai in the south of Nepal is inhabited by the indo-Nepali, i.e., Danuwar, Mooshar and Tharu communities (Manandhar 2002, Gurung 1996). Tharu people are a native ethnic group who have lived in the lowlands for centuries and developed a distinct and self-sufficient society with their own language, religion and culture. Tharu are the largest and oldest ethnic group of Nepal Tarai. There are different views about their origin (Gurung 1996, Skar 1999, McLean 1999, Bista 2004). The Rana Tharus claim to be of Rajput origin and have migrated from Far Western Tarai, while farther easterns claim to be the descendants of the Śākya and Koliya peoples of Kapilvastu (Hagen 1960, Gurung 1996, Muller-Boker 1999a, b, Manandhar 1985, Lall 1983, Rajaure 1975).

Tharu community has worked under the system of bonded labor known as Kamaiya since 18th century but Nepal Government banned this practice in 2000 (Muller-Boker 1999a, McDonaugh 1999, Gurung 1996). Tharus from the west have been practicing the Badghar system. Badghar is the elected chief of a village who works for the village welfare and also manages the cultural traditions. Tharus have Mongoloid features with dark and semi-dark colors, and are believed to have come to Nepal from India during the Muslim invasion in 12th and 13th century (Pyakuryal 1982, Gurung 1996). They are rich in culture, traditions, rites and rituals. In 1854, the Mulkiain categorized Nepal's legal system and divided society into a system of castes. Tharus were placed between the touchable and untouchables social hierarchy (Muller-Boker 1999a,b, McDonaugh 1999, Gurung 1996).

Even when the Tarai was avoided by others communities because of the influence of malaria, only the Tharus managed to live there (Pyakuryal 1982, Manandhar 2002, Gurung 1996). Their illiteracy and lack of representation in government made it difficult for them to struggle against the new landlords (Pyakuryal 1982). Many fell into debt and were forced to become bonded laborers-Kamaiyas (men) and Kamlaharis (women) (McDonaugh 1999, Patterson 1982, Gurung 1996). Tharus also celebrate Hindu festivals. Each of the Tharu family worships its personal tutelary deity who is represented by a lump of earth mixed with multi-colour cotton threads, crude sugarcane and a gold coin in the centre (Pyakuryal 1982, Muller-Boker 1999a,b).

Food, Language and Religion

Tharu's unique food items are popular throughout the country (Bhattachan 2016). The special food items are *Dhikri* and *Ghonghi*. *Dhikri* is made of rice flour. The dough is cooked over steam and eaten together with pickle and or curry. *Ghonghi* is an edible snail collected in nearby water bodies and left overnight so that all the gooey material inside them comes out. Their tail end is cut so that it is easier to suck out the meat from the shell. They are boiled and later cooked like curry (Krauskopff 1995). Tharu grow barley, wheat, maize, and rice, as well as raise animals such as chickens, ducks, pigs, etc. They are also popular in using large nets (Figs. 15.2 and 15.3) for fishing (Muller-Boker 1999a,b, Manandhar 2002).

Tharus have their indigenous dialect, but speak a mixture of local dialects, such as Prakriti, Bhojpuri, Mughali and Nepali. Tharu communities in different parts of Nepal do not share the same language but speak various endemic Tharu languages. As, for example, in western Nepal, they speak variants of Hindi, Urdu and Awadhi whereas in and near central Nepal, they speak a variant of Bhojpuri (Pyakuryal 1982). Likewise, in eastern Nepal, they speak a variant of Maithili. Tharus already lived in the Terai before Indo-Europeans arrived. Hinduism is the main religion of the Tharus and includes belief in spirits and devils (Manandhar 2002). Small numbers have converted to Buddhism in the recent years and few have converted to Christianity.

Festivals, Marriage, Birth and Death

Tharu people are fond of festivals and celebrate *jitiya*, *chhath*, *maghi*, *fagu* and *dashain*. *Jitiya* is a ten days festival celebrated by Tharu women before dashain (in

Fig. 15.2. Tharu women carrying fishing nets.

Fig. 15.3. Tharu men knitting fishing net and carrying *Bauhinia vahlii* leaves.

the aastami-8th days of Dashain). During *jitiya*, Mothers keep fasting even without drinking a drop of water with full dedication and faith for the long life and well-being of her son. The fasting completes after ten days from the sunrise (Pyakuryal 1982). In this day, Tharu women perform their traditional dance in the various programs. Tharu people also celebrate Dashain wearing new clothes and eating delicious foods. It is celebrated for ten days which falls after two days of *jitiya*.

Chhath is another important festival. On this day, people worship the sun and keep fasting without drinking a drop of water. *Maghi* is the most important festival which is considered as the New Year for the Tharu community. People take bath at the nearest water sources. On this day, delicious foods like sesame laddus, rice pudding, fish and meat are made and served (Pyakuryal 1982). *Fagu* is another important festival known as *Holi* in Nepal which is celebrated by sprinkling colors on each other and singing songs (Pyakuryal 1982, Manandhar 2002). The author has been working with the Tharu communities of Nepal since long time and has gathered a rich knowledge of the Tharu cultures and traditions (Fig. 15.4).

Monogamous marriage system has been practiced among the Tharu people. Traditionally, marriages were often arranged during the pregnancies of two women.

Fig. 15.4. Interaction with the local Tharu people about the medicinal plants in Bardiya.

Most Tharus now practice conventional arranged marriages and also practice love marriages, inter caste marriage, marriage after courtship and eloping (Muller-Boker 1999a,b). The birth ceremony is celebrated at the age of five months for the girl child and at the age of six-seven months for boys. On this day, rice pudding is eaten by the child and the name of the baby is also kept. Most of the Tharu cremate their deceased; others, however, bury them. There is a strange custom of keeping men face down and women face up during the burial and the mourning period is about thirteen days (Pyakuryal 1982).

Medicinal Plants

A total of 324 plant species belonging to 100 families used by the Tharu community in Nepal were presented in this chapter (Table 15.1). The plant species were arranged alphabetically according to scientific name, followed by family names, parts used, Nepali and Tharu names, medicinal usage and reference citations (Table 15.1).

The top ten largest families are: Fabaceae (29 species), followed by Lamiaceae (21 species), Asteraceae (18 species), Solanaceae (13), Poaceae, Euphorbiaceae and Moraceae (11 species each), Malvaceae and Apocynaceae (10 species each) and Cucurbitaceae (9 species) (Fig. 15.5).

Similarly, Amaranthaceae and Rubiaceae are represented by 7 species each followed by Convolvulaceae and Cucurbitaceae (6 species each). Four families Anacardiaceae, Boraginaceae, Myrtaceae and Zingiberaceae are represented by 5 species each (Table 15.2). Likewise, Acanthaceae, Combretaceae, Lythraceae, Polygonaceae and Rutaceae were represented by 4 species each. Eleven families, Amaryllidaceae, Annonaceae, Apiaceae, Cyperaceae, Dioscoraceae, Phyllanthaceae, Plantaginaceae, Pteridaceae, Sapotaceae, Urticaceae and Verbenaceae were represented by 3 species each. Twelve families, Bigioniaceae, Equisetaceae, Lauraceae, Linaceae, Meliaceae, Nyctaginaceae, Orchidaceae, Oxalidaceae,

Table 15.1. Medicinal plants of Tharu communities of Nepal.

Plant's Scientific Name (family name); Parts Used	Nepali and Tharu names	Medicinal Usage	References
Abelmoschus moschatus Moench. (Malvaceae); Rt	Bankapas	Cuts, wounds	Dangol and Gurung 1991
Abelmoscus esculentus (L.) Moench. (Malvaceae); Rt	Bhindi (N)	Constipation	Chapagain et al. 2004
Abrus precatorius L. (Fabaceae); Sd, Rt, Lvs	Titihar, Chilahariyak thond (T)	Fever, stomach and eye diseases, asthma, uterus problem, boils, cough, sore throat, headache, cold, blood purifier, malaria, paralysis, nerve diseases, skin diseases, sciatica, rheumatism, asthma, dental caries, urinary infection	Acharya and Acharya 2009, Manandhar 2002, Chapagain et al. 2004
Abutilon indicum (L.) St Sweet (Malvaceae); Lvs	Baliyari (T)	Body swellings, wounds, blisters, boils	Dangol and Gurung 1991
Acacia catechu (L.f.) Willd. (Fabaceae); St, Lvs, Sap	Khayar (N)	Diarrhea, dysentery, bodyache, blood accumulation, bone fracture, cough	Bhattarai et al. 2009, Chapagain et al. 2004, Manandhar 1985
Acacia nilotica (L.) Del. (Fabaceae); Br, Fl	Babul	Dysentery, diarrhea, quench thirst	Dangol and Gurung 1991
Achyranthes aspera L. (Amaranthaceae); Rt, Lvs, St, WP	Naksirka, Uthaanna, Ultakur (T)	Diarrhea, dysentery, constipation, blisters, stomachache, headache, toothache, snake scorpion sting, delivery problems, vomiting, fever, fatigue, itchy skin, indigestion, dental trouble, bleeding, menstrual disorders	Bhattarai et al. 2009, Chapagain et al. 2004, Manandhar 1985, 1988, 1990, 1989a,b, Muller-Boker 1999a, Dangol and Gurung 1991, Mandar and Chaudhary 1993
Achyranthes bidentata Blume (Amaranthaceae); WP	Ultakur (T)	Wounds, blisters	Chapagain et al. 2004
Acorus calamus L. (Acoraceae); Rt, Rh	Bojho (N); Katara, Bajh (T)	Roundworm, hookworm cough, cold, sore throat, tonsillitis, bronchitis, anthelmintic	Bhattarai et al. 2009, Acharya and Acharya 2009, Chapagain et al. 2004, Dangol and Gurung 1991
Adiantum caudatum L. (Pteridaceae); Sht	Ratijari	Fever	Dangol and Gurung 1991

Table 15.1 contd. ...

...Table 15.1 contd.

Plant's Scientific Name (family name); Parts Used	Nepali and Tharu names	Medicinal Usage	References
Adiantum lunulatum Burm.f. (Pteridaceae); Rh	Ratamur (T)	Fever, dysentery, glandular swelling	Manandhar 1985, 2002
Aegle marmelos (L.) Correa (Rutaceae); Fr, Br, Lvs, Rt	Bel (N); Jogchmuda (T)	Diarrhea, dysentery, constipation ringworm, scabies, snake scorpion sting, diabetes, inflammation of sex organs of child, fever, weakness, wounds, headache, fever, bowel complaints, cooling, piles	Bhattarai et al. 2009, Acharya and Acharya 2009, Chapagain et al. 2004, Taylor et al. 1996, Muller-Boker 1999a, Manandhar 1989b, 1990, Mandar and Chaudhary 1993
Agave cantala var. cantala (Asparagaceae); Lvs	Ketu	Bodyache	Chapagain et al. 2004
Agave sisalana Perrine (Asparagaceae); Br	Kedli	Paralysis	Mandar and Chaudhary 1993
Ageratum conyzoides (L.) L. (Asteraceae); Rt, Lvs	Gandhejhar (N); Raunne (T)	Typhoid, cuts	Bhattarai et al. 2009, Chapagain et al. 2004, Dangol and Gurung 1991
Allium cepa L. (Amaryllidaceae); Bu	Piyaj (N)	Vomiting, diarrhea, dysentery	Chapagain et al. 2004, Mandar and Chaudhary 1993
Allium sativum L. (Amaryllidaceae); Bu	Lasun (N)	Headache	Mandar and Chaudhary 1993
Aloe vera (L.) Burm.f. (Xanthorrheaceae); Ltx, Lvs	Ghauekumari (N); Ghyukuwanr (T)	Cuts, wounds, burns, boils, rheumatic pains, skin irritations, fever, constipation, jaundice, gonorrhea, kidney pains, indigestion, peptic ulcers, cough, cold, dropsy	Bhattarai et al. 2009, Chapagain et al. 2004, Manandhar 2002
Alstonia scholaris (L.) R.Br. (Apocynaceae); Br, Ltx, WP	Chhativan (N)	Diarrhea, dysentery, bodyache, breast pain, tonic, fever, great thrist, menstrual disorders	Bhattarai et al. 2009, Chapagain et al. 2004, Muller-Boker 1999a, Mandar and Chaudhary 1993
Alternanthera sessilis (L.) R.Br. ex DC. (Amaranthaceae); WP, Aep, Rt	Saranchi, gantha phula, garri (T)	Fever, cough, cold, scabies, cuts, wounds, venereal diseases, dysuria, bloody dysentery, mental disorder, heat stroke	Taylor et al. 1996, Manandhar 1985, 2002, Mandar and Chaudhary 1993, Manandhar 1985
Amaranthus spinosus L. (Amaranthaceae); Rt, Lvs	Setolode; Kattia maattia, Kantiya, Makhan (N,T)	Diuretic, cooling agent, constipation, fever, diarrhea, dysentery, indigestion, vomiting, gonorrhea, menorrhagia, boils, antidote	Bhattarai et al. 2009, Chapagain et al. 2004, Manandhar 2002, Dangol and Gurung 1991

Botanical name (Family); Parts used	Local name	Uses	References
Ananas comosus (L.) Merr. (Bromeliaceae); Fr, Lvs	Anar (T)	Arthritis/rheumatism	Chapagain et al. 2004
Annona reticulata L. (Annonaceae); Lvs, Fr	Sarifa (N)	Cuts, wounds, diabetes	Mandar and Chaudhary 1993
Annona squamosa L. (Annonaceae); Lvs	Sarifa (N)	Boils	Chapagain et al. 2004
Antidesma acidum Retz. (Phyllanthaceae); Lvs, Fr, Br	Dakhi, Dakhee (T)	Snake scorpion sting, bone fracture	Chapagain et al. 2004
Areca catechu L. (Arecaceae); Br	Supari (N)	Arthritis	Mandar and Chaudhary 1993
Argemone mexicana L. (Papaveraceae); Fl, St, Sd	Bharbanda (T)	Boils, eye complaints, itches	Chapagain et al. 2004, Mandar and Chaudhary 1993
Artemisia indica Willd. (Asteraceae); Rt, Lvs	Titepati (N); Pati (T)	Gastritis, paralysis, scabies, anthelmintic, ringworm	Bhattarai et al. 2009, Manandhar 1985, 1994
Artemisia vulgaris L. (Asteraceae); WP	Kurza	Repellent for bugs, fleas	Dangol and Gurung 1991
Artocarpus heterophyllus Lam. (Moraceae); Fr	Kathar (N)	Swelling part of scrotum	Bhattarai et al. 2009
Artocarpus integer (Thunb.) Merr. (Moraceae); Sd	Rukh katahar (N)	Tuberculosis	Chapagain et al. 2004
Artocarpus lacucha Buch.-Ham. (Moraceae); WP, St, Br, Rt	Barhar (N)	Headache, constipation, diarrhea, dysentery	Chapagain et al. 2004, Mandar and Chaudhary 1993
Asparagus racemosus Willd. (Asparagaceae); Rt	Santapsatauri, Kurela (T)	Tonic in lactating a postpartum mother, anthelmintic, delivery, vomiting, burning sensation in urination	Bhattarai et al. 2009, Chapagain et al. 2004, Muller-Boker 1999a, Mandar and Chaudhary 1993, Manandhar 1985
Asparagus racemosus Willd. var. *subacerosus* Baker (Asparagaceae); Rt	Santawar (T)		Acharya and Acharya 2009
Azadirachta indica A. Juss. (Meliaceae); Lvs, Br	Neem (N)	Anthelmentic, fever, gastritis, stomachache, cough, reduce sugar level, cooling agent, diabetes, wounds, blisters, itches, skin diseases, tootache	Bhattarai et al. 2009, Acharya and Acharya 2009, Chapagain et al. 2004, Muller-Boker 1999a, Mandar and Chaudhary 1993
Bauhinia vahlii Wight and Arn. (Fabaceae); Rt, Lvs, Sd	Bhorla (N); Malu, Moharain (T)	Pulmonary tuberculosis, abrasions, urinary infection, boils	Bhattarai et al. 2009, Chapagain et al. 2004, Manandhar 1985

Table 15.1 contd. ...

...Table 15.1 contd.

Plant's Scientific Name (family name); Parts Used	Nepali and Tharu names	Medicinal Usage	References
Bauhinia variegata L. (Fabaceae); Br, St	Koilara (T)	Antidote in snake bite, dysentery	Acharya and Acharya 2009
Berberis asiatica DC. (Berberidaceae); Rt	Chutro (N)	Boils	Chapagain et al. 2004
Biophytum sensitivum (L.) DC. (Oxalidaceae); WP	Lajmohani (T)	Insomnia, fever	Chapagain et al. 2004, Dangol and Gurung 1991
Boehmeria platyphylla D.Don (Urticaceae); Rt, Lvs	Khasreti (N)	Cut, wounds	Bhattarai et al. 2009, Manandhar 1985
Boerhavia diffusa L. (Nyctaginaceae); WP	Churchuriya (T)	Headache	Manandhar 1985
Bombax ceiba L. (Malvaceae); Ltx, Fr, Br, Rt, Sht, Sap	Simal; Simra (N, T)	Diarrhea, urinary disorder, vaginal and intestinal bleeding, bone fracture, constipation, wounds, cuts, blisters, worms, fever, dysentery, cough	Bhattarai et al. 2009, Acharya and Acharya 2009, Chapagain et al. 2004, Muller-Boker 1999a, Manandhar 1989a,b, 1990, Mandar and Chaudhary 1993
Brucea javanica (L.) Merr. (Simaroubaceae); Fr	Bhakimlo (N)	Diarrhea	Chapagain et al. 2004
Butea monosperma (Lam.) Taub. Kuntz (Fabaceae); Br	Parasin (T)	Appetizer, weakness	Chapagain et al. 2004
Caesalpinia bonduc (L.) Roxb. (Fabaceae); Lvs, Bd	Kathgarel	Fever	Mandar and Chaudhary 1993
Caesulia axillaris Roxb. (Asteraceae); WP	Gerguj (T)	Wounds	Muller-Boker 1999a
Calamus tenuis Roxb. (Arecaceae); Rt	Bet (N)	Delivery problems, menstrual problems, miscarriages	Chapagain et al. 2004
Callicarpa macrophylla Vahl. (Lamiaceae); St, Rt, Fr, Lvs	Dahigola, Dahigona (T)	Headache, diarrhea, wounds, blisters, fever, chickenpox, boils, rash, indigestion, rheumatic troubles, ear ache	Chapagain et al. 2004, Muller-Boker 1999a, Manandhar 1980, 1987a, 1989b, 1990, Shrestha 1985, Dangol and Gurung 1991

Species (Family); Parts	Local names	Uses	References
Calotropis gigantea (L.) Dryand. (Apocynaceae); Ltx, Twg, Lvs, Fr, St	Aank (N); Madar, Yank (T)	Fractured bone, boils, gingivitis, sinusitis, swelling of testes, body pain, pimples, headache, blood accumulation, body swelling, cuts, delivery, wounds, alleviate pain, cough, sprain, whitlow, injurious to eye, hastening suppuration, arthritis	Bhattarai et al. 2009, Acharya and Acharya 2009, Chapagain et al. 2004, Manandhar 1985, 1990, 1986b, 2002, Muller-Boker 1999a, Dangol and Gurung 1991
Calotropis procera (Aiton.) Dryand. (Apocynaceae); Br	Akon	Diarrhea, dysentery	Mandar and Chaudhary 1993
Cannabis sativa L. (Cannabaceae); Sd, Fl, Lvs	Gajha (N)	Diarrhea, dysentery, constipation, cough, cold, cuts, indigestion	Bhattarai et al. 2009, Chapagain et al. 2004, Dangol and Gurung 1991
Careya arborea Roxb. (Lecythidaceae); Br, Fr	Kumbhi (T)	Wound, diarrhea, fever, snake bite, cuts	Acharya and Acharya 2009, Chapagain et al. 2004
Carica papaya L. (Caricaceae); Sd, Fr	Mewa (N)	Insecticidal, Jaundice	Mandar and Chaudhary 1993
Carissa carandas L. (Apocynaceae); Rt	Karaundath, Karonda (T)	Diarrhea, dysentery, bloody dysentery	Taylor et al. 1996, Manandhar 1985
Cassia fistula L. (Fabaceae); Sd, Fr, Rt, St	Rajbriksha (N), Aairogha (T)	Stomachache, diarrhea, dysentery, diuretic, blisters, extreme thirst, blindness, throat infection, skin diseases, syphilis, diabetes, asthma, laxative, headache, arthritis	Bhattarai et al. 2009, Chapagain et al. 2004, Manandhar 2002, Dangol and Gurung 1991, Mandar and Chaudhary 1993
Cassia mimosoides L. (Fabaceae); WP	Chotaki (T)	Leprosy	Manandhar 1985
Cautleya spicata (Sm.) Baker (Zingiberaceae); Rh	Bayada (T)	Hand/leg ache, indigestion	Chapagain et al. 2004
Celosia argentea L. (Amaranthaceae); Fl, Lvs, Fr	Murga kesar (T)	Nasal bleeding	Chapagain et al. 2004

Table 15.1 contd. ...

...Table 15.1 contd.

Plant's Scientific Name (family name); Parts Used	Nepali and Tharu names	Medicinal Usage	References
Centella asiatica (L.) Urb. (Apiaceae); WP, Lvs	Ghod tapre (N); Dhoataiyae, Bhatbhate, Barmaruwa (T)	Fever, tonic, enhances memory, skin diseases, nerve troubles, cooling agent, cuts, diabetes, dysentery, mental disorders, weakness, heartburn, syphilis	Bhattarai et al. 2009, Acharya and Acharya 2009, Chapagain et al. 2004, Muller-Boker 1999a, Dangol and Gurung 1991, Mandar and Chaudhary 1993
Chamaerops humilis L. (Arecaceae); Sd	Khajuri (T)	Itchy skin	Muller-Boker 1999a
Cheilocostus speciosus (J. Koenig) C.D. Specht (Costaceae); Rt, St	Joghidangha (T)	Headache, diuretic, wounds, swollen feet	Bhattarai et al. 2009, Muller-Boker 1999a
Cheilanthes farinosa (Forsk.) Kaulf. (Pteridaceae); Rt	Dubai sinki (N)	Wounds	Manandhar 1985
Chenopodium album L. (Amaranthaceae); Lvs, St, Rt, WP	Bethuwa (N)	Constipation, clean bronchi, lungs, joint pain, blood pressure	Muller-Boker 1999a, Dangol and Gurung 1991, Mandar and Chaudhary 1993
Chenopodium murale L. (Amaranthaceae); WP	Pahadia Bethuwa (N); Bangain (T)	Gastritis, indigestion	Dangol and Gurung 1991
Chlorophytum nepalense (Lindl.) Baker (Asparagaceae); Tu	Dhud kutri, Banpyajia (T)	Milk secretion, hydroccle, joint pain	Chapagain et al. 2004, Dangol and Gurung 1991, Manandhar 1985
Chrozophora rottleri (Geisel.) A. Juss ex Spreng. (Euphorbiaceae); Fr	Chotaki hunkatath (T)	Cold	Manandhar 1985
Chrysopogon aciculatus (Retz.) Trin. (Poaceae); Rt	Sarauth (T)	Boils, wounds	Manandhar 1985
Cinnamomum camphora (L.) J. Presl (Lauraceae); Lvs	Kapur (N)	Cough, cold	Bhattarai et al. 2009
Cirsium wallichii DC. (Asteraceae); Rt	Markatiya (T)	Stomach inflammation	Manandhar 1985

Plant name (Family); Parts	Local names	Uses	References
Cissampelos pareira L. (Menispermaceae); Rt, Aep	Gohman, Batulia (T)	Chronic diseases of children, headache, appetizer, diarrhea, weakness, fever, induce abortion in human	Bhattarai et al. 2009, Chapagain et al. 2004, Dangol and Gurung 1991
Cissus repens Lam. (Vitaceae); Br	Rechu (T)	Cough, cold, bronchitis	Bhattarai et al. 2009
Cissus sp. (Vitaceae); St	Charcharia	Extreme thirst	Chapagain et al. 2004
Citrus medica L. (Rutaceae); Fr, Lvs	Nemo	Diarrhea, dysentery, insecticidal, itches	Mandar and Chaudhary 1993
Cleome viscosa L. (Cleomaceae); Fr, Lvs, Sd, Rt	Harhur, Hurhuwa (T)	Earache, headache, arthritis/rheumatism, loss of weight, mental disorders, cuts, wounds	Chapagain et al. 2004, Dangol and Gurung 1991, Manandhar 1985
Clerodendrum indicum (L.) Kuntze (Lamiaceae); Aer, WP	Chuchure (T)	Fever, cuts, wounds	Taylor et al. 1996
Clerodendrum infortunatum L. (Verbenaceae); Lvs, Rt, Sht, Bd	Bhanthi, Dhus (T)	Fever, delivery problems, boils, blood purification, eliminate lice	Dangol and Gurung 1991, Chapagain et al. 2004, Muller-Boker 1999a, Mandar and Chaudhary 1993, Manandhar 1985
Coffea benghalensis B. Heyne ex Schult. (Rubiaceae); Bd	Bharemase phul (N)	Delivery	Chapagain et al. 2004
Colebrookea oppositifolia Sm. (Lamiaceae); Lvs, Rt, St	Dhurseli (N); Dhurseta, Bhogate (T)	Eye pain, anthelmintic, premature ejaculation, rheumatic troubles, ear-ache	Bhattarai et al. 2009, Dangol and Gurung 1991, Manandhar 1994
Crateva unilocularis Buch.-Ham. (Capparaceae); Sd, Br, Rt	Sipligan (N)	Sinusitis, stomachache, cuts, wounds, boils	Bhattarai et al. 2009
Crinum amoenum Ker Gawl. Ex Roxb. (Amaryllidaceae); Bu	Ban piyaju (T)	To swollen testicles	Muller-Boker 1999a
Crotalaria prostrata Rottb. ex Willd. (Fabaceae); WP	Bansan (T)	Wounds, blisters	Chapagain et al. 2004, Manandhar 1985
Crypsinus hastatus (Thunb.) Copel. (Polypodiaceae); Rt	Harjor (N)	Fever	Manandhar 1985
Cucumis melo L. (Cucurbitaceae); Rt, WP	Gurmi, Goima (T)	Dysentery, difficulty in urination	Dangol and Gurung 1991, Manandhar 1985

Table 15.1 contd. ...

...Table 15.1 contd.

Plant's Scientific Name (family name); Parts Used	Nepali and Tharu names	Medicinal Usage	References
Cucumis sativus L. (Cucurbitaceae); Sd	Khira (T)	Heartburn	Muller-Boker 1999a
Curculigo orchioides Gaertn. (Hypoxidaceae); Rt, Rh	Musaleri, Mussar (T)	Headache, lactation in postpartum mother, snake scorpion sting, diarrhea, dysentery, milk secretion, rheumatism	Bhattarai et al. 2009, Chapagain et al. 2004, Mandar and Chaudhary 1993, Manandhar 1985
Curcuma angustifolia Roxb. (Zingiberaceae); Rt	Haldi (N)	Cough, cold, fever, fracture and dislocated bones	Mandar and Chaudhary 1993, Manandhar 1985
Cuscuta reflexa Roxb. (Convolvulaceae); WP	Amarlathi, Budhbaula, Saraksewal (T)	Hookworm, roundworm, jaundice, rabies infection, depression, body swelling, bone fracture, cuts, indigestion, dislocated parts	Bhattarai et al. 2009, Chapagain et al. 2004, Dangol and Gurung 1991, Manandhar 1985
Cuscuta reflexa var. *anguina* (Edgew.) C.B. Clarke (Convolvulaceae); WP	Baora (T)	Stomachache, headache, body pain, jaundice	Acharya and Acharya 2009
Cymbidium madidum Lindl. (Orchidaceae); Rt	Harjor (T)	Fracture and dislocated bone	Manandhar 1988, 1985
Cymbopogon flexuosus (Nees ex Steud.) W.Waston (Poaceae); Lvs		Cough, cold	Chapagain et al. 2004
Cymbopogon jwarancusa (Jones) Schult. (Poaceae); Rt	Jarakus (T)	Snake scorpion sting, asthma	Chapagain et al. 2004
Cynodon dactylon (L.) Pers. (Poaceae); WP, Rt, Lvs	Dubho (N)	Tonic, vertigo/dizziness, anthelmintic, stomachache, defect in vision, toothache, cuts, boils, nasal bleeding, white leprosy	Bhattarai et al. 2009, Acharya and Acharya 2009, Chapagain et al. 2004
Cynoglossum lanceolatum Forssk. (Boraginaceae); Rt, Fl		Cataracts	Bhattarai et al. 2009
Cynoglossum zeylanicum (Vahl.) Brand (Boraginaceae); Lvs	Chakchira (T)	Headache, wounds, blisters	Chapagain et al. 2004
Cyperus compressus L. (Cyperaceae); WP	Jhusuna (T)	Cuts, scabies	Dangol and Gurung 1991
Cyperus difformis L. (Cyperaceae); WP	Ghanaune chhatia (T)	Earache	Chapagain et al. 2004
Cyperus rotundus L. (Cyperaceae); Rt, Lvs	Motha (N); Bhada (T)	Stomach diseases, indigestion, diarrhea, vomiting, cough, bronchitis, fever	Bhattarai et al. 2009, Acharya and Acharya 2009

Dalbergia latifolia Roxb. (Fabaceae); Rt, Br	Satisal (N)	Bodyache	Acharya and Acharya 2009
Dalbergia sissoo Roxb. ex DC. (Fabaceae); Br, Lvs, Sht	Sisso (N); Sisawa, Sisava (T)	Feeling of higher level of heat inside the body, impotency, cooling effect, anthelmintic, gout, fever	Bhattarai et al. 2009, Chapagain et al. 2004, Dangol and Gurung 1991, Manandhar 1985, 1994
Datura metel L. (Solanaceae); Sd, Fr	Dhatur (T)	Mental disorders, diarrhea, dysentery, filaria	Chapagain et al. 2004, Mandar and Chaudhary 1993
Datura stramonium L. (Solanaceae); Sd, Lvs	Dhaturo (N); Dhatur (T)	Premature ejaculation, menstrual disorders, bronchitis, arthritis	Bhattarai et al. 2009, Muller-Boker 1999a, Mandar and Chaudhary 1993
Dendrocalamus hamiltonii Nees and Arn. ex Munro (Poaceae); Sht	Tama (N)	Purgative	Dangol and Gurung 1991
Desmodium gangeticum (L.) DC. (Fabaceae); Lvs, St, Fl	Gatkosiya (T)	Headache	Bhattarai et al. 2009
Desmodium laxiflorum DC. (Fabaceae); St	Tangari (N)	Abortification, fish poison	Mandar and Chaudhary 1993
Desmostachya bipinnata (L.) Stapf (Poaceae); Rt	Kush (N), Kusli (T)	Headache, toothache	Chapagain et al. 2004
Dillenia pentagyana Roxb. (Dilleniaceae); Sd	Tetari (N); Agai (T)	Headache, to tick bites	Chapagain et al. 2004, Muller-Boker 1999a
Dioscorea bulbifera L. (Dioscoreaceae); Tu	Gittha tarul (N)	Earache	Chapagain et al. 2004
Dioscorea deltoidea Wall. ex Griseb. (Dioscoreaceae); Fr	Gittha (Nep)	Stomach pain	Acharya and Acharya 2009
Dioscorea sp. (Dioscoreaceae); Rt	Hardgohi (T)	Fever	Muller-Boker 1999a
Diospyros malabarica (Desr.) Kostel. (Ebenaceae); Fr	Tendu (N)	Boils	Chapagain et al. 2004
Diplocyclos palmatus (L.) C. Jeffrey (Cucurbitaceae); Lvs	Kundru (T)	Blindness	Chapagain et al. 2004
Diploknema butyracea (Roxb.) H.J. Lam (Sapotaceae); Fr, Sd	Chyuri (N)	Wounds, muscle pain, astringent	Acharya and Acharya 2009, Manandhar 1985
Drynaria quercifolia (L.) J. Sm. (Polypodiaceae); WP	Hathajori (T)	Arheya (a disease killing affected livestock)	Muller-Boker 1999a

Table 15.1 contd. ...

...Table 15.1 contd.

Plant's Scientific Name (family name); Parts Used	Nepali and Tharu names	Medicinal Usage	References
Dryopteris sp. (Dryopteridaceae); Lvs	Damsinki (N)	Gastric, ulcer, worms	Acharya and Acharya 2009
Eclipta prostrata (L.) L. (Asteraceae); Lvs, WP, Rt, Sd	Bhangarella, Bhegruna, Bhangarail (T)	Infection, to treat white spots in eye, athlete's foot, cuts, wounds, ulcer, headache	Chapagain et al. 2004, Muller-Boker 1999a, Dangol and Gurung 1991, Mandar and Chaudhary 1993, Manandhar 1985
Ehretia laevis Roxb. (Boraginaceae); Br	Khatkhajuwa (T)	Blisters, diarrhea	Chapagain et al. 2004
Elephantopus scaber L. (Asteraceae); Rt, WP	Dadri, Khasuriya (T)	Headache and sinusitis, tonic, wounds, blisters, aphrodisiac	Bhattarai et al. 2009, Chapagain et al. 2004, Manandhar 1985
Engelhardia spicata Lesch. Ex Blume (Juglandaceae); Fl	Mahuwa (N)	Cough, cold	Chapagain et al. 2004
Equisetum diffusum D.Don (Equisetaceae); Lvs, WP, Rt	Akchomka, Ankhachiukan (T)	Headache, dysentery, indigestion, chest complaints, fracture and dislocated bones	Chapagain et al. 2004, Manandhar 1985, 1987b, 1990
Erigeron trilobus (Decne.) Boiss. (Asteraceae); Lvs	Undhmunte (T)	Wounds, blisters	Chapagain et al. 2004
Erythrina arborescens Roxb. (Fabaceae); Lvs	Pharhed (N)	Diarrhea, dysentery, constipation, ear pain	Bhattarai et al. 2009
Eucalyptus alba Reinw. ex Blume (Myrtaceae); Br	Kalapti (T)	Body-ache	Chapagain et al. 2004
Eulaliopsis binata (Retz.) C.E. Hubb. (Poaceae); WP	Bankas (T)	Blood purifier	Chapagain et al. 2004
Euphorbia fusiformis Buch.-Ham. ex D.Don (Euphorbiaceae); WP, Lvs, Rt	Banmurai (T)	Body swelling, milk secretion	Chapagain et al. 2004
Euphorbia hirta L. (Euphorbiaceae); WP, Lvs, Rt	Duddhi (N); Dudhiya jhyang (T)	Constipation, cuts, milk secretion, wounds, blisters, numb legs	Chapagain et al. 2004, Dangol and Gurung 1991
Euphorbia parviflora L. (Euphorbiaceae); WP	Sano duddhi (N)	Menstrual problems	Chapagain et al. 2004
Euphorbia prostrata Ait. (Euphorbiaceae); WP	Dundhi (T)	Snake-bite	Manandhar 1985

Scientific name (Family); Parts used	Local name	Uses	References
Euphorbia royleana Boiss. (Euphorbiaceae); Lvs, Rt, St	Sihundisighe, Sihur (T)	Cough, cold, blisters, wounds, constipation, pneumonia, anthelmintic	Bhattarai et al. 2009, Chapagain et al. 2004, Manandhar 1994
Evolvulus nummularius (L.) L. (Convolvulaceae); WP	Dinghumni phul (T)	Scabies	Manandhar 1985
Ficus benghalensis L. (Moraceae); Br, Lvs	Bargat, pipra (T)	Diarrhea, dysentery, diabetes	Acharya and Acharya 2009, Chapagain et al. 2004, Mandar and Chaudhary 1993
Ficus hispida L.f. (Moraceae); Lvs, Rt	Kothaiya dumari (T)	Diuretic, ear pain, ear wound, weakness of hearing	Bhattarai et al. 2009, Muller-Boker 1999a, Dangol and Gurung 1991
Ficus lacor Buch.-Ham. (Moraceae); Ltx	Gular (T)	Boils	Manandhar 1985
Ficus racemosa L. (Moraceae); Ltx	Gullar (T)	Wounds, blisters	Chapagain et al. 2004
Ficus religiosa L. (Moraceae); St, Lvs	Pipal (N); Pipra (T)	Cuts, wounds, body swelling, aching ear, regularize a woman's monthly period, migraine	Bhattarai et al. 2009, Chapagain et al. 2004, Muller-Boker 1999a, Dangol and Gurung 1991, Mandar and Chaudhary 1993
Ficus semicordata Buch.-Ham. ex Sm. (Moraceae); Rt, Lvs, Twg	Khurhuri (T)	Headache, fever, aching ear	Acharya and Acharya 2009, Muller-Boker 1999a
Flemingia chappar Benth. (Fabaceae); Lvs, Rt	Bansapti	Cuts, diarrhea	Chapagain et al. 2004
Flemingia macrophylla (Willd.) Merr. (Fabaceae); St	Majilauta (T)	Toothbrush	Muller-Boker 1999a
Flemingia strobilifera (L.) W.T.Aiton (Fabaceae); Lvs	Banasapti (T)	Cuts	Chapagain et al. 2004
Fragaria indica Wall. (Rosaceae); WP	Chauranchata (T)	Fever, urinary disorders	Dangol and Gurung 1991
Gardenia jasminoides J.Ellis (Rubiaceae); Fl	Indrakaul	Quench thirst	Dangol and Gurung 1991
Garuga pinnata Roxb. (Burseraceae); Br, Sht	Jhengra (T)	Snake scorpion sting, cuts, diarrhea, boils, malaria, wounds	Chapagain et al. 2004, Manandhar 1985
Gaultheria fragrantissima Wall. (Ericaceae); Lvs	Pakbhemi (T)	To get relief from par of head, legs and hands	Acharya and Acharya 2009
Gmelina arborea Roxb. (Lamiaceae); Lvs	Gamhar (T)	Swelling	Dangol and Gurung 1991

Table 15.1 contd. ...

...Table 15.1 contd.

Plant's Scientific Name (family name); Parts Used	Nepali and Tharu names	Medicinal Usage	References
Grewia sapida Roxb. ex DC. (Malvaceae); WP	Farsa (T)	Dysentery	Dangol and Gurung 1991
Haldina cordifolia (Roxb.) Ridsdale (Rubiaceae); Lvs	Haldu (N)	Wounds, blisters	Chapagain et al. 2004, Dangol and Gurung 1991
Hedychium spicatum Sm. (Zingiberaceae); Rt		Menstrual problems	Bhattarai et al. 2009
Helianthus annuus L. (Asteraceae); Fr, Fl, Lvs, Sd	Suryamandal phul (N)	Headache, deafness, pneumonia, wounds, blisters	Chapagain et al. 2004
Helicteres isora L. (Malvaceae); Rt	Patuha, Karatha (T)	Worm infestations, Fever	Muller-Boker 1999a, Dangol and Gurung 1991
Heliotropium strigosum Willd. (Boraginaceae); WP	Chiraigoar, Darbahi (T)	Appetizer, Infection	Chapagain et al. 2004
Helminthostachys zeylanica Hook.(L.) (Ophioglossaceae); Rh	Majurgoda (T)	Impotency	Chapagain et al. 2004
Hemigraphis hirta (Vahl) T. Anders. (Anacardiaceae); WP	Banpan (T)	Throat problems	Manandhar 1985
Hibiscus rosa-sinensis L. (Malvaceae); Lvs, Fl	Barhamase phul (N)	Boils, diuretic, wounds, blisters	Bhattarai et al. 2009, Chapagain et al. 2004
Hippochaete debilis (Roxb. ex Vaucher) Ching (Equisetaceae); WP	Sanohadchure (N)	Bone fractures	Bhattarai et al. 2009, Chapagain et al. 2004
Hiptage benghalensis (L.) Kurz (Malpighiaceae); St	Madhhulata (T)	Toothache	Chapagain et al. 2004
Holarrhena pubescens Wall. ex. G.Don. (Apocynaceae); Br, Fr, Ltx, Rt, Fl	Dudhakoria, Kacheri (T)	Bloody dysentery, diarrhea, worms, anthelmintic, snake scorpion sting, body swelling, cough, cold, constipation, tuberculosis, urinary infection, menstrual disorders	Bhattarai et al. 2009, Chapagain et al. 2004, Muller-Boker 1999a, Mandar and Chaudhary 1993
Hydrocotyle rotundifolia Roxb. (Apiaceae); WP	Tarpurin (T)	Fever, urinary disorders	Dangol and Gurung 1991

Hygrophila auriculata (Schumach.) Heine (Acanthaceae); WP, Rt, Lvs	Makhana (T)	Toothache, body swelling, vomiting, wounds, blisters, invigorating	Chapagain et al. 2004, Manandhar 1985
Hymenodictyon orixense (Roxb.) Mabb. (Rubiaceae); Sht, Fr	Bhudkul, Bhurkun (T)	Body swelling, miscarriages	Chapagain et al. 2004
Ichnocarpus frutescens (L.) W.T.Aiton (Apocynaceae); Rt, Lvs	Chhekar duddhi (T)	Diarrhea, milk secrection	Chapagain et al. 2004
Imperata cylindrica (L.) Raeusch. (Poaceae); Rt	Siru (N); Churki (T)	Anthelmintic, cooling agent, vomiting, anthelmintic	Chapagain et al. 2004, Muller-Boker 1999a, Manandhar 1994
Indigofera linifolia (L.f.) Retz. (Fabaceae); Rt	Muscamna (T)	Milk secrection	Chapagain et al. 2004
Inula cappa (Buch.-Ham. ex D.Don) DC. (Asteraceae); Rt	Pataya (T)	Headache, to reduce disorders arising due to intake of meat	Bhattarai et al. 2009
Ipomoea aquatic Forssk (Convolvulaceae); Lvs	Dhodi (T)	Wounds, blisters	Chapagain et al. 2004
Ipomoea carnea subsp. *fistulosa* (Mart. ex Choisg) D.F. Austin (Convolvulaceae); Ltx	Besarma, Behaya (T)	Cuts, wounds	Chapagain et al. 2004, Manandhar 1985
Ipomoea quamoclit L. (Convolvulaceae); WP	Chhotaki gurubans (T)	Blood vomiting	Manandhar 1985
Isodon sp. (Lamiaceae); WP	Mutmuhari (T)	Infection	Chapagain et al. 2004
Jatropha curcas L. (Euphorbiaceae); Ltx, St	Sajiwan (N); Nijot, Ramjeevanmam jyoti (T)	Wounds, blisters, burns, boils, gum infection, athlete's foot, tootache	Bhattarai et al. 2009, Chapagain et al. 2004, Muller-Boker 1999a, Dangol and Gurung 1991, Mandar and Chaudhary 1993
Justicia adhatoda L. (Acanthaceae); Lvs, Sht, Rt	Asuro (N); Rus, Asur (T)	Body swelling, cough, cold, bronchitis, fever, asthma, bronchitis, typhoid, stomach pain	Chapagain et al. 2004, Dangol and Gurung 1991, Bhattarai et al. 2009, Muller-Boker 1999a, Mandar and Chaudhary 1993, Manandhar 1985
Justicia sp. (Acanthaceae); Lvs, WP	Gorkatla, Panchuiya (T)	Cooling agent, cuts, delivery	Chapagain et al. 2004
Kalanchoe integra (Medik.) Kuntze (Crassulaceae); Lvs	Ajambari (N)	Ear pain, ear wounds	Bhattarai et al. 2009

Table 15.1 contd. ...

...Table 15.1 contd.

Plant's Scientific Name (family name); Parts Used	Nepali and Tharu names	Medicinal Usage	References
Lablab purpureus (L.) Sweet (Fabaceae); WP, Rt	Simi (N)	Headache, bone fracture	Chapagain et al. 2004
Lagenaria siceraria (Molina) Standl. (Cucurbitaceae); Lvs	Louka (N)	Burns, boils	Chapagain et al. 2004
Lagerstroemia parviflora Roxb. (Lythraceae); Fl	Dhayaro (N)	Diarrhea, dysentery, typhoid fever	Bhattarai et al. 2009
Lannea coromandelica (Houtt.) Merr. (Anacardiaceae); St	Gingad (T)	Nosebleeding	Dangol and Gurung 1991
Launaea aspleniifolia (Willd.) Hook.f. (Asteraceae); WP	Chikini dudhi (T)	Skin irritation	Manandhar 1985
Lawsonia inermis L. (Lythraceae); Lvs, Sd	Mehandi (N)	Abrasion, backbone pain, menstrual problems and disorders, cooling agent, infection	Bhattarai et al. 2009, Chapagain et al. 2004
Leucas aspera (Willd.) Link (Lamiaceae); Lvs, Rt	Danefu (T)	Ringworm	Dangol and Gurung 1991
Leucas cephalotes (Roth) Spreng. (Lamiaceae); Rt, WP, Lvs	Sano gum (N); Makhan (T)	Snake scorpion sting, asthma, ringworm, boils, urinary complaints	Chapagain et al. 2004, Dangol and Gurung 1991, Manandhar 1985
Leucas zeylanica (L.) W.T.Aiton (Lamiaceae); Lvs	Dulphi	Jaundice	Mandar and Chaudhary 1993
Linum usitatissimum L. (Linaceae); WP, Sd	Arasi (T)	Wounds, boils	Manandhar 2002
Lippia nodiflora (L.) Rich. (Verbenaceae); WP	Kokan (T)	Headache	Manandhar 1985
Litchi chinensis Sonn. (Sapindaceae); Ltx	Litchi (N)	Dysentery	Bhattarai et al. 2009
Litsea monopetala (Roxb.) Pers. (Lauraceae); Sd	Kutmero (N)	Stomachache	Bhattarai et al. 2009
Ludwigia octovalvis (Jack.) Raven (Onagraceae); WP	Bihi (N)	Wounds	Manandhar 1985

Plant name (family); parts	Local name	Uses	References
Luffa cylindrica (L.) M. Roem. (Cucurbitaceae); Sd	Ghira (T)	Chicken pox	Müller-Böker 1999a
Lygodium flexosum (L.) Sw. (Lygodiaceae); Rt	Bhalunia	Wounds, blisters	Chapagain et al. 2004
Macrotyloma uniflorum (Lam.) Verdc. (Fabaceae); Sd	Gahat (T)	Arthritis/Rheumatism	Chapagain et al. 2004
Madhuca longifolia (J.Koenig ex L.) J.F.Macbr. (Sapotaceae); Fl	Mahuwa (N)	Tierdness	Mandar and Chaudhary 1993
Mallotus philippensis (Lam.) Mull. Arg. (Euphorbiaceae); Br, Rt, Fr	Sindure (N); Rohini (T)	Diarrhea, dysentery, asthma, bronchitis, urinary tract infection, constipation, skin diseases, typhoid, gastratis	Bhattarai et al. 2009, Chapagain et al. 2004, Dangol and Gurung 1991
Mangifera indica L. (Anacardiaceae); Br, Ltx, Rt, Fr	Aamp (N)	Chronic diarrhea, gastritis, dysentery, loss of weight, vitamin deficiency, menstrual disorders	Bhattarai et al. 2009, Chapagain et al. 2004, Mandar and Chaudhary 1993
Marsdenia tinctoria R.Br. (Apocynaceae); Fl	Dudhe laharo (N)	Gout, disorder of thymus, to purify blood	Acharya and Acharya 2009
Melia azedarach L. (Meliaceae); Br, Rt, Fr, Lvs	Bakaino, Nimpata (N); Bakain (T)	Diarrhea, cholera, constipation, anthelmintic, vomiting, blood impurities, urinary discharges, fever, paralysis	Bhattarai et al. 2009, Acharya and Acharya 2009, Dangol and Gurung 1991, Mandar and Chaudhary 1993
Mentha arvensis L. (Lamiaceae); Lvs	Patina (N)	Cough, cold, dehydration, gastritis	Bhattarai et al. 2009
Mentha longifolia (L.) L. (Lamiaceae); Lvs	Pudina (N)	Headache, cooling agent, dysentery, skin diseases	Chapagain et al. 2004
Mentha spicata L. (Lamiaceae); Lvs	Bawari, Pudina (N)	Throat infection, indigestion, boils	Acharya and Acharya 2009, Manandhar 1985
Mimosa pudica L. (Fabaceae); Rt	Lajjawati (N); Lajmohani (T)	Cuts, wounds, insomnia, fever, paralysis	Bhattarai et al. 2009, Chapagain et al. 2004, Mandar and Chaudhary 1993
Mirabilis jalapa L. (Nyctaginaceae); Rt	Malati phul (N); Nakesari, Barka gurubans (T)	Delivery problems, menstrual problems, milk seccretion, nasal bleeding, scabies	Chapagain et al. 2004, Manandhar 1985

Table 15.1 contd. ...

...Table 15.1 contd.

Plant's Scientific Name (family name); Parts Used	Nepali and Tharu names	Medicinal Usage	References
Momordica charantia L. (Cucurbitaceae); WP, Lvs	Kareli jhar (N)	Body swelling, blindness	Chapagain et al. 2004
Momordica dioica Roxb. ex Willd (Cucurbitaceae); Rt	Kheksi (T)	Diabetes, urinary infection	Chapagain et al. 2004
Moringa oleifera Lam. (Moringaceae); Fr, Lvs, Br	Munga (N)	Blood pressure	Mandar and Chaudhary 1993
Morus serrata Roxb. (Moraceae); Rt, Br	Tathimashlari (T)	Diarrhea, anthelmintic, typhoid	Bhattarai et al. 2009
Murraya koenigii (L.) Spreng. (Rutaceae); Fr, Lvs, Rt	Ban neem, binbinbheria (T)	Diarrhea, dysentery	Acharya and Acharya 2009, Chapagain et al. 2004
Musa paradisiac L. (Musaceae); Fr, St	Kera (N)	Loss of weight, diarrhea, dysentery	Chapagain et al. 2004, Mandar and Chaudhary 1993
Myrica esculenta Buch.Ham ex D.Don (Myricaceae); Br	Kafal (N)	Dysentery	Chapagain et al. 2004
Nelumbo nucifera Gaertn. (Nelumbonaceae); St	Kamalko phul (N)	Pneumonia	Chapagain et al. 2004
Neopicrorhiza scrophulariiflora (Pennell) D.Y.Hong (Plantaginaceae); Rt	Kutki (N)	Cuts	Chapagain et al. 2004
Nepeta leucophylla Benth. (Lamiaceae); WP	Gandheli gharra (T)	Boils	Manandhar 1985
Nicotiana tabacum L. (Solanaceae); Lvs	Surti (N)	Swelling of scrotum and testes	Bhattarai et al. 2009
Nyctanthes arbortristis L. (Oleaceae); Fl, Lvs	Parijat (N)	Headache	Chapagain et al. 2004
Ocimum americanum L. (Lamiaceae); Rt	Bawari (N)	Fever	Manandhar 1985
Ocimum basilicum L. (Lamiaceae); Sd	Gathiwan (T)	Cough, cold, feeling of higher level of heat inside the body	Bhattarai et al. 2009
Ocimum tenuiflorum L. (Lamiaceae); Lvs	Tulsi (N)	Blood pressure, wounds, ear pain, cough, cold, bronchitis, sinusitis, typhoid, fever	Bhattarai et al. 2009, Chapagain et al. 2004, Mandar and Chaudhary 1993

Plant name	Local name	Uses	References
Opuntia monacantha (Willd.) Haw. (Cactaceae); St, Sap	Pate siundi (N)	Burns, boils, wounds, blisters	Chapagain et al. 2004
Oroxylum indicum (L.) Kurz (Bignoniaceae); Rt, Fr, Br	Sauna tata (T)	Bone fracture, dysentery, pneumonia, vomiting, wounds, menstrual disorders	Chapagain et al. 2004, Muller-Boker 1999a, Mandar and Chaudhary 1993
Oxalis corniculata L. (Oxalidaceae); WP, Lvs, St, Rt	Chariamilo (N), Chamchama, Amchocha (T)	Fever, diarrhea, boils, dysentery, fever, pneumonia, wounds, blisters, nasal bleeding, relieves pains, cold	Bhattarai et al. 2009, Acharya and Acharya 2009, Chapagain et al. 2004, Muller-Boker 1999a, Dangol and Gurung 1991
Parmelia nepalensis Tayl. (Parmeliaceae); WP	Jhyau (N)	Diseases of gum, throat, scabies, piles, leprosy	Acharya and Acharya 2009
Pericampylus glaucus (Lam.) Merr. (Menispermaceae); Rt	Khuranijhar (N)	Diuretic, dysuria	Bhattarai et al. 2009
Peristrophe bicalyculata (Retz.) Nees (Acanthaceae); WP	ChuChure (T)	Fever, cuts, wounds	Taylor et al. 1996
Persicaria barbata (L.) H.Hara (Polygonaceae); WP	Miriya bikh (T)	Swelling	Dangol and Gurung 1991
Phyla nodiflora (L.) Greene (Verbenaceae); Rt, Lvs, WP	Jalnim, Jhyangrin (T)	Headache, appetizer, cooling agent, stomachache, ulcers	Chapagain et al. 2004, Dangol and Gurung 1991
Phyllanthus amarus Schumach. and Thonn. (Phyllanthaceae); WP, Rt, Lvs	Bhuiamala (N); Chhotaki dahigola (T)	Asthma, bronchitis, urinary tract infection, diuretic, stomachic, febrifuge, pimples, cuts, wounds	Bhattarai et al. 2009, Manandhar 1985, 2002
Phyllanthus emblica L. (Phyllanthaceae); Fr, Sd, St, Lvs	Amala(N); Aura, Auraha, Yawara (T)	Heart pain, constipation, cough, cold, sinusitis, diarrhea, dysentery, gastritis, diuretic, laxative, vitamin, burns, hair tonic	Bhattarai et al. 2009, Acharya and Acharya 2009, Chapagain et al. 2004, Muller-Boker 1999a, Mandar and Chaudhary 1993
Phyllanthus urinaria L. (Phyllanthaceae); Rt, Lvs	Sano dahigola (T)	Wounds, blisters	Chapagain et al. 2004
Physalis divaricata D.Don (Solanaceae); WP	Galbhomara (T)	Body thin and feeble	Manandhar 1985

Table 15.1 contd. ...

...*Table 15.1 contd.*

Plant's Scientific Name (family name); Parts Used	Nepali and Tharu names	Medicinal Usage	References
Physalis minima L. (Solanaceae); Lvs	Budak yadur (T)	Menstrual disorders, eye infection	Dangol and Gurung 1991
Pinus roxburghii Sarg. (Pinaceae); Oil	Sallo (N)	Arthritis/rheumatism	Chapagain et al. 2004
Piper cubeba L.f. (Piperaceae); Fr	Marich (N)	Boils, cough, cold, tonsillitis, bronchitis, wounds	Bhattarai et al. 2009
Piper longum L. (Piperaceae); Rt, Fr, St	Kharipipar, Pharipipra (T)	Menstrual disorders, cough, digestion, asthma, cold	Bhattarai et al. 2009, Acharya and Acharya 2009, Chapagain et al. 2004, Muller-Boker 1999a, Dangol and Gurung 1991
Plantago major L. (Plantaginaceae); Sd	Bhatbhadwa (T)	Fever, diarrhea, dysentery, laxative demulcent, expectorant, diuretic	Acharya and Acharya 2009
Plumbago zeylanica L. (Plumbaginaceae); Ltx, WP, Rt	Abijale kuro (N); Kalamnath (T)	Blister, wart, ringworm, diarrhea, dysentery, fever, stomachache, boils, stomach disorders	Bhattarai et al. 2009, Manandhar 1985, 1987b, 1990
Plumeria rubra L. (Apocynaceae); St, Ltx, Br	Galaini (N)	Stomachache, rabies infection toothache, anthelmintic	Bhattarai et al. 2009, Chapagain et al. 2004
Pogostemon benghalensis (Burm.f.) Kuntze (Lamiaceae); WP, Rt, Sd, Lvs	Rudhilo (Nep); Bhati, Udhara, Kohabar (T)	Fever, cough, cold, typhoid, bodyache, diarrhea, pneumonia, dysentery	Bhattarai et al. 2009, Chapagain et al. 2004, Manandhar 1985, 1988
Pogostemon parviflorus Benth. (Lamiaceae); Lvs	Utajari (T)	Fever	Dangol and Gurung 1991
Polyalthia longifolia (Sonn.) Thw. (Annonaceae); Br	Ashoka (N)	Delivery	Chapagain et al. 2004
Polygala arillata Buch.-Ham. ex D.Don (Polygalaceae); WP	Thanak (T)	Tonsillitis, stomachache, gastritis, cough, cold, wounds	Bhattarai et al. 2009
Polygonatum verticillatum (L.) All. (Asparagaceae); Rt	Kaituwa (T)	Tonic	Manandhar 1988

Polygonum plebejum R.Br. (Polygonaceae); WP	Chiraik gor (T)	Wounds	Manandhar 1985
Polygonum viscosum (Buch.-Ham ex D.Don) (Polygonaceae); Lvs	Biria (T)	Inflammation of sex organs of child	Chapagain et al. 2004
Pouzolzia zeylanica (L.) Benn. (Urticaceae); WP	Kakara khopta, Kakarakucha	Ringworm, wounds	Dangol and Gurung 1991
Premna barbata Wall. ex Schauer (Lamiaceae); St Br	Giliyar (T)	Fever, quench thirst	Dangol and Gurung 1991
Premna serratifolia L. (Lamiaceae); Rt, Lvs	Gineri (N)	Body swelling, wounds, blisters	Chapagain et al. 2004
Prunus persica (L.) Batsch (Rosaceae); Lvs	Aru (N)	Wounds	Bhattarai et al. 2009, Chapagain et al. 2004
Psidium guajava L. (Myrtaceae); Br, Fr, St, Lvs, Rt	Amba, Belauti (N); Runi, amrud (T)	Diarrhea, dysentery, constipation, fever, cough, cold, anthelmintic, hoarseness of voice, vomiting, tootache, stomach complaints	Bhattarai et al. 2009, Chapagain et al. 2004, Muller-Boker 1999a, Mandar and Chaudhary 1993, Manandhar 1985
Pterocarpus marsupium Roxb. (Fabaceae); St, Br	Bijaya sal (N)	Neurological problems, pneumonia	Acharya and Acharya 2009
Pulicaria dysenterica (L.) Gaertn. (Asteraceae); WP	Gandhaiya (T)	Wounds	Manandhar 1985
Punica granatum L. (Lythraceae); Fr, Br	Darim, Anar (Nep)	Diarrhea, dysentery, skin abrasions, control abortion, miscarriage	Bhattarai et al. 2009, Mandar and Chaudhary 1993
Ranunculus sceleratus L. (Ranunculaceae); WP	Nakapolba (T)	Gastric inflammation	Manandhar 1985
Raphanus sativus var. *hortensis* Backer (Brassicaceae); Lvs, Tu	Mula (N)	Indigestion, skin diseases	Chapagain et al. 2004
Rauvolfia serpentina (L.) Benth. ex Kurz (Apocynaceae); Rt, Fr	Sarpagandha (N); Dhaldhaliya, Dharmarua, Dhambarbiruwa (T)	Cuts, wounds, boils, fever, stomachache, menstrual disorders, snake bite, blood pressure, hypotonic, dysentery, indigestion, mental disorders, diarrhea	Bhattarai et al. 2009, Acharya and Acharya 2009, Chapagain et al. 2004, Manandhar 1985, 1988, Muller-Boker 1999a, Dangol and Gurung 1991

Table 15.1 contd. ...

...*Table 15.1 contd.*

Plant's Scientific Name (family name); Parts Used	Nepali and Tharu names	Medicinal Usage	References
Reinwardtia indica Dumort. (Linaceae); Rt	Dauthi Phul (N)	Lactation in postpartum mothers	Bhattarai et al. 2009
Remusatia vivipara (Roxb.) Schott (Araceae); St	Kachchu (N)	Boils, blisters	Mandar and Chaudhary 1993
Ricinus communis L. (Euphorbiaceae); Sd, Lvs, Fr, St, Rt	Andir (N); Reyar, Reru (T)	Skin abrasions, to relieve pain on fractured bone, body swelling, deafness, burn, constipation, boils	Bhattarai et al. 2009, Chapagain et al. 2004, Muller-Boker 1999a, Mandar and Chaudhary 1993, Manandhar 1985
Rumex nepalensis Spreng. (Polygonaceae); Sd	Dhaldhaliya (T)	Chickenpox	Manandhar 1985
Sagittaria guayanensis Kunth (Alismataceae); WP	Banarbhega (T)	Fever	Manandhar 1985
Sapium insigne (Royle) Benth. (Euphorbiaceae); Ltx	Khirro (N)	Diarrhea	Manandhar 1985
Saraca asoca (Roxb.) Willd. (Fabaceae); Br	Aasho	Diarrhea, dysentery, pimples, stomachic	Mandar and Chaudhary 1993
Schefflera venulosa (Wight & Arn.) Harms (Araliaceae); Twg	Simarlati (T)	Irregular menstrual cycle	Muller-Boker 1999a
Schima wallichii Choisy (Theaceae); Br	Chilaunee (N)	Wounds, diarrhea, dysentery	Bhattarai et al. 2009
Scleichera oleosa (Lour.) Mers. (Sapindaceae); Br	Kusum (N)	Ulcer	Chapagain et al. 2004
Scoparia dulcis L. (Plantaginaceae); Lvs, WP	Chiniya jhyang (N); Gurikijhani, Bhera chachura (T)	Diabetes, fever, great thirst, cooling, diarrhea, dysentery, mental disorders, boils	Chapagain et al. 2004, Muller-Boker 1999a, Dangol and Gurung 1991, Mandar and Chaudhary 1993, Manandhar 1985
Scurrula elata (Edgew.) Danser (Loranthaceae); Lvs	Aainjeru (N)	Joint pain	Acharya and Acharya 2009
Semecarpus anacardium L.f. (Anacardiaceae); Fr	Bhela (T)	Dysentery, asthma, acute rheumatism, cuts, chapped feet	Acharya and Acharya 2009, Chapagain et al. 2004, Manandhar 1985

Plant	Local name	Uses	References
Senna occidentalis (L.) Link (Fabaceae); Rt	Bangain (T)	Ringworm infection	Dangol and Gurung 1991
Senna tora (L.) Roxb. (Fabaceae); Lvs, St, Fr, Rt, Sd	Tapra (N), Chakon, Chilbile (T)	Headache, toothache, snake scorpion sting, skin diseases, cough, leprosy, fever, anthelmintic	Bhattarai et al. 2009, Chapagain et al. 2004, Manandhar 1985, 1986b, 1987b, 1990, 1994
Sesamum indicum L. (Pedaliaceae); Sd, Rt	Til (N)	Arthritis/rheumatism	Chapagain et al. 2004
Shorea robusta Gaertn. (Dipterocarpaceae); Ltx, Sd, Res, Lvs	Sal (N); Sakuwa (T)	Diarrhea, bloody dysentery, cuts, appetizer	Bhattarai et al. 2009, Chapagain et al. 2004, Muller-Boker 1999a, Manandhar 1985
Sida cordata (Burm.f.) Borss Waalk. (Malvaceae); Lvs, St, Rt	Biskhopra, Baliyari (T)	Snake scorpion sting, body swellings, wounds, blisters, boils, tonic	Chapagain et al. 2004, Dangol and Gurung 1991, Manandhar 1985
Sida rhombifolia L. (Malvaceae); Lvs, Rt, WP	Barchar, Biskhopra, Balu jhar (T)	Snake scorpion sting, cuts, loss of weight, bruises, wounds, boils	Chapagain et al. 2004, Muller-Boker 1999a, Manandhar 1985
Smilax ovalifolia Roxb. ex D.Don (Smilacaceae); Rt, St	Bagnucha (T)	Dysentery, blindness	Chapagain et al. 2004
Solanum aculeatissimum Jacq. (Solanaceae); Fr	Ghorhyenta (T)	Toothache, fever	Dangol and Gurung 1991
Solanum americanum L. (Solanaceae) Fr, Rt	Kanthakari, bihi (N); Bhutka	Headache, asthma, cooling agent, eye complaints	Bhattarai et al. 2009, Chapagain et al. 2004, Mandar and Chaudhary 1993
Solanum anguivi Lam. (Solanaceae); Fr, WP	Sanobihi (Nep)	Headache, insomnia	Bhattarai et al. 2009, Manandhar 1985
Solanum erianthum D.Don (Solanaceae); Fr	Bhurakath (T)	Cough, cold	Chapagain et al. 2004
Solanum surattense Brum.f. (Solanaceae); Fr	Kacharehata (T)	Decaying teeth	Manandhar 1985
Solanum torvum Swartz (Solanaceae); Fr	Bihi (N)	Headache	Manandhar 1985
Solanum tuberosum L. (Solanaceae); Tu	Alu (N)	Snake scorpion sting, boils	Chapagain et al. 2004, Mandar and Chaudhary 1993
Solanum virginianum Dunal (Solanaceae); Lvs, Sd	Khasretha (T)	Toothache	Chapagain et al. 2004

Table 15.1 contd. ...

...Table 15.1 contd.

Plant's Scientific Name (family name); Parts Used	Nepali and Tharu names	Medicinal Usage	References
Solena heterophylla Lour. (Cucurbitaceae); WP	Tilkor (N)	Asthma	Mandar and Chaudhary 1993
Spatholobus parviflorus (DC.) Kuntze (Fabaceae); Rt, Twg	Praslati (T)	Menstrual disorders, fever	Muller-Boker 1999a
Spermacoce alata Aubl. (Rubiaceae); WP	Paundhi (T)	Dislocated bone	Manandhar 1988
Spilanthes calva DC. (Asteraceae); Fl, Fr	Kawakchimi, Gorakhpan (T)	Cough, cold, gingivitis	Bhattarai et al. 2009, Chapagain et al. 2004
Spondias pinnata (L.f.) Kurz (Anacardiaceae); Fr	Amar (T)	Cough, cold	Chapagain et al. 2004
Sterospermum cheloniodes DC. (Bigioniaceae); St Br	Dudh khiri (T)	Leprosy	Dangol and Gurung 1991
Streblus asper Lour. (Moraceae); Br, St	Sehor (T)	Diarrhea, dysentery	Taylor et al. 1996, Muller-Boker 1999a
Syzygium cumini (L.) Skeels (Myrtaceae); Br, Fr	Jamuna (N)	Diarrhea, miscarriages, stomachache	Bhattarai et al. 2009, Chapagain et al. 2004, Muller-Boker 1999a
Syzygium nervosum A.Cunn. ex DC. (Myrtaceae); Br	Kathjamune (T)	Sinusitis, headache, cough, cold, sinusitis	Bhattarai et al. 2009
Syzygium operculata (Roxb.) Neid. (Myrtaceae); Br, Rt, Sht, Lvs	Bhadra jam (T)	Cuts, diarrhea, hoarseness of voice, measles	Chapagain et al. 2004
Tagetes erecta L. (Asteraceae); Lvs	Gendaphul	Diarrhea, dysentery	Mandar and Chaudhary 1993
Tamarindus indica L. (Fabaceae); Lvs	Yemili (T)	Cuts, wounds, diarrhea, dysentery	Dangol and Gurung 1991, Mandar and Chaudhary 1993
Tectona grandis L.f. (Lamiaceae); Br	Sagun (T)	Menstrual disorders	Chapagain et al. 2004
Terminalia alata Heyne ex Roth (Combretaceae); Br	Saj (N); Asna (T)	Diarrhea, dysentery, stomachache	Bhattarai et al. 2009
Terminalia arjuna (Roxb. ex DC) W. and A. (Combretaceae); Ltx	Arjun Kaath	Bodyache	Chapagain et al. 2004

Species (Family); Part	Tharu name	Uses	References
Terminalia bellirica (Gaertn.) Roxb. (Combretaceae); Fr	Barro (N), Asida, Baheri (T)	Cough, cold, appetizer, gastritis, diarrhea, dysentery, Stomachache, constipation, eye diseases, bronchitis	Bhattarai et al. 2009, Chapagain et al. 2004, Muller-Boker 1999a, Dangol and Gurung 1991, Mandar and Chaudhary 1993
Terminalia chebula Retz. (Combretaceae); Fr	Harro (N); Harai (T)	Cough and cold, gastritis, bronchitis, chest pain, diarrhea, dysentery	Bhattarai et al. 2009, Chapagain et al. 2004, Muller-Boker 1999a, Mandar and Chaudhary 1993
Tetrastigma serrulatum (Roxb.) Planch. (Vitaceae); Lvs	Tinpatiya (T)	Wounds, blisters	Chapagain et al. 2004
Thevetia peruviana (Pers.) K. Schum. (Apocynaceae); Lvs, Fr	Karbir (N	Ear pain, wounds	Bhattarai et al. 2009, Dangol and Gurung 1991
Thysanolaena maxima (Roxb.) Kuntze (Poaceae); Rt	Bashadi, Bankucho (T)	Boils, anthelmintic	Bhattarai et al. 2009
Tiliacora sp. (Menispermaceae); Rt	Karot (T)	Anthelmintic	Chapagain et al. 2004
Tinospora sinensis (Lour.) Merr. (Menispermaceae); Rt	Gurjo (N)	Diarrhea, dysentery, stomachache, diuretic	Bhattarai et al. 2009
Torenia sp. (Linderniaceae); WP	Sauna phul (T)	Wounds, blisters	Chapagain et al. 2004
Trachyspermum ammi (L.) Sprague (Apiaceae); Sd, WP	Jawain (T)	Colds, sore throat, wounds, pimples, rashes, pneumonia, diarrhea	Muller-Boker 1999a
Trianthema protulacastrum L. (Aizoaceae); Rt	Gaspurna (T)	Body swelling	Chapagain et al. 2004
Trichodesma indicum (L.) Lehm. (Boraginaceae); WP	Oonmodia jhyang (T)	Headache	Dangol and Gurung 1991
Trichosanthes dioica Roxb. (Cucurbitaceae); Rt	Parwar (N)	Hand/leg ache	Chapagain et al. 2004
Tridax procumbens L. (Asteraceae); Fl, WP	Dhusere (N)	Cough, cold, boils	Manandhar 1985
Triumfetta rhomboids Jacq. (Tiliaceae): WP	Bishakhopra (T)	Boils, pimples	Manandhar 1985

Table 15.1 contd. ...

...Table 15.1 contd.

Plant's Scientific Name (family name); Parts Used	Nepali and Tharu names	Medicinal Usage	References
Uraria legapodioides (L.) Desv. (Fabaceae); WP	Odarbau (T)	Boils	Manandhar 1985
Urena Lobata L. (Malvaceae); Rt	Bariyar	Ringworm infection	Mandar and Chaudhary 1993
Urtica dioica L. (Urticaceae); Rt	Sisna (T)	Wounds	Manandhar 1985
Vanda roxburghii R.Br. (Orchidaceae); WP	Rukh hadjwar (N)	Bone fracture	Chapagain et al. 2004
Vernonia cinerea (L.) Less. (Asteraceae); Sd, Rt, WP, Lvs, Fl	Marcha jhar (N); Sahadeya (T)	Cough, cold, cuts, wounds, skin diseases, fever, anthelmintic, conjunctivitis, anthelmintic	Manandhar 2002, Manandhar 1994
Vetiveria zizanoides (L.) Nash (Poaceae); Rt	Sikuhal (T)	Toothache	Chapagain et al. 2004
Viscum articulatum Burm.f. (Santalaceae), WP	Bhuin hadjwar (N)	Bone fracture	Chapagain et al. 2004
Vitex negundo L. (Verbenaceae); Lvs, Br	Simali (N)	Cough, cold, fever, ear pain, wounds, headache, toothache, nasal bleeding	Chapagain et al. 2004, Bhattarai et al. 2009
Woodfordia fruticosa (L.) Kurz (Lythraceae); Fl, Br, Lvs, Rt	Dhayaro (N); Dhawatha, Dhaira (T)	Stomachache, diarrhea, dysentery, typhoid, labor pain, control bleeding, maintain healthy condition of new-born child, boils, burns, pneumonia, sprains, pains	Bhattarai et al. 2009, Acharya and Acharya 2009, Chapagain et al. 2004, Manandhar 1985
Xanthium strumarium L. (Asteraceae); Fr	Kuchakuchiya (T)	Conjunctivitis	Manandhar 1985
Xeromphis spinosa (Thunb.) Keay (Rubiaceae); Br, Fr	Mainkada (N); Piralu (T)	Gastritis, stomachache, arthritis/rheumatism, diarrhea, dysentery, fish poison, leprosy	Bhattarai et al. 2009, Chapagain et al. 2004, Muller-Boker 1999a, Manandhar 1989a, 1990, Dangol and Gurung 1991
Xeromphis uliginosa (Retz.) Maheshw. (Rubiaceae); Fr	Perar (T)	Diarrhea	Chapagain et al. 2004
Zanthoxylum armatum DC. (Rutaceae); Fr	Timur (N)	Skin diseases	Chapagain et al. 2004
Zea mays L. (Poaceae); Fl	Makai (N)	Delivery	Chapagain et al. 2004

Zingiber officinale (Willd.) (Zingiberaceae); Lvs, Rh	Aduwa (N); Adrakh (T)	Diarrhea, sinusitis, cough, cold	Bhattarai et al. 2009, Chapagain et al. 2004, Muller-Boker 1999a, Mandar and Chaudhary 1993
Zizyphus mauritiana Lam. (Rhamnaceae); Fr, Sht, Sd, Lvs	Bayar (N); Bairi (T)	Measles, breast pain, diarrhea, chicken pox, dysentery	Bhattarai et al. 2009, Chapagain et al. 2004, Muller-Boker 1999a, Manandhar 1985
Zizyphus nummularia (Burm.f.) Wight and Am. (Rhamnaceae); Fr, Lvs, Rt	Jangali bayar (N)	Fever, earache	Dangol and Gurung 1991

Keys: N: Nepali; T: Tharu; Lvs: Leaves; Fr: Fruits; Sd: Seeds; Rt: Roots; Rh: Rhizomes; Br: Barks; WP: Whole plant; Ltx: Latex; Sht: Shoots; Fl: Flowers; Tu: Tubers; Psb: Pseudobulb; Arp: Aerial parts; Bd: Buds; SP: Spines; Twg: Twigs; Bu: Bulbs.

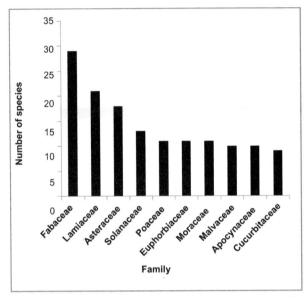

Fig. 15.5. Top ten largest families used for medicine by the Tharu communities in Nepal.

Table 15.2. Number of medicinal plants represented by each family.

Name of Family	No. of Species
Fabaceae	29
Lamiaceae	21
Asteraceae	18
Solanaceae	13
Euphorbiaceae, Poaceae and Moraceae	11 species each
Apocynaceae and Malvaceae	10 species each
Cucurbitaceae	9
Amaranthaceae and Rubiaceae	7 species each
Convolvulaceae and Cucurbitaceae	6 species each
Anacardiaceae, Boraginaceae, Myrtaceae and Zingiberaceae	5 species each
Acanthaceae, Combretaceae, Lythraceae, Polygonaceae, Rutaceae	4 species each
Amaryllidaceae, Annonaceae, Apiaceae, Cyperaceae, Dioscoraceae, Phyllanthaceae, Plantaginaceae, Pteridaceae, Sapotaceae, Urticaceae and Verbenaceae	3 species each
Bigioniaceae, Equisetaceae, Lauraceae, Linaceae, Meliaceae, Nyctaginaceae, Orchidaceae, Oxalidaceae, Piperaceae, Rosaceae, Sapindaceae and Vitaceae	2 species each
Acoraceae, Arecaceae, Aizoaceae, Alismataceae, Araceae, Berberidaceae, Brassicaceae, Bromeliaceae, Burseraceae, Cactaceae, Cannabaceae, Capparaceae, Caricaceae, Cleomaceae, Costaceae, Crassulaceae, Dilleniaceae, Dipterocarpaceae, Dryopteridaceae, Ebenaceae, Ericaceae, Hypoxidaceae, Juglandaceae, Lecythidaceae, Linderniaceae, Loranthaceae, Lygodiaceae, Malphighinaceae, Moringaceae, Musaceae, Myricaceae, Nelumbonaceae, Oleaceae, Ophioglossaceae, Papaveraceae, Parmeliaceae, Pedaliaceae, Phyllanthaceae, Pinaceae, Plumbaginaceae, Rhamnaceae, Santalaceae, Simaroubaceae, Smilaceae, Theaceae, Tiliaceae and Xanthorrheaceae	1 species each

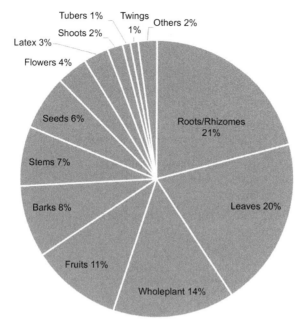

Fig. 15.6. Plant parts used in medicine by the Tharu communities in Nepal.

Piperaceae, Rosaceae, Sapindaceae and Vitaceae were represented by 2 species each. The remaining 56 families were represented by one species each (Table 15.2).

The Tharu communities of Nepal have been using different parts of plants as medicine. Various plant parts used in medicine were roots/rhizomes (114 species), followed by leaves (108 species), whole plant (78 species), fruits (58 species), barks (46 species), stems (38 species), seeds (35 species), flowers (20 species), latex (16 species), shoots (10 species), tubers (5 species), twigs (5 species), bulbs (3 species), aeriel parts (3 species), buds (3 species), oils/resins (2 species) and spines (1 species). The percentage of plant parts used was also given (Fig. 15.6). The listed plants species' lifeform were herbs, shrubs and trees.

Conclusion

The *Guruwas* (traditional healers) of the Tharu community have a vast knowledge of the medicinal value of the plants. Even today medicinal plants are widely used in the Tharu communities to treat common ailments. But the social survival of those people is disappearing day by day. Tharu people are highly linked with the bioresources and the preservation and protection of those resources are essential to preserve their cultural and ethnobotanical knowledge. Through a Nepalese perspective, community participation has long been tested and considered successful for the protection and conservation of bioresources. Hence, the government should update such policy which will be helpful to preserve the traditional healing practices of the several ethnic communities, including the Tharus of Nepal.

References

Acharya, R. and Acharya, K.P. 2009. Ethnobotanical study of medicinal plants used by tharu community of parroha VDC, Rupandehi district, Nepal. Sci. Word 7(7): 80–84.

Baral, S.R. and Kurmi, P.P. 2006. A Compendium of Medicinal Plants in Nepal. Mass Printing Press, Chhauni, Kathmandu.

Bhattachan, K. 2016. Post-shifting cultivation: struggles for livelihood and food security among Tharu people displaced by the Chitwan national park in Nepal. pp. 291–325. *In*: Christian, E. (ed.). Shifting Cultivation, Livelihood and Food Security: New and Old Challenges for Indigenous Peoples in Asia. Bangkok: The Food and Agriculture Organization of the United Nations, International Work Group for Indigenous Affairs and Asia Indigenous Peoples Pact.

Bhattarai, S. 2009. Ethnobotany and antibacterial activities of selected medicinal plants of Nepal Himalaya. PhD. Thesis, Central Department of Botany, Tribhuvan University, Kirtipur, Kathmandu, Nepal.

Bhattarai, S.B., Chaudhary, R.P. and Taylor, R.S.L. 2009. Ethnomedicinal plants used by the people of Nawalparasi District, Central Nepal. Our Nat. 7: 82–99.

Bista, D.B. 2004. People of Nepal. Ratna Pustak Bhandar, Kathmandu, Nepal.

Chapagain, D.J., Joshi, S.D. and Jnawali, S.R. 2004. Use of medicinal plants by Tharu community in the southwestern bufferzone of Royal Bardiya National park. Sci. Word. 2(2): 50–62.

Chaudhary, R.P. 1998. Biodiversity in Nepal: Status and Conservation. S. Devi, Saharanpur (U.P.), India and Tecpress Books Bangkok, Thailand.

Dangol, D.R. and Gurung, S.B. 1991. Ethnobotany of the Tharu tribe of Chitwan District, Nepal. Int. Journ. Pharmacog. 29(3): 203–209.

Ghimire, K. and Bastakoti, R. 2009. Ethnomedicinal knowledge and healthcare practices among the Tharus of Nawalparasi district in central Nepal. For. Ecol. Manage. 257(10): 2066–2072.

GON. 2006. Plants of Nepal: Fact Sheet. Government of Nepal, Ministry of Forest and Soil Conservation, Department of Plant Resources, Thapathali, Kathmandu, Nepal.

Gurung, H. 1996. Ethnic Demography of Nepal. Kathmandu: Nepal Foundation Paper 5, Nepal Foundation for Advanced Studies.

Hagen, T. 1960. Nepal, Konigreich am Himalaya, Bern, Kummerly and Frey.

Krauskopff, G. 1995. The anthropology of the Tharus: an annoted bibliography. Kail. 17(3/4): 185–213.

Katewa, S.S., Chaudhary, B.L. and Jain, A. 2004. Folk herbal medicines from tribal area of Rajasthan, India. Jour. Ethnophar. 92: 41–46.

Lall K. 1983. The Tharus. The Rising Nepal 20(233): 3

Lewis, M.P., Simons, G.F. and Fennig, C.D. (eds.). 2014. Tharu, Chitwania: a language of Nepal. Ethnologue: Languages of the World, Seventeenth edition. Dallas, Texas: SIL International, online version.

Mahishi, P., Srinivasa, B.H. and Shivanna, M.B. 2005. Medicinal plant wealth of local communities in some villages in Shimoga district of Karnataka, India. Journ. Ethnopharmo. 98: 307–312.

Manandhar, N.P. 1985. Ethnobotanical notes on certain medicinal plants used by Tharus of Dang-deokhuri district, Nepal. Int. J. Crude Drug Res. 23(4): 153–159.

Manandhar, N.P. 1986b. Ethnobotany of Jumla district, Nepal. Int. J. Crude Drug Res. 24(2): 81–89.

Manandhar, N.P. 1987a. Traditional medicinal plants used by tribals of Lamjung district, Nepal. Int. J. Crude Drug Res. 25(4): 236–240.

Manandhar NP. 1988. Ethno Veterinary medicinal drugs of Central development region of Nepal. Bull. of Med. Ethno. Bot. Res. 10(3-4): 93–99.

Manandhar, N.P. 1989a. Medicinal plants used by Chepang tribes of Makawanpur district, Nepal. Fitoter. 60(1): 61–68.

Manandhar, N.P. 1989b. Useful wild plants of Nepal. Franz Steiner Verlag Wiesbaden GMBH, Stuttgart, W Germany.

Manandhar, N.P. 1990. Folk-lore Medicine of Chitwan District, Nepal. Ethnob. 2: 31–38.

Manandhar, N.P. 1994. Medicinal Folk-lore about the plants used as anthelmintic agents in Nepal. Fitother. 16(2): 149–155.

Manandhar, N.P. 2002. Plants and People of Nepal. Timber Press, Portland, Oregon.

Mandar, L.N. and Chaudhary, R.P. 1993. Medicinal Plants and their traditional use by tribal people of Saptari district, Nepal. *In*: Proceedings of First National Botanical Conference (1992). Nepal Botanical Society, Kathmandu.

McDonaugh, C. 2007. Spirit, substance, vehicle. Kinship and cosmology among the Dangaura Tharu, Nepal.

McLean, J. 1999. Conservation and the Impact of Relocation on the Tharus of Chitwan, Nepal, Himalaya. The Journ. Assoc. Himalay. Studi. 19(2): 38–44.

Meyer, K.W. and Deuel, P. 1999. Who are the Tham? National Minority and Identity as manifested in Housing forms and Practices. *In*: Skar, H.O. (ed.). Nepal: Tharu and Tarai Neighbours, Kathmandu: Bibliotheca Himalayica.

Muller-Boker, U. 1991. Wild Animals and Poor people: Conflicts between Conservation and Human needs in Citawan (Nepal). Europ. Bullet. Himalay. Res. 2: 28–31.

Muller-Boker, U. 1993. Ethnobotanical Studies among the Citawan Tharus. Journ. Nep. Rese. Cent. 9: 17–56.

Müller-Böker, U. 1999a. The Chitawan Tharus in Southern Nepal. An Ethnoecological Approach. Nepal Re-search Centre Publications, No. 21. Kathmandu, Stuttgart.

Müller-Böker, U. 1999b. The Chitwall Tharus ill Southern Nepal: All Ethnoecological Approach. Franz Stiner Verlag Stuttgart.

Pyakuryal, K. 1982. Ethnicity and Rural Development: A Sociological Study of Four Tharu Villages in Chitwan, Nepal. PhD. Thesis, Michigan State University.

Rajaure, D.P. 1975. Tatooing among the Tharus of Dang-Deokhuri, Far western Nepal. Contri. Nepal. Stud. 2(1): 91–98.

Sapkota, M. 2014. Contested Identity Politics in Nepal: Implications from Tharu Movement. IOSR Journ. Humanit. Soci. Sci. 19(7): 16–25.

Shrestha, P. 1985. Research note: Contribution to the ethnobotany of the Palpa area. Contr. Nepal. Stud. 12(2): 63–74.

Skar, H.O. (ed.). 1999. Nepal: Tharu and Tarai Neighbours. Kathmandu: Bibliotheca Himalayica.

Taylor, R.S.L., Hudson, J.B., Manandhar, N.P. and Towers, G.H.N. 1996. Antiviral activities of medicinal plants of southern Nepal. Journ. Ethnophar. 53: 97–104.

Thapa, S. 2001. Documentation of Traditional Uses of Plants by Tharu Community around Royal Shuklaphanta Wildlife Reserve, Far-Western Nepal. MSc Dissertation, Central Department of Botany, Tribhuvan University, Kathmandu, Nepal.

Verma, S.C. 2010. The eco-friendly Tharu tribe: A study in socio-cultural dynamics. Journ. Asi. Pac. Stud. 1(2): 177–187.

Index

Editors Biography

Professor José Luis Martínez S.

Professor José Luis Martínez S. is a chemical biologist and did his MSc from the Pontifical Catholic University of Chile. He works at the Vice-Rectory for Research, Development and Innovation of the University of Santiago, Chile. He is the author and co-author of scientific articles in pharmacology and ethnobotany and has a hundred presentations in conferences at national and international scientific events. He is the editor and founder of the Latin American and Caribbean Bulletin of Medicinal and Aromatic Plants (BLACPMA) and is also the Editor of other journals and scientific research books.

Professor Amner Muñoz-Acevedo

Professor Amner is PhD in Chemistry and teaches in the Department of Chemistry and Biology at Universidad of the North Colombia, with wide experience in studies of natural products (plants/microorganisms—volatile/non-volatile secondary metabolites) based on ethnomedicine, organic synthesis, *in vitro* biological assays (antioxidant capacity, cito-toxicity/ anticancer, antimicrobial, repellency/insecticidal, antiviral activities), functionalized beverages and food, sustainable uses of agroindustrial wastes, sample preparation techniques (HD, SPME, S-HS, SDE, SFE, SPE, Soxhlet, US, LLE) and instrumental methods of chemical analysis (TLC, GC, LC, MS, NMR, IR, UV/Vis). He has authored/co-authored several papers and delivered lectures at national/international scientific events.

Professor Mahendra Rai

 Mahendra Rai is a senior professor and Basic Science Research Faculty (UGC) at the Department of Biotechnology, Sant Gadge Baba Amravati University, Maharashtra, India. He has been Visiting Scientist in the University of Geneva, Debrecen University, Hungary; University of Campinas, Brazil; Nicolaus Copernicus University, Poland. VSB Technical University of Ostrava, Czech Republic, and National University of Rosario, Argentina. He has published more than 380 research papers in national and international journals. In addition, he has edited/authored more than 45 books and 6 patents.

Printed and bound by CPI Group (UK) Ltd, Croydon, CR0 4YY

24/10/2024

01778304-0009